Topological Analysis
and Synthesis of
Communication Networks

Topological

Analysis and Synthesis

IN LIT TERIS
LIBER TAS ∴

NEW YORK AND LONDON   1962

# of Communication Networks

WAN HEE KIM

Associate Professor of Electrical Engineering, Columbia University

ROBERT TIEN-WEN CHIEN

Member of Research Staff, IBM

Adjunct Assistant Professor of Electrical Engineering, Columbia University

Columbia University Press

To Chung Sook

and

Sophie

Preface

THE WORD "NETWORK" implies, intrinsically, a collection of line-segments which interconnect points with a specified topological configuration. If each line-segment of a network represents an electrical element such as a resistor, an inductor, a capacitor, and a voltage or a current source, then the network is an "electrical network." On the other hand, if each line-segment of a network characterizes a contact of an electrical switch, then it is called a "switching network" and each line-segment is associated with two states of the contact, whether it is open or closed. One more important category of networks which has had a significant engineering application in recent years is that of the so-called "communication nets," often called "flow networks." Each line-segment of a communication net is associated with a fixed or random varying capacity of information flow or material flow. If the reliability of a communication net is studied and tested, each line-segment of the net may be associated with a probability of reliability (or of failure).

For an electrical network, we consider the form of the voltage or current response in time or in frequency through the network. However, the sequence of closing contacts decides the characteristics of a switching network. The major objective of a communication net is the design of a network which will achieve maximum flow or minimum cost. Thus, each category of network described has its own techniques of analysis and synthesis. However, they are all still networks; e.g., they are basically the so-called "connected linear graphs."

This book presents a uniform approach to various engineering problems, such as analysis and synthesis of electrical networks, sampled-data feedback control systems, switching networks, and communication nets.

Chapter 1 of Part I studies the fundamentals of linear graph theory, always with an eye to their applications. The review of the theory of linear graphs is, however, minimized by including only what is essential for comprehension of the materials presented in the book. The book is designed to study the solutions of engineering problems rather than the theory of linear graphs itself. Topological analysis of *RLC* networks is the first subject discussed, since it is the most direct application of the theory. We then extend the analysis techniques to multipoles, i.e., networks containing multiterminal linear devices, such as vacuum tubes and transistors, transformers, gyrators, and circulators.

Part II contains rigorous proof and derivation of the various properties of the so-called "signal-flow graphs" and of their applications to the analysis of feedback control systems and single-rate as well as multirate sampled-data systems. Part III discusses the analysis and synthesis of *n*-port networks. Then, in Part IV, the realization of loop and cut-set matrices is investigated, and a simple algorithm for the realization is proposed as an extension of some of the analytic techniques developed for the synthesis of *n*-port resistive networks. The realization of loop and cut-set matrices is, however, presented in terms of the synthesis of contact-switching networks. Part V deals with basic contributions already made in the area of communication nets. A complete theory is presented for the analysis and synthesis of one-terminal-pair flow networks. Also discussed are topics such as *n*-terminal-pair simultaneous-flow problems and applications of linear programming. Parts II and V are more or less self-contained, except that an understanding of a very few fundamental notions of linear graph theory is necessary for their comprehension.

Part IV may not be comprehensible without the background of most of the materials presented in Chapter 1 of Part I and also of some developments in Part III. This book is written for use in a course on advanced network theory or as a reference book for those who desire to familiarize themselves with this subject or to pursue further research in this area.

For the most part, this book is based on the senior author's class notes for a one-year graduate course which has been given at the Department of Electrical Engineering, Columbia University, since 1957. These notes were based on materials developed by doctoral students and staff members in the Department under research grants from the National Science Foundation. Part V is based mainly on contributions made by Dr. D. T. Tang, Dr. W. Mayeda, Dr. R. E. Gomory, and Dr. T. C. Hu, of the IBM Research Center, Yorktown Heights, New York. The authors acknowledge the sponsorship of the National Science Foundation and International Business Machines Corporation which has made this research possible.

The senior author wishes to express his thanks to former students who have made his manuscript possible, including Dr. Omar Wing,

Dr. C. V. Freiman, and Dr. J. A. Bernstein, particularly to Dr. R. T. Ash, Dr. I. T. Frisch, and also to a number of students in the Columbia University courses EE 305–306 (EEE 8201x and E 8202y) who have made numerous constructive suggestions for the improvement of the presentation of the materials. A number of constructive criticisms by Professor Unger, particularly for Part IV, were greatly helpful. We are grateful to Dr. C. C. Halkias for his careful reading of the manuscript and the proofs.

The authors are indebted to the Circuit Theory Group at the University of Illinois, and especially to Professor M. E. Van Valkenburg and Professor S. Seshu (both authors were members of the group at one time), for their numerous contributions to much of the material included in this book. We owe a special debt to many authors who have made significant contributions in the field which we have incorporated into the materials presented here. Finally, we express our special thanks to Mrs. Susan McCauley for typing the manuscript.

WAN H. KIM

June, 1962   ROBERT T. CHIEN

Contents

PART I

Topological Analysis
of Electrical Networks

Introduction

THE HIGHLY SYSTEMATIZED development of the modern theory of electrical networks which is the basis of the study of almost all subjects in electrical engineering may be attributed to a number of reasons. One of the most important of these is the abundant supply of available analytic tools. In particular, the formulation of conventional network theory, which is based on ordinary two-terminal devices such as resistors, inductors, and capacitors, has been greatly advanced by the application of linear graphs. Moreover, this particular method is suitable for digital computers. Thus, the topological study of electrical networks has become indispensable to modern engineering.

We shall therefore begin with the study of the fundamentals of linear graph theory. We will then formulate the various functions of ordinary networks in topological terms, and then extend the analysis techniques applied to an ordinary network so that they can also be used for the topological study of networks containing mutually coupled devices, as well as linear active devices.

CHAPTER 1

Fundamentals of the Theory
of Linear Graphs

IT IS BELIEVED that the work of Euler[1] in 1736 on the well-known "Königsberg bridge problem" (see Reference 2 for discussion of the problem) was the first formal study of linear graphs. In 1847 Kirchhoff[3] introduced the topological study of electrical networks and laid the groundwork for present network theory. Since 1847 numerous contributions have been made concerning the analytic properties of linear graphs as well as their broad applications to engineering problems. We shall, however, review only those basic concepts and properties of linear graphs which are necessary for subsequent discussions in this book, since the purpose of this book is to show the applications of graph theory to engineering problems, particularly in electrical engineering; the book is not designed for the study of graph theory itself. We will therefore study some of the fundamental concepts of linear graphs, always keeping their applications in mind. For a more extensive and rigorous study of the theory of linear graphs, other books[3-8] and papers[9] are recommended. The symbols and terminology used in this chapter follow, for the most part, that of References 5-7, 16, 20a, 32, and 33e.

Definition 1. An "edge" (or arc, element, 1-cell), is a line segment with two end points. An end point of an edge is called a "node" (or vertex, 0-cell). If two nodes of an edge are identified with each other, the edge will be called a "self-looped edge." A node and an edge are "incident" with each other if the node is an end point of the edge.

Definition 2. A "linear graph" (or graph, linear complex, 1-complex) is a collection of edges with no self-looped edges. An "oriented graph" (or directed graph) is a graph in which the direction of each edge is specified. Otherwise, the graph is "nonoriented" (or nondirected). In the definitions which follow, we shall not specify whether a graph under consideration is oriented or nonoriented unless necessary.

According to Definition 2, a linear graph, or simply a graph, should not contain any isolated points. However, on many occasions, we will consider a single node of a graph to be a subgraph of the graph. If a graphical representation of a network contains self-looped edge(s) it will be referred to as a "signal-flow graph" (see Part II) so that we will be able to distinguish it from a linear graph.

Definition 3. The "degree" of a node is the number of edges incident at that node.

Definition 4. A graph in which every node is of even degree is called an "Euler graph." If every node of a graph is of the same degree, it is referred to as a "regular graph," and a regular graph in which every node is of degree k is a "regular graph of degree k."

Definition 5. A "path" between nodes i and j in a graph is an ordered sequence of edges in the graph in which every node is of degree two but nodes i and j are of degree one. If a direction of a path is specified, then the path is called a "directed path."

Definition 6. If in a graph there exists a path between every pair of nodes in the graph it is "connected." It is therefore clear that a connected graph *cannot* contain any isolated edge or point. If there exists exactly one edge between any two nodes of a graph, the graph is a "completely connected" graph or, simply, a "complete graph," since there exists a direct path from one node to any other nodes in the graph.

Definition 7. A connected graph is "separable" if it contains at least one subgraph which has only one node in common with its complement. This common node is called a "cut node" or "cut vertex."

Definition 8. A "loop" (or circuit) is a connected regular graph or subgraph of *degree two*. If a loop of a connected graph contains all nodes of the graph, it has a special name: "Hamilton loop" (or Hamilton circuit). A collection of loops is called a "union of loops" and if no two loops in the union have an edge (or a node) in common it is a "union of edge-disjoint loops" (or union of node-disjoint loops). A set of edges which form a single loop is referred to as "prime loop-edges" or "prime loop-set" while the term "loop-edges" or "loop set" is used for a set of edges which form a union of edge-disjoint loops.[15]

Definition 9. A "tree" is a connected graph or subgraph containing all nodes of the graph but no loops. A "branch" (link) is an edge of a tree, and a "chord" (tie) is an edge of the complement of a tree. The complement of a tree of a connected graph is sometimes called a "co-tree." A k-tree of a connected graph of n nodes is a collection of k disjoint subgraphs of the graph containing n nodes but containing no loop. A k-chord, or "k-co-tree" is the complement of a k-tree of a connected graph.

The concept of a k-tree of a graph is illustrated in the following example.

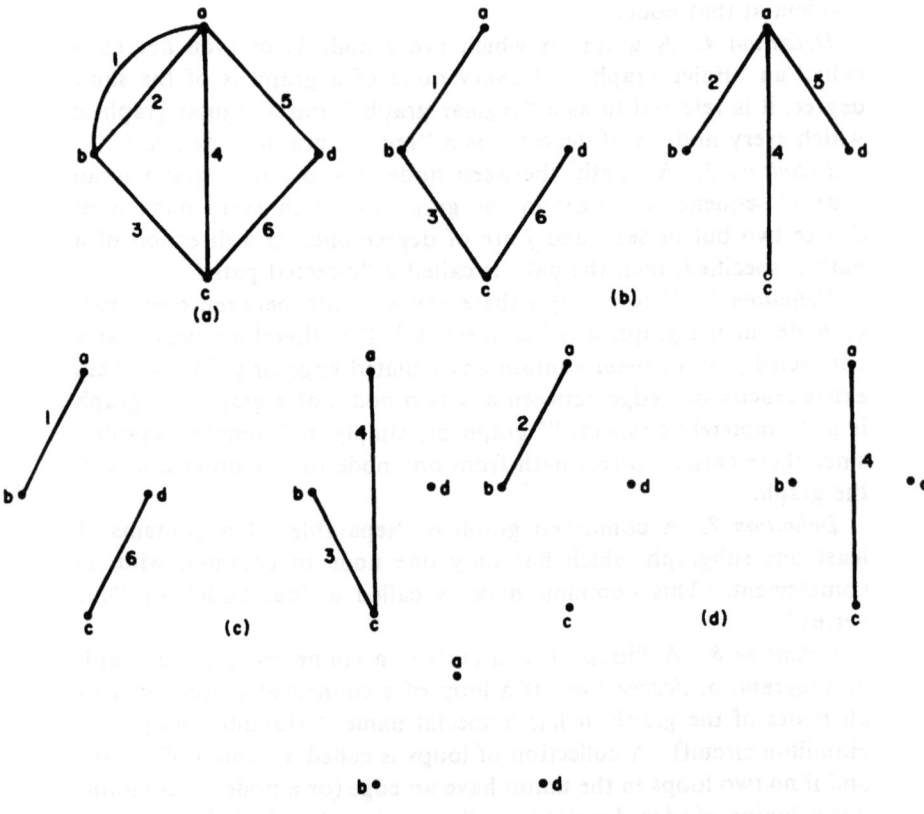

Fig. 1.1. Illustration of a k-tree

(a) Graph G; (b) some of the trees of G; (c) some of the 2-trees of G; (d) some of the 3-trees of G; (e) a 4-tree of G.

Example 1. Let us consider a graph G as shown in Fig. 1.1a, where the number of nodes is 4 and the number of edges 6. Some trees, 2-trees, 3-trees, and a 4-tree of G, are given respectively in Figs. 1.1b, 1.1c, 1.1d, and 1.1e. Note here that an isolated node is considered to be a subgraph of G.

Definition 10. A "maximally connected subgraph" G_m of a graph G is a subgraph of G or the graph itself such that the addition of an edge in the complement of G_m to G_m makes the resultant subgraph no longer connected.

Therefore, if G is a connected graph, the maximally connected subgraph of G is the graph itself. If G is a collection of disjoint subgraphs and each subgraph is connected, then each subgraph is a maximally connected subgraph of G. This is, therefore, a more elegant name for a so-called "connected piece."

Property 1. Let G be a connected graph of b edges and n nodes. Then:

A k-tree of G has $(n-k)$ edges, where $n \geqslant k$;

A subgraph of G, which contains n nodes and $(n-k)$ edges but no loop, is a k-tree of G.

Proof: We shall prove the first property by the induction method. For $n = 2$, a tree of the graph contains one edge, and a 2-tree of the graph is a collection of two isolated points containing no edges. It is thus clear that the property is true for $n = 2$. Let the property be true for $n = r$, i.e. a k-tree of a graph of r nodes contains $(r-k)$ edges and $r \geqslant k$. Since there should exist at least one node of degree one in a k-tree of r nodes, because it contains no loop, we add an edge at the node of degree one of the k-tree such that no loop is formed. Then, the resulting graph contains $(r+1)$ nodes and $(r-k+1)$ edges but no loop. That is, a k-tree of $(r+1)$ nodes has $(r-k+1)$ edges. This completes the proof of the first property.

For the proof of the second property, let us denote a subgraph of G which satisfies the conditions of the second property by G_1. Let us assume that G_1 consists of m maximally connected subgraphs $g_1, g_2, \ldots, g_m$ and that the number of nodes in each subgraph is $n_1, n_2, \ldots, n_m$, respectively. Since G_1 has no loop, each subgraph is its own tree. Thus,

$$\text{(number of nodes in } G_1) = \sum_{i=1}^{m} n_i = n, \qquad (1.1)$$

and from the first property it is clear that

$$(\text{number of edges in } G_1) = \sum_{i=1}^{m} (n_i - 1) = n - m. \qquad (1.2)$$

Thus, if $m = 1$, G_1 is a connected subgraph of G and contains n nodes but no loop; G_1 is then a tree of G and has $(n-1)$ edges. If $m = 2$, G_1 consists of two disjoint subgraphs of G and contains n nodes but no loop; G_1 is therefore a 2-tree of G and has $(n-2)$ edges. If $m = k$, G_1 is a collection of k disjoint subgraphs of G and contains n nodes but no loop. Then G_1 is a k-tree of G and has $(n-k)$ edges. Thus, the proof of the second property is completed.

Definition 11. A "basic loop" (or fundamental loop, f-circuit or c-circuit) is a loop formed by each chord and some or all branches of a tree of a connected graph.

It is therefore clear that a connected graph with n nodes and b edges contains $(b-n+1)$ basic loops.

Definition 12. The "incidence matrix" (or incidence cut-set, or vertex cut-set matrix), denoted by $\mathbf{A} = [a_{ij}]$, of a graph with n nodes and b edges, is the matrix with n rows and b columns. Each row corresponds to a node, and each column corresponds to an edge, such that

$a_{ij} = 1$,   if edge j is incident at node i and directed away from node i;

$a_{ij} = -1$,   if edge j is incident at node i and directed toward node i;

$a_{ij} = 0$,   if edge j is not incident at node i.

If a graph under consideration is nonoriented, then each element of the incidence matrix of the graph will take values of 1 or 0.

Property 2. The incidence matrix of a connected graph of n nodes and b edges, $\mathbf{A}$, has the following properties:

The rank of $\mathbf{A}$, called the "rank of the graph," is $(n-1)$, and the "nullity" (or cyclomatic number or connectivity) of the graph is $(b-n+1)$;

The determinant of every square submatrix of $\mathbf{A}$ is 1, -1, or 0, i.e., $\mathbf{A}$ is an "E-matrix"[10] or a "unimodular matrix";

If the graph consists of P maximally connected subgraphs, the rank of the incidence matrix, or of the graph, is $(n-P)$, and the nullity of the graph is $(b-n+P)$.

Proof: The rank of the incidence matrix $\mathbf{A}$ can be at most $(n-1)$, since the sum of all n rows in each column yields the sequence of zeros, because each column has exactly one 1 and one -1. Furthermore, the sum of $(n-1)$ rows contains at least one nonzero element

(1 or -1). Otherwise, the row which was not included must have all zeros so that the two nonzero elements, 1 and -1, must have been included to yield the sum of all zeros. This implies that the node corresponding to the row which was left out is isolated from the rest of the graph represented by **A**, which contradicts our assumption that the graph is *connected*. Therefore **A** contains at least $(n-1)$ independent rows. Thus, the rank of **A** (or the number of independent rows in A) must be $(n-1)$. Thus, the first property is proved.

The second property can be directly proved by expanding the determinant of a square submatrix of **A** by the Laplace expansion method. Let us consider the determinant of a square submatrix of order $k \leqslant n$, Δ, of $\mathbf{A} = [a_{ij}]$. Then, the determinant vanishes if each column of the determinant contains exactly one 1 and one -1, or if any of the columns or rows contains all zeros. It is therefore clear that there must be at least one column in the determinant which contains only one nonzero element so that the determinant will be nonvanishing. Let us assume that the ith column of the determinant has only one nonzero element in the jth row. We therefore have

$$\Delta = (-1)^{i+j} a_{ij} M_{ij}, \tag{1.3}$$

where $a_{ij} = 1$ or -1, and M_{ij} is the minor of Δ for (i,j)-position.

We again try to expand M_{ij} and meet the same situation as before. That is, the minor may vanish, and if not, it should have at least one column containing only one nonzero element. Let this nonzero element be at the mth column and kth row in the minor. Thus, we have

$$\Delta = (-1)^{i+j+k+m} a_{ij} a_{km} M_{ik,jm}, \tag{1.4}$$

where, again, $a_{km} = 1$ or -1, and $M_{ij,km}$ is the minor of M_{ij} for (k, m)-position.

By repeating this expansion, we will reach a single element, i.e., a first-order subdeterminant. It is thus clear that a subdeterminant of any order of the incidence matrix is either 0, 1, or -1.

Definition 13. The "principal node" of an edge *of a tree* of a connected graph is the one of the two nodes of the edge that is located farthest from a fixed reference node in a path of the tree containing the edge and the reference node. The other node of the edge is called the "minor node."

Note that the principal and minor nodes of an edge are determined only with respect to a particular tree of a connected graph. Therefore, if there exists another tree of the graph which contains the same edge, then the principal and minor nodes of the edge in this tree may not be the same as found in the first tree. It should be obvious that a reference node of a graph cannot be the principal node of any edge of the graph. Thus, the following property is evident.

Property 3. In a graph or subgraph of a graph which contains no loop, no node can be the principal node of more than one edge of the graph or the subgraph of a graph.

Example 2. Given a graph G in Fig. 1.2a, where node 4 is chosen to be the reference node, let us choose a pair of trees of G, t_1, and

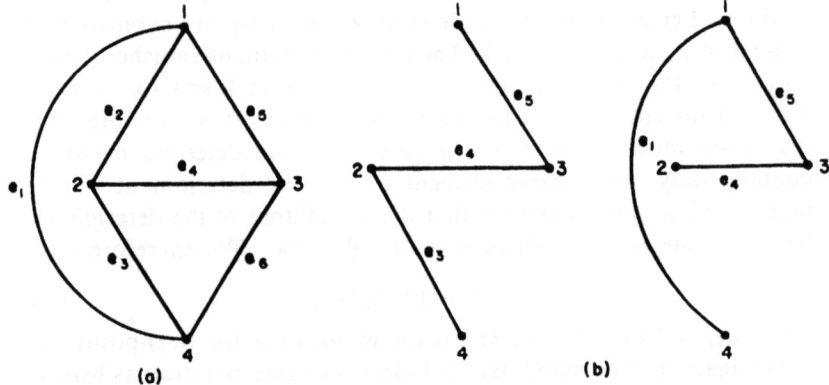

(a)   (b)

Fig. 1.2. Illustration of the principal and minor nodes
(a) Graph G; (b) trees t_1 and t_2 of G.

t_2, as shown in Fig. 1.2b. Then, in t_1, the principal nodes of edges e_4 and e_5 are nodes 3 and 1, respectively. However, in t_2 nodes 2 and 3 are the principal nodes of e_4 and e_5. It is noted that the principal and minor nodes of e_4 in t_1 are interchanged in t_2 as are the principal and minor nodes of e_5.

Definition 14. The loop matrix of an oriented graph with b edges, $\mathbf{B}_l = [b_{ij}]$, is the matrix in which each row corresponds to each loop for all possible loops; each column corresponds to an edge for all edges of the graph:

$b_{ij} = 1$,   if edge j is included in ith loop and the orientation of the edge coincides with the orientation of loop i;

$b_{ij} = -1$,   if edge j is included in ith loop but the orientation of the edge and the loop are reverse with each other;

$b_{ij} = 0$,   if edge j is not included in ith loop.

Note here that the orientation of a loop is chosen arbitrarily, whether it is clockwise or counterclockwise.

Definition 15. A basic loop matrix (or fundamental loop matrix, or c-circuit matrix) with respect to a tree of a connected graph $\mathbf{B}_f$ is defined to be a submatrix of the loop matrix of the graph which contains only basic loops of the graph. That is, each row of $\mathbf{B}_f$ corresponds to each basic loop and each column to each edge. The orientation of each basic loop is determined by the orientation of each chord by which the loop is formed. This rectangular representation is called a "tie set schedule" by Guillemin.[11]

As an illustration of loop matrices, let us consider a graph given in Fig. 1.3a. It is clear that there exist three loops in the graph, and

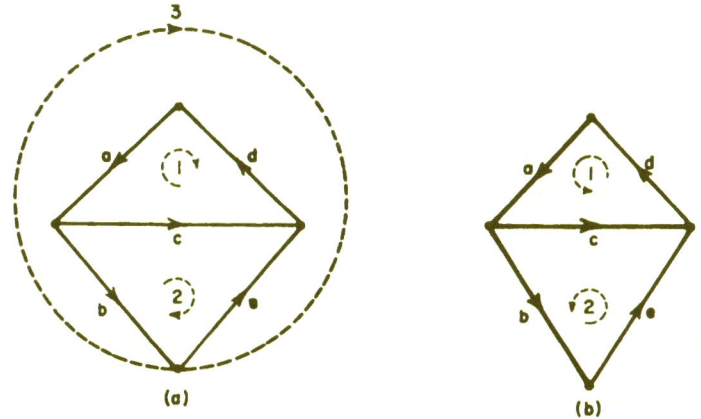

Fig. 1.3. *(a) All possible loops and (b) only basic loops of a graph*

we assign them the orientation clockwise as indicated in Fig. 1.3a. Then, the loop matrix $\mathbf{B}_l$ corresponding to the loops is given by

$$\mathbf{B}_l = \begin{array}{c} \\ \text{loop 1} \\ \text{loop 2} \\ \text{loop 3} \end{array} \begin{array}{ccccc} a & b & c & d & e \\ \left[\begin{array}{ccccc} -1 & 0 & -1 & -1 & 0 \\ 0 & -1 & 1 & 0 & -1 \\ -1 & -1 & 0 & -1 & -1 \end{array} \right]. \end{array} \qquad (1.5)$$

If we choose a tree of the graph, consisting of edges a, b, and c, and if each basic loop is formed by chords d and e as indicated in Fig. 1.3b, then the basic loop matrix corresponding to the loops defined, $\mathbf{B}_f$, is obtained by

$$\mathbf{B}_f = \begin{array}{c} \\ \text{loop 1} \\ \text{loop 2} \end{array} \begin{array}{ccccc} a & b & c & d & e \\ \left[\begin{array}{ccccc} 1 & 0 & 1 & 1 & 0 \\ 0 & 1 & -1 & 0 & 1 \end{array} \right]. \end{array} \qquad (1.6)$$

Now, let us rearrange the columns of $\mathbf{B}_f$ so that the columns corresponding to chords d and e are moved into the left side of the matrix and the columns corresponding to branches a, b, and c into the right side. Then we have

$$\begin{array}{ccccc} d & e & a & b & c \end{array}$$

$$\mathbf{B}_f = \begin{bmatrix} 1 & 0 & 1 & 0 & 1 \\ 0 & 1 & 0 & 1 & -1 \end{bmatrix} = [\mathbf{U} \quad \mathbf{B}_{f_{11}}], \qquad (1.7)$$

where

$$\mathbf{U} = \begin{bmatrix} 1 & 0 \\ 0 & 1 \end{bmatrix} \qquad \mathbf{B}_{f_{11}} = \begin{bmatrix} 1 & 0 & 1 \\ 0 & 1 & -1 \end{bmatrix}. \qquad (1.8)$$

In general, for a connected graph of n nodes and b edges a basic loop matrix of the graph with respect to a tree of the graph $\mathbf{B}_f$ is always decomposed into two submatrices such as

$$\mathbf{B}_f = [\mathbf{U}_2 \quad \mathbf{B}_{f_{11}}]. \qquad (1.9)$$

Here $\mathbf{U}_2$ is a unit matrix of order $(b-n+1)$ in which each column corresponds to a chord of a tree and each column of $\mathbf{B}_{f_{11}}$ to a branch of the tree. This is clear because each basic loop is defined by exactly one chord of a tree. Thus, the following property is evident.

Property 4. A basic loop matrix of a connected graph with n nodes and b edges, $\mathbf{B}_f$, has the following properties:

It is always decomposable into a "basic form" $\mathbf{B}_f = [\mathbf{U}_2 \quad \mathbf{B}_{f_{11}}]$ where $\mathbf{U}_2$ is a unit matrix of order $(b-n+1)$;

The rank of $\mathbf{B}_f$, i.e., the number of independent basic loops, is $(b-n+1)$;

$\mathbf{B}_f$ is a unimodular matrix.

In order to obtain a simple proof for the last property, it may be convenient for us to rely upon some physical aspects of an electrical network corresponding to a graph. It is therefore given in the next chapter (see also References 10 and 12).

Property 5. Let $\mathbf{A}$ be the incidence matrix of a graph of n nodes and b edges. We also denote a loop matrix of a graph which contains all possible loops or a part of them by $\mathbf{B}$. Then, if the columns of both $\mathbf{A}$ and $\mathbf{B}$ are arranged in the same edge order, we have:

$\mathbf{AB}^t = \mathbf{O}$ and $\mathbf{BA}^t = \mathbf{O}$;

If a matrix $\mathbf{Q}$, of order k by b, whose elements are 0, 1, or -1, satisfies that $\mathbf{Q}^t\mathbf{A} = \mathbf{AQ}^t = \mathbf{O}$, where $\mathbf{A}$ is the incidence matrix of a connected graph of n nodes and b edges, then each row of $\mathbf{Q}$ corresponds to a loop or an edge-disjoint union of loops, where the superscript t indicates the transpose of a matrix. For nonoriented graphs, the property is true in modulo 2 arithmetic (mod 2).

Proof: Let us prove the first property. Take ith row of $\mathbf{A}$ and mth row of $\mathbf{B}$ (that is, mth column of $\mathbf{B}^t$). The ith row of $\mathbf{A}$ and mth row of $\mathbf{B}$ will contain nonzero elements if and only if an edge of a graph is incident at node i and also included in loop m. Otherwise, the product is zero. If node i is in loop m, then, by the definition of a loop, there should be two edges incident at node i and both of them are included in loop m. Therefore, the product of ith row of $\mathbf{A}$ and mth column of $\mathbf{B}^t$ will contain one 1 and one -1, so that their sum is zero.

We now prove the second property. That is, we assume that a matrix $\mathbf{Q}$ satisfies the condition that $\mathbf{QA}^t = \mathbf{O}$. Let us choose ith row of $\mathbf{Q}$ which contains nonzero elements in the first, second, ..., kth columns. Then, if edge b_1 corresponding to the first column of the ith row of $\mathbf{Q}$ is incident at a node, say node α, there should exist another edge, say b_2, incident at node α, because the product of the ith row of $\mathbf{Q}$ and the αth column of $\mathbf{A}$ must produce a zero. Edge b_2 must be incident at another node, say node β. Then, there should exist an edge, say edge b_3, also incident at node β. If we continue this process of reasoning until we have exhausted all nodes of the graph, then it is clear that the set of the edges corresponding to the nonzero entries in the ith row of $\mathbf{Q}$ must correspond to a loop or edge-disjoint union of loops in order to satisfy our assumption. This proves the second property.

Since Property 5 is of primary importance in the study of network

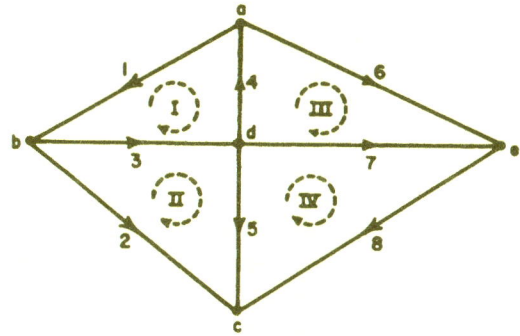

Fig. 1.4. A graph for the illustration of Property 5

topology, it needs further investigation. Let us consider a graph given in Fig. 1.4, with four loops and their orientations chosen arbitrarily. Then, the incidence matrix **A** and the loop matrix of the four chosen loops **B** are found to be

$$
\mathbf{A} = \begin{array}{c} \\ a \\ b \\ c \\ d \\ e \end{array}
\begin{array}{cccccccc}
1 & 2 & 3 & 4 & 5 & 6 & 7 & 8 \\
\end{array}
\left[\begin{array}{rrrrrrrr}
1 & 0 & 0 & -1 & 0 & 1 & 0 & 0 \\
-1 & 1 & 1 & 0 & 0 & 0 & 0 & 0 \\
0 & -1 & 0 & 0 & -1 & 0 & 0 & -1 \\
0 & 0 & -1 & 1 & 1 & 0 & 1 & 0 \\
0 & 0 & 0 & 0 & 0 & -1 & -1 & 1
\end{array}\right] \quad (1.10a)
$$

$$
\mathbf{B} = \begin{array}{c} \text{I} \\ \text{II} \\ \text{III} \\ \text{IV} \end{array}
\left[\begin{array}{rrrrrrrr}
-1 & 0 & -1 & -1 & 0 & 0 & 0 & 0 \\
0 & -1 & 1 & 0 & 1 & 0 & 0 & 0 \\
0 & 0 & 0 & 1 & 0 & 1 & -1 & 0 \\
0 & 0 & 0 & 0 & -1 & 0 & 1 & 1
\end{array}\right] \quad (1.10b)
$$

Now let us consider the second, third, and fifth columns of **A**. We multiply $(+1)$ by all the elements of the first column, and (-1) by the elements of both the second and fifth columns, and add the columns for each row. The resulting column then consists of all zeros. That is, the three columns are linearly dependent, because the edges corresponding to the columns form a loop.

It is thus clear that a submatrix of the incidence matrix of a graph is singular if the set of edges corresponding to the columns of the submatrix contains a loop or loops. Bearing this in mind, let us study the following property.

Property 6. A square submatrix of order $(n-1)$ of the incidence matrix **A** of a connected graph of n nodes is nonsingular if and only if the edges corresponding to the columns of the submatrix constitute a tree of the graph.

Proof: We know that there always exists a nonsingular submatrix of order $(n-1)$ since the rank of **A** is $(n-1)$ due to Property 2. We also know that a submatrix of order $(n-1)$ of **A** is nonsingular if the columns of the submatrix correspond to a set of branches of a tree, because the columns are linearly independent (due to Property 5).

For the proof of the sufficient part, we should show that the edges corresponding to the columns of any nonsingular submatrix of order

$(n-1)$ of **A** constitute a tree. Let us consider a square submatrix of **A** of order $(n-1)$ which is nonsingular, i.e., all the columns of the submatrix are linearly independent. Then, due to Property 5, the set of the edges corresponding to the columns of the submatrix contain no loop. Since the set of $(n-1)$ edges forms a subgraph of a connected graph but contains no loop, the edges therefore constitute a tree (because of Property 1).

The proof of the property can also be obtained without using Property 5 at all. That is, first show that the sum of any k rows of the incidence matrix of a connected graph of b edges and n nodes, $k < n$, produces a row containing at least one nonzero element. Using this fact, we rearrange the rows and columns and apply the elementary transformation so that one can always reduce the incidence matrix to a triangular matrix with the last row of all zeros (see Reference 7).

Property 6 is essential to the topological analysis as well as synthesis of the various types of networks. We shall study an interesting application of Property 6.

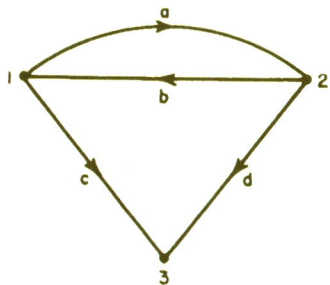

Fig. 1.5. A connected graph

Let us assume that we are given a connected graph in Fig. 1.5. Then **A**, the incidence matrix of the graph of Fig. 1.5, is found to be

$$\mathbf{A} = \begin{array}{c} \\ 1 \\ 2 \\ 3 \end{array} \begin{array}{cccc} a & b & c & d \\ \left[\begin{array}{cccc} 1 & -1 & 1 & 0 \\ -1 & 1 & 0 & 1 \\ 0 & 0 & -1 & -1 \end{array} \right] \end{array}. \tag{1.11}$$

However, we would not need all three rows of **A** of Eq. (1.11) to completely characterize the graph. Since each column of the incidence matrix contains exactly one -1 and one 1 we can construct in general the whole matrix of n rows by knowing only $(n-1)$ rows. This was actually implied by Property 2. This submatrix is denoted by A_1. Thus, for the graph given in Fig. 1.5, the incidence matrix which we will consider would be

$$A_1 = \begin{array}{c} 1 \\ 2 \end{array} \begin{bmatrix} \overset{a}{1} & \overset{b}{-1} & \overset{c}{1} & \overset{d}{0} \\ -1 & 1 & 0 & 1 \end{bmatrix}, \qquad (1.12)$$

which is obtained from A of Eq. (1.11) by deleting the last row.

Now, applying the second condition listed in Property 2 and Property 6, we know that the determinant of every submatrix of A_1 is either 1 or -1 if the columns of each submatrix correspond to a set of edges which form a tree. This property of the incidence matrix of a connected graph actually enables us to compute the number of all possible trees of the graph.

Let us evaluate the determinant of the product of A_1 and $A_1{}^t$ by using the Binet–Cauchy theorem.† That is,

$|A_1A_1{}^t| = \Sigma$ product of every corresponding pair of major determinants of A_1 and $A_1{}^t$, $\qquad (1.13)$

where the summation is over all products of the corresponding majors.

Referring to the incidence matrix of the graph of Fig. 1.5 we shall evaluate the determinant of $(A_1A_1{}^t)$ using the Binet–Cauchy theorem:

$$|A_1A_1{}^t| = \begin{vmatrix} \begin{bmatrix} \overset{a}{1} & \overset{b}{-1} & \overset{c}{1} & \overset{d}{0} \\ -1 & 1 & 0 & 1 \end{bmatrix} \begin{bmatrix} 1 & -1 \\ -1 & 1 \\ 1 & 0 \\ 0 & 1 \end{bmatrix} \begin{matrix} a \\ b \\ c \\ d \end{matrix} \end{vmatrix}$$

---

† Binet–Cauchy theorem: Let matrices C and D be of order (p,q) and (q,p), respectively, and $q \geqslant p$. Then

$\det(CD) =$ sum of the products of the corresponding pairs of major determinants of C and D,

where a major is a determinant of the highest order formed in a matrix. Or, in terms of the corresponding compound matrices, we have

$$(CD)^{(k)} = C^{(k)}D^{(k)},$$

where k indicates the order of a compound and the elements of a kth order compound of a matrix are subdeterminants of order k of the matrix arranged in lexicographical order. See Reference 13.

$$
= \begin{vmatrix} 1 & -1 \\ -1 & 1 \end{vmatrix} \begin{vmatrix} 1 & 1 \\ -1 & 1 \end{vmatrix} \begin{matrix} a \\ b \end{matrix} + \begin{vmatrix} 1 & 1 \\ -1 & 0 \end{vmatrix} \begin{vmatrix} 1 & -1 \\ 1 & 0 \end{vmatrix} \begin{matrix} a \\ c \end{matrix} + \begin{vmatrix} 1 & 0 \\ -1 & 1 \end{vmatrix} \begin{vmatrix} 1 & -1 \\ 0 & -1 \end{vmatrix} \begin{matrix} a \\ d \end{matrix}
$$

$$
+ \begin{vmatrix} -1 & 1 \\ 1 & 0 \end{vmatrix} \begin{vmatrix} -1 & 1 \\ 1 & 0 \end{vmatrix} \begin{matrix} b \\ c \end{matrix} + \begin{vmatrix} -1 & 0 \\ 1 & 1 \end{vmatrix} \begin{vmatrix} -1 & 1 \\ 0 & 1 \end{vmatrix} \begin{matrix} b \\ d \end{matrix} + \begin{vmatrix} 1 & 0 \\ 0 & 1 \end{vmatrix} \begin{vmatrix} 1 & 0 \\ 0 & 1 \end{vmatrix} \begin{matrix} c \\ d \end{matrix}
$$

$$
= 0+1+1+1+1+1 = 5. \tag{1.14}
$$

That is, the total number of trees of the graph of Fig. (1.5) is five.

The product of the first corresponding pair of major determinants consists of two subdeterminants; one of which is formed from A_1 taking the columns corresponding to edges a and b and the other from A_1^t with the rows corresponding to edges a and b. It is, therefore, clear that these determinants both have the same value. Since edges a and b form a loop the majors should both vanish, according to Property 5. For the other product terms in Eq. (1.14) each set of edges corresponding to the columns of a major of A_1 (or the rows of a major of A_1^t) forms a tree. Therefore, both majors of each pair in the last five products will have the same values of 1 or -1. Thus each product of the corresponding majors in the evaluation of $(A_1A_1^t)$ will take the value of 0 if the edges corresponding to the columns (or rows) of the major contain a loop(s), or unconnected and of 1 if the set of edges constitute a tree. Thus, the determinant of $(A_1A_1^t)$ is equal to the number of total trees in a connected graph. Hence, the following property.[11]

Property 7. The determinant of $(A_1A_1^t)$ is equal to the number of total trees of a connected graph. A_1 is the incidence matrix of the graph with one row deleted.

Let us denote the incidence matrix of a connected and oriented graph G by A and a square submatrix of the incidence matrix corresponding to a tree of the graph or called a "tree matrix"† by A_t. The matrix $A_{t_{ij}}$ is obtained by deleting row i and column j from A_t, and $|A_t|_{ij}$ is the minor of the determinant of A_t of (i, j)-position.

Property 8. The minor of the determinant of a tree matrix A_t has the following property (Frisch [20a]):

$|A_t|_{ij} =$ nonzero, if ith node is the principal node of jth edge in t;
$|A_t|_{ij} =$ zero, if ith node is the minor node of jth edge in t.

---

† The tree matrix here is formed on a single tree. On the other hand, a tree matrix proposed by Hakimi[14] contains all possible trees.

Proof: Consider an edge j of a tree t of a connected graph. Denote the principal and minor nodes of edge j by nodes i and k, respectively. Since a tree is a loopless graph or subgraph of a connected graph, it is possible to cut tree t by removing edge j into two disjoint subgraphs t' and t'' such that t' contains the reference node of t, as shown in

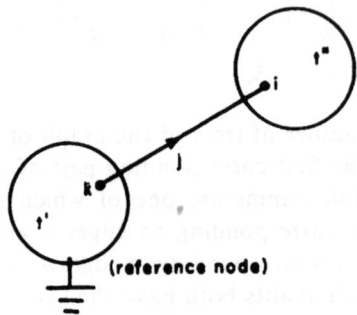

Fig. 1.6. Division of tree t into disjoint subgraphs t' and t''

Fig. 1.6. Now expand the determinant of $A_t = [a_{pq}]$ about column j. Then, one gets

$$|A_t| = (-1)^{i+j}a_{ij}|A_t|_{ij} + (-1)^{k+j}a_{kj}|A_t|_{kj} \neq 0, \qquad (1.15)$$

where a_{ij} and a_{kj} are nonzero elements of A_t.

The submatrix $A_{t_{ij}}$ is the incidence matrix of the subgraph of t obtained by removing edge j and identifying node i with node k in t as shown in Fig. 1.7a, while $A_{t_{jk}}$ represents the subgraph of t obtained

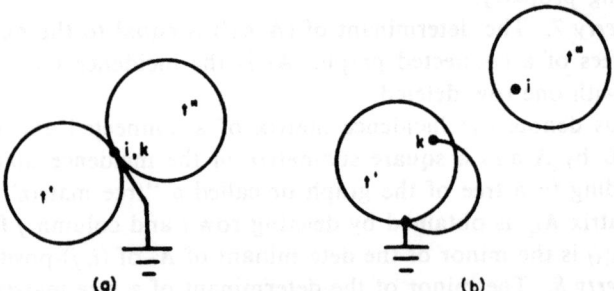

Fig. 1.7. (a) Subgraph corresponding to $A_{t_{ij}}$; (b) subgraph corresponding to $A_{t_{kj}}$

by removing edge j and identifying node k with the reference node as shown in Fig. 1.7b. Since the subgraph corresponding to $A_{t_{kj}}$

contains a loop and is not connected, $|A_t|_{kj} = 0$ due to Property 6. Thus, from Eq. (1.15), $|A_t|_{ij} \neq 0$. It is also clear that the subgraph corresponding to $|A_t|_{ij}$ is a connected subgraph of t with no loop. This completes the proof of the property.

Property 9. The incidence matrix of a connected graph A of b edges and n nodes can always be partitioned as:

$$A = \begin{bmatrix} A_{11} & A_{12} \\ A_{21} & A_{22} \end{bmatrix},$$

where the columns of A_{12} correspond to a set of branches of a tree, and so is a nonsingular matrix of order $(n-1)$. Then, the basic loop matrix of the graph, B_f, with respect to the tree is given by

$$B_f = [U_2 \quad -(A_{12}^{-1}A_{11})^t],$$

where U_2 is a unit matrix of order $(b-n+1)$ and A_{12}^{-1} is the inverse of A_{12}.[7]

Proof: Let us denote the loop matrix which includes all possible loops of the graph by B_l. Next permute the rows of B_l, if necessary, so that the matrix is written as:

$$B_l = \begin{bmatrix} B_f \\ B_{l_{21}} \end{bmatrix}. \tag{1.16}$$

In other words, we arrange the rows of B_l so that the first $(b-n+1)$ rows correspond to the basic loops with respect to a tree of the graph. We now use Property 4 and rewrite B_f as:

$$B_f = [U_2 \quad B_{f_{12}}]. \tag{1.17a}$$

Again partitioning $B_{l_{21}}$ into two submatrices:

$$B_{l_{21}} = [\alpha_{11} \quad \alpha_{12}]. \tag{1.17b}$$

Thus, we will have

$$B_l = \begin{bmatrix} U_2 & B_{f_{12}} \\ \alpha_{11} & \alpha_{12} \end{bmatrix}. \tag{1.18}$$

We then apply Property 5 so that

$$\begin{aligned} AB_l^t &= \begin{bmatrix} A_{11} & A_{12} \\ A_{21} & A_{22} \end{bmatrix} \begin{bmatrix} U_2 & \alpha_{11}^t \\ B_{f_{12}}^t & \alpha_{12}^t \end{bmatrix} \\ &= \begin{bmatrix} A_{11} + A_{12}B_{f_{12}}^t & A_{11}\alpha_{11}^t + A_{22}\alpha_{12}^t \\ A_{21} + A_{12}B_{f_{12}}^t & A_{21}\alpha_{11}^t + A_{22}\alpha_{12}^t \end{bmatrix} = O. \end{aligned} \tag{1.19}$$

Thus,

$$\mathbf{A}_{11} + \mathbf{A}_{12}\mathbf{B}_{f_{12}}{}^t = \mathbf{O}$$

and

$$\mathbf{B}_{f_{12}}{}^t = -\mathbf{A}_{12}{}^{-1}\mathbf{A}_{11}. \qquad (1.20)$$

If we substitute Eq. (1.20) to Eq. (1.17a) we have

$$\mathbf{B}_f = [\mathbf{U}_2 \quad -(\mathbf{A}_{12}{}^{-1}\mathbf{A}_{11})^t]. \qquad (1.21)$$

Hence, the property.

Property 10. The rank of the loop matrix $\mathbf{B}_l$ of a connected graph, with b edges and n nodes, i.e., the number of independent loops of the graph, is $b-n+1$. If a graph has n nodes and b edges but consists of P isolated subgraphs it has $(b-n+P)$ independent loops.[7]

Proof: Instead of proving the property directly, we shall find the upper and lower bounds of the rank of $\mathbf{B}_l$. That is, the rank of $\mathbf{B}_l$ is at least $(b-n+1)$ but cannot exceed $(b-n+1)$.

Since a basic loop matrix $\mathbf{B}_f$ with respect to a tree of the graph is a submatrix of $\mathbf{B}_l$, we know that the rank of $\mathbf{B}_l \geqslant b-n+1$. If we apply Sylvester's law of nullity[†] to Property 5, we have

$$\text{(the rank of } \mathbf{A} = n-1) + \text{(the rank of } \mathbf{B}_l{}^t) \leqslant b \qquad (1.22a)$$

or

$$\text{the rank of } \mathbf{B}_l \leqslant b-n+1, \qquad (1.22b)$$

since the rank of a matrix is identical to that of the transpose of the matrix. Thus,

$$\text{the rank of } \mathbf{B}_l = b-n+1. \qquad (1.23)$$

If a graph under consideration consists of P separated subgraphs, then the foregoing discussion is valid for each subgraph. Therefore, for the whole graph we have $(b-n+P)$ independent loops.

Property 11. A square submatrix of order $(b-n+1)$ of the loop matrix $\mathbf{B}_l$ of a connected graph with b edges and n nodes is non-singular, if and only if the edges corresponding to the columns of the submatrix constitute the set of all chords of a tree or a co-tree of the graph.[32]

Proof: First we know that there should always exist a nonsingular submatrix of order $(b-n+1)$ of $\mathbf{B}_l$ because the rank of $\mathbf{B}_l$ is $(b-n+1)$ due to Property 11. Furthermore, it has already been found that

---

† Sylvester's law of nullity: Let matrices $\mathbf{P}$ and $\mathbf{Q}$ be of order $(n \times b)$ and $(b \times m)$, respectively. We also let $\mathbf{PQ} = \mathbf{C}$. Then, the rank of $\mathbf{C} \geqslant$ (the rank of $\mathbf{P}$) + (the rank of $\mathbf{Q}$) $- b$.

the submatrix of order $(b-n+1)$ of a basic loop matrix is non-singular, actually a unit matrix, if the edges corresponding to the columns of the submatrix constitute the set of chords of a tree.

Next, let us arrange the columns of $\mathbf{B}_l$ so that the leading square submatrix of order $(b-n+1)$ is nonsingular. We therefore have

$$\mathbf{B}_l = \begin{bmatrix} \mathbf{B}_{l_{11}} & \mathbf{B}_{l_{12}} \\ \mathbf{B}_{l_{21}} & \mathbf{B}_{l_{22}} \end{bmatrix}, \tag{1.24}$$

where $\mathbf{B}_{l_{11}}$ is a nonsingular submatrix of order $(b-n+1)$. Suppose there exists a loop formed solely by the edges corresponding to some of the columns of $\mathbf{B}_{l_{12}}$. Then, this would produce a row of zeros in $\mathbf{B}_{l_{11}}$. However, $\mathbf{B}_{l_{11}}$ is nonsingular; this is therefore impossible. Now consider $\mathbf{B}_{l_{22}}$, and assume again that the edges corresponding to some of the columns of the submatrix form a loop. Then, this will produce a row of zeros in $\mathbf{B}_{l_{12}}$. Consequently the rank of $\mathbf{B}_l$ must be greater than $(b-n+1)$ and equal to at least $(b-n+2)$. This contradicts Property 10. It is therefore clear that the edges corresponding to the last $(n-1)$ columns of $\mathbf{B}_l$ of Eq. (1.24) must constitute a tree. Thus, the edges corresponding to the columns of $\mathbf{B}_{l_{11}}$ must constitute a set of chords of a tree.

Property 12. Let $\mathbf{A}_1$ and $\mathbf{B}$ be the incidence and loop matrices of a connected graph with n nodes and b edges so that $\mathbf{A}_1$ contains $(n-1)$ rows and $\mathbf{B}$ a set of independent $(b-n+1)$ loops. If the columns of both $\mathbf{A}_1$ and $\mathbf{B}$ are arranged in the same order so that

$$\mathbf{A}_1 = [\mathbf{A}_{11} \quad \mathbf{A}_{12}]$$
$$\mathbf{B} = [\mathbf{B}_{11} \quad \mathbf{B}_{12}],$$

where the first $(b-n+1)$ columns of both matrices correspond to the co-tree of a tree and the last $(n-1)$ columns to a set of the branches. Then,

$$\mathbf{A}_1 = [-\mathbf{A}_{12}\mathbf{B}_{12}{}^t(\mathbf{B}_{11}{}^t)^{-1} \quad \mathbf{A}_{12}]$$
$$= \mathbf{A}_{12}[-\mathbf{B}_{12}(\mathbf{B}_{11}{}^t)^{-1} \quad \mathbf{U}_1] \tag{1.25a}$$

$$\mathbf{B} = [\mathbf{B}_{11} \quad -\mathbf{B}_{11}\mathbf{A}_{11}{}^t(\mathbf{A}_{12}{}^t)^{-1}]$$
$$= \mathbf{B}_{11}[\mathbf{U}_2 \quad -\mathbf{A}_{11}{}^t(\mathbf{A}_{12}{}^t)^{-1}], \tag{1.25b}$$

where $\mathbf{U}_1$ and $\mathbf{U}_2$ are identity matrices of order $(n-1)$ and $(b-n+1)$, respectively.[7]

Proof: Due to Property 5, we have

$$\mathbf{BA_1}^t = [\mathbf{B}_{11} \quad \mathbf{B}_{12}] \begin{bmatrix} \mathbf{A}_{11}^t \\ \mathbf{A}_{12}^t \end{bmatrix}$$

$$= \mathbf{B}_{11}\mathbf{A}_{11}^t + \mathbf{B}_{12}\mathbf{A}_{12}^t = \mathbf{O} \qquad (1.26a)$$

and

$$\mathbf{A_1B}^t = [\mathbf{A}_{11} \quad \mathbf{A}_{12}] \begin{bmatrix} \mathbf{B}_{11}^t \\ \mathbf{B}_{12}^t \end{bmatrix}$$

$$= \mathbf{A}_{11}\mathbf{B}_{11}^t + \mathbf{A}_{12}\mathbf{B}_{12}^t = \mathbf{O}. \qquad (1.26b)$$

Thus, from Eq. (1.26a)

$$\mathbf{B}_{12} = -\mathbf{B}_{11}\mathbf{A}_{11}^t(\mathbf{A}_{12}^t)^{-1}, \qquad (1.27a)$$

and from Eq. (1.26b)

$$\mathbf{A}_{11} = -\mathbf{A}_{12}\mathbf{B}_{12}^t(\mathbf{B}_{11}^t)^{-1}, \qquad (1.27b)$$

where $\mathbf{A}_{12}$ and $\mathbf{B}_{11}$ are nonsingular submatrices due to Properties 6 and 11. Hence, the property.

From Properties 9 and 12, one may deduce the following property.

Property 13. Let $\mathbf{B}_f$ be a basic loop matrix of a connected graph with n nodes and b edges. We also denote a loop matrix containing exactly $(b-n+1)$ independent loops of the same graph by $\mathbf{B}$. If we arrange the columns of both matrices such that the first $(b-n+1)$ columns correspond to the co-tree of a tree, then

$$\mathbf{B}_f = [\mathbf{U}_2 \quad \mathbf{B}_{f_{12}}] \qquad (1.28a)$$

$$\mathbf{B} = [\mathbf{B}_{11} \quad \mathbf{B}_{12}]. \qquad (1.28b)$$

Then $\mathbf{B}_f$ and $\mathbf{B}$ are related to each other by

$$\mathbf{B} = \mathbf{B}_{11}\mathbf{B}_f \qquad (1.29a)$$

$$\mathbf{B}_f = \mathbf{B}_{11}^{-1}\mathbf{B} \qquad (1.29b)$$

$$\mathbf{B}_{f_{12}} = \mathbf{B}_{11}^{-1}\mathbf{B}_{12}. \qquad (1.29c)$$

Definition 16. A "simple cut-set" (or a prime cut-set) of a graph is a set of edges such that the removal of the set of edges from the graph reduces the rank of the graph by one and no proper subset of the set of edges has the same property, and a "cut-set" implies a simple cut-set or an edge-disjoint union of simple cut-sets.[†]

From the definition of a cut-set it is clear that the set of edges incident at a node in a connected, *nonseparable* graph is a simple

[†] As an extension of the concept of a cut-set, we will define later a "seg" and a "cut." See Part III and Part V.

cut-set, and if the graph is separable, the set of edges incident at a cut-node is a cut-set, i.e., an edge-disjoint union of simple cut-sets, since every nonloop edge is again a simple cut-set.

Let us consider a connected and separable graph as shown in Fig. 1.8. In Fig. 1.8a, let us take the set of edges incident at node 2, edges (a, b, d, e), and remove the edges from the graph which will yield the subgraph of Fig. 1.8b. The rank of the subgraph is one, while the

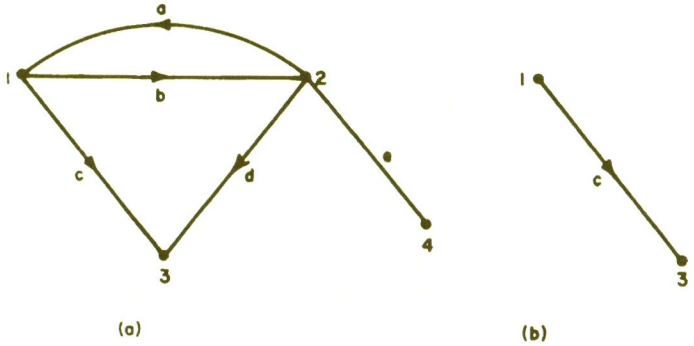

(a)   (b)

Fig. 1.8. Illustration of a cut-set of a separable graph
(a) A separable graph; (b) edges a, b, d, and e removed.

rank of the original graph is three. The rank of the graph, therefore, is reduced by two by removing the set of edges. This violates the first property of the definition of a simple cut-set. Furthermore, it violates also the second property of the definition since there exists a proper subset of the set of edges, namely, edges (a, b, c, d) or edge e, which satisfies the first property.

Definition 17. The "basic set of cut-sets" (or basic cut-sets†) is the $(n-1)$ simple cut-sets in a connected graph of n nodes which are formed by *each branch* of a tree and some or all chords included in the basic loops (with respect to the same tree) containing the branch. Sometimes a basic cut-set is called a "fundamental cut-set" or an "*f*-cut-set."

Let a connected graph G be shown in Fig. 1.9a. Then, choose a tree as indicated in Fig. 1.9b, and form a basic cut-set with respect to edge b_2, that is, the set of edges (b_1, b_2, b_3). The removal of this basic

---

† The word "basic cut-set" is sometimes used to imply a simple cut-set with respect to a specified pair of nodes i and j such that the removal of the set of edges will separate the graph into two disjoint subgraphs, one of which contains node i and the other of which contains node j.

cut-set from G separates G into two subgraphs G_1 and G_2 as shown in Fig. 1.9c. The graph G has the rank of five, and the rank of the subgraph consisting of G_1 and G_2 (shown in Fig. 1.9c) is $4 = 2+2 = =$ (the rank of G_1)+(the rank of G_2). The removal of the set of edges (b_1, b_2, b_3), which is a basic cut-set, reduces the rank of G by one, thereby satisfying the first property of Definition 16. If we take one of the edges b_1, b_2, and b_3 back to the subgraph of Fig. 1.9c, then the resultant graph becomes connected and its rank is the same as the rank of G, that is, five. This shows that the basic cut-set (b_1, b_2, b_3) satisfies the second property of the definition of a simple cut-set.

If we wish to assign an orientation to a cut-set it can be assigned arbitrarily as it was for a loop. Thus we have the following definition.

Definition 18. The simple cut-set matrix $\mathbf{C} = [c_{ij}]$, of a connected graph, has one row for each simple cut-set including all possible simple cut-sets and one column for each edge in the graph, such that:

$c_{ij} =$ 1, if edge j is in simple cut-set i and their orientations coincide;

$c_{ij} = -1$, if edge j is in simple cut-set i and their orientations are opposite;

$c_{ij} =$ 0, if edge j is not in simple cut-set i.

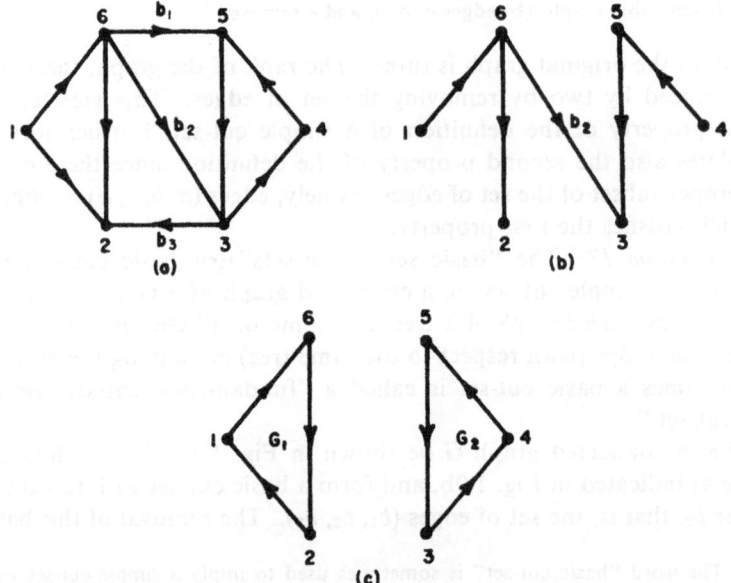

Fig. 1.9. Illustration of a basic cut-set
(a) A graph G; (b) a tree of G; (c) removal of a basic cut-set.

When we refer to a "cut-set matrix," each row of the matrix may include a simple cut-set or an edge-disjoint union of simple cut-sets.

A "basic cut-set matrix" C_f is a submatrix of a cut-set matrix whose rows are defined only for the basic cut-sets with respect to a tree, and the orientation of each basic cut-set is defined by the branch contained in the basic cut-set.

As the readers may have already realized at this point, each row of the incidence matrix corresponds to a simple cut-set or a cut-set. It is therefore sometimes referred to as an "incident cut-set" or "node cut-set." We thus have the following property as the generalization of Property 5.

Property 14. If all the columns are arranged in the same order, we have, for loop and cut-set matrices of a connected graph,†[32]

$$B_l C = O, \; CB_l{}^t = O$$
$$B_f C_f{}^t = O, \; C_f B_f{}^t = O.$$

Property 15. A basic cut-set matrix C_f of a connected graph with b edges and n nodes has the following properties:

C_f is a unimodular matrix;

C_f is always partitioned to a basic form $C_f = [C_{f_{11}} \quad U_1]$, U_1 is a unit matrix of order $(n-1)$ and the columns of the unit matrix correspond to branches of a tree;

The rank of $C_f = n-1$;

$C_f = [-B_{f_{11}} \quad U_1]$ and $B_{f_{11}}$ is a submatrix of the basic loop matrix with respect to the tree, such that $B_f = [U_2 \quad B_{f_{11}}]$ and U_2 is a unit matrix of order $(b-n+1)$;

$C_f = [A_{12}{}^{-1}A_{11} \quad U_1]$ and A_{12} and A_{11} are the submatrices of the incidence matrix A_1 such that $A_1 = [A_{11} \quad A_{12}]$ and the columns of A_{12} correspond to all the branches of the same tree which was considered in forming the basic cut-set matrix.[7, 10]

Proof: The second property is obvious from the definition of a basic cut-set, and the third one follows directly from the second. The fourth one can be easily derived from Property 14. The last property again follows from the fourth one and Property 9. The first property follows from the fourth; from the fact that B_f is a unimodular matrix and so is its submatrix $B_{f_{11}}$. An alternate proof of the first property will be found in the next chapter.

† This property is a necessary and sufficient condition for a matrix B to be a loop matrix, provided that C is a cut-set matrix of a connected graph and vice versa (see Reference 15).

Thus far, the relationships between the incidence, cut-set, and loop matrices have been investigated. We shall now study the properties of paths of a connected graph since they play an important role in the application of linear graph theory to the various engineering problems. (See Part IV and Reference 16.)

Definition 19. A "path-isolated subgraph," with respect to the paths between nodes i and j of a connected graph, is a connected subgraph none of whose edges are contained in any path between the nodes i and j. If a connected graph is separable, there always exist some path-isolated subgraphs in the graph with respect to the paths between a particular pair of nodes.

Let us consider a separable graph G as shown in Fig. 1.10a. Since G is separable at node b and c into three subgraphs G_1, G_2, and G_3

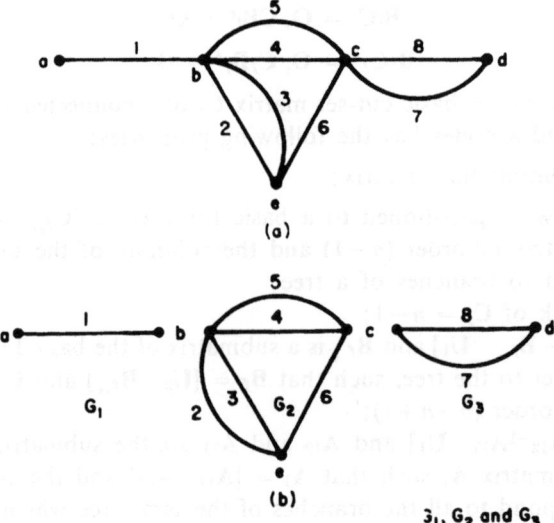

Fig. 1.10. *Illustration of path-isolated subgraphs*
(a) A separable graph G; (b) G is decomposed into three subgraphs G_1, G_2, and G_3.

as shown in Fig. 1.10b, the paths with respect to nodes a and b will not contain any edges in subgraphs G_2 and G_3. Therefore, G_2 and G_3 together become a path-isolated subgraph with respect to nodes a and b. Similarly, the subgraph consisting of G_1 and G_3 is a path-isolated subgraph with respect to the paths between nodes b and c, and G_1 and G_2 together become a path-isolated subgraph of G with respect to the paths between nodes c and d.

Definition 20. A path matrix of a connected and nonoriented graph of b edges and n nodes with respect to all possible paths between nodes α and β, $\mathbf{P}_{\alpha\beta} = [p_{ij}]$ is defined by

$p_{ij} = 1$, if edge j is contained in path i;
$p_{ij} = 0$, otherwise.

The subscripts of the path matrix may be dropped if no possible confusion arises.

An expression of the so-called "configuration matrix" introduced by Gould[15] is actually identical to the concept of a path matrix even though the two expressions were defined differently.

Then, the relationship between the incidence matrix and path matrix of a connected graph is given by the following.

Property 16. If the edges of a nonoriented, connected graph are arranged in the same order for the columns of the incidence matrix $\mathbf{A}$ and of a path matrix $\mathbf{P}_{\alpha\beta}$, then

$$\mathbf{A}\mathbf{P}_{\alpha\beta} = \mathbf{H} = [h_{km}], \qquad \text{mod } 2$$

and $h_{km} = 1$ only for $k = \alpha, \beta$, and $h_{km} = 0$ otherwise (Wing[16]).

Proof: $\mathbf{P}_{\alpha\beta}$ contains all paths between nodes α and β; that is, the terminal nodes of all the paths are nodes α and β; h_{km} is nonzero if an edge is in path m and is incident at node k. Therefore, $h_{\alpha m} = h_{\beta m} = 1$ for all the paths; $m = 1, 2, \ldots$, since every path between nodes α and β terminates at nodes α and β. However, all internal nodes of any path between nodes α and β are of degree two. Thus, $h_{km} = 0$ mod 2 for $k \neq \alpha, \beta$.

Example 3. The incidence matrix $\mathbf{A}$ and the path matrix $\mathbf{P}_{bc}$ of the graph of Fig. 1.10 are found, respectively, by

$$\mathbf{A} = \begin{array}{c} a \\ b \\ c \\ d \\ e \end{array} \begin{array}{cccccccc} 1 & 2 & 3 & 4 & 5 & 6 & 7 & 8 \\ \left[\begin{array}{cccccccc} 1 & 0 & 0 & 0 & 0 & 0 & 0 & 0 \\ 1 & 1 & 1 & 1 & 1 & 0 & 0 & 0 \\ 0 & 0 & 0 & 1 & 1 & 1 & 1 & 1 \\ 0 & 0 & 0 & 0 & 0 & 0 & 1 & 1 \\ 0 & 1 & 1 & 0 & 0 & 1 & 0 & 0 \end{array} \right] \end{array} \qquad (1.30a)$$

$$\mathbf{P}_{bc} = \left[\begin{array}{cccccccc} 0 & 1 & 0 & 0 & 0 & 1 & 0 & 0 \\ 0 & 0 & 1 & 0 & 0 & 1 & 0 & 0 \\ 0 & 0 & 0 & 1 & 0 & 0 & 0 & 0 \\ 0 & 0 & 0 & 0 & 1 & 0 & 0 & 0 \end{array} \right] \qquad (1.30b)$$

and

$$\mathbf{AP}_{bc}{}^t = \begin{bmatrix} 0 & 0 & 0 & 0 \\ 1 & 1 & 1 & 1 \\ 1 & 1 & 1 & 1 \\ 0 & 0 & 0 & 0 \\ 0 & 0 & 0 & 0 \end{bmatrix}. \qquad (1.30c)$$

Property 17. The path matrix $\mathbf{P}_{ij}$ with respect to nodes i and j of a connected but nonoriented graph with b edges and n nodes has the following properties:

Every loop or edge-disjoint of loops in a graph is the ring sum of some two paths between i and j, where the ring sum of two paths p_1 and p_2 is a set of edges which belong to either path p_1 or path p_2 but not to both†;

The rank of $\mathbf{P}_{ij}$ is $(b-n+2)$ if the graph contains no path-isolated subgraphs with respect to nodes i and j;

The rank of $\mathbf{P}_{ij}$ is $(b-n+2-S)$ if the graph contains path-isolated subgraphs and S is the number of independent loops in the path-isolated subgraphs.

Proof: The first property is obvious because the ring sum of any two paths between nodes i and j is a loop if the graph is nonseparable and may be an edge-disjoint union of loops if the graph is separable.

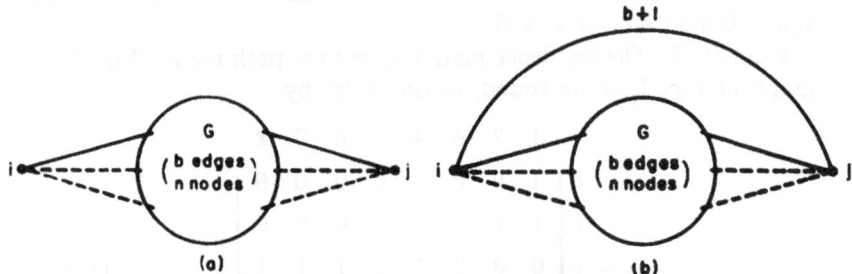

Fig. 1.11. Proof of Property 17

(a) A graph with no path-isolated subgraphs; (b) an edge is introduced between nodes i and j.

To prove the second property, consider all paths of a graph G with respect to nodes i and j as shown in Fig. 1.11a.

Let the path matrix $\mathbf{P}_{ij}$ of G be written as:

† This property was proposed by Ashenhurst for a nonseparable graph.[17]

$$\mathbf{P}_{ij} = \begin{bmatrix} p_{11} & p_{12} \dots p_{1b} \\ p_{21} & p_{22} \dots p_{2b} \\ \cdot & \cdot \quad \cdot \quad \cdot \quad \cdot \\ p_{m1} & p_{m2} \dots p_{mb} \end{bmatrix}, \tag{1.31}$$

where m is the number of all the paths between nodes i and j.

Now, let us introduce an extra edge, say $(b+1)$th edge, between nodes i and j. Then, all the paths between i and j defined by the rows of $\mathbf{P}_{ij}$ become loops, and each of the loops includes the $(b+1)$th edge and also nodes i and j. The loop matrix $\mathbf{B}$ for the loops defined by the $(b+1)$th edge will be

$$\mathbf{B} = \begin{bmatrix} p_{11} \dots p_{1b} & 1 \\ p_{12} \dots p_{2b} & 1 \\ \cdot \quad \cdot \quad \cdot \quad \cdot & \cdot \\ p_{m1} \dots p_{mb} & 1 \end{bmatrix}. \tag{1.32}$$

That is, $\mathbf{B}$ will have the $(b+1)$th column of all ones.

Since there exists no path-isolated subgraph in G, and the ring sum of two distinct paths corresponding to the rows of $\mathbf{B}$ yields a loop or an edge-disjoint union of loops and the path matrix $\mathbf{P}_{ij}$ contains all possible paths between nodes i and j, every loop or every edge-disjoint union of loops of the original graph will therefore be generated by modulo 2 sum of any two rows of $\mathbf{B}$. It is thus clear that the rank of $\mathbf{P}_{ij}$ is $(b-n+2)$ because the rank of $\mathbf{B}$ is $(b+1)-n+1$ due to Property 10. Note here that an additional column of all ones will not alter the rank of the matrix. The third property follows directly from the second. Alternate proofs for the second and third properties are given by Wing by showing an algorithm for finding all the independent paths in a connected graph (see Reference 16).

Thus far, the basic definitions and properties of a linear graph have been studied. However, all the concepts were defined in terms of a single graph, whether a connected graph or one which consisted of several isolated subgraphs. So far no study has been done on the interrelationship among or comparison of more than one graph. On many occasions, however, it will be necessary for us to investigate the relationships between more than one graph, particularly for the applications of the theory of linear graphs to engineering problems. This is because the graphical representations of some physical systems require more than one graph to completely characterize the systems (see Chapter 3). Furthermore, we are always interested in finding a class of physically equivalent systems. (See Parts III, IV, and V.)

In terms of graph theory Whitney[9c,d] has defined some classification of graphs and proposed operational procedures on linear graphs. We shall extend the concepts proposed by Whitney such that they can be applied to the various problems in the area of the analysis of electrical networks and communication systems as well as their synthesis.[18-20]

Let us now define the operations to be performed on a graph G_1 containing b edges and n nodes.

OPERATIONS

1. Break G_1 at a single node into two disjoint subgraphs or join two disjoint subgraphs at a node.

2. Cut G_1 at two nodes into two disjoint subgraphs and turn one of the subgraphs around at the two nodes.

3. Remove one or more edges in G_1 and put them back between any two nodes in G_1 or change the orientations of any member edges in G. Operation 3 is proposed[20a] as an addition to operations 1 and 2 defined by Whitney.

If we let the graph resulting from any of the defined operations, or combination of the three operations, be denoted by G_2, then Table 1 gives the relationships between G_1 and G_2.

Table 1. Operations and classification of graphs

Operations on G_1	Relationships between G_1 and G_2
none	isomorphic
1	1–isomorphic
1 and/or 2	2–isomorphic
1 and/or 2, and/or 3	2–semi-isomorphic

The following examples illustrate these relationships between G_1 and G_2.

Example 4. One-isomorphic graphs: If we are given a graph G_1 in Fig. 1.12a, cutting G_1 at node c results in G_2 as shown in Fig. 1.12b. Then G_1 and G_2 are 1-isomorphic with each other.

Two-isomorphic graphs: We are given a graph G_1 in Fig. 1.13a. Break G_1 at nodes b and e into subgraphs G'_1 and G''_1, G'_1 containing edges 1 and 5, and G''_1, edges 2, 3, 4, and 6. Next turn G'_1 around nodes b and e to obtain G_2 as shown in Fig. 1.13b. Then G_1 and G_2 are said to be 2-isomorphic with each other.

This concept of 2-isomorphism has been extensively used by Gould[15] for the realization of contact switching networks (see Part IV). Gould defined 2-isomorphic graphs as an "equivalent class of

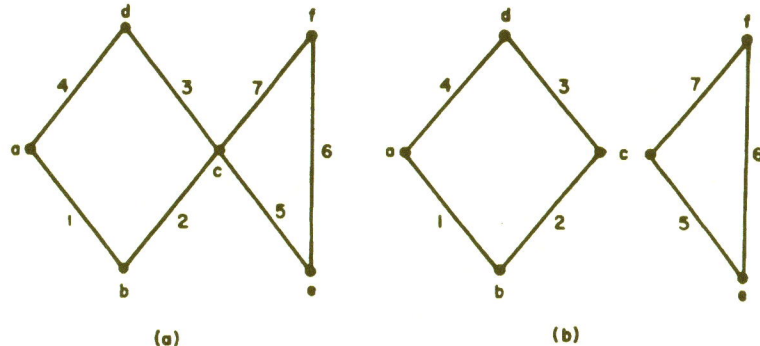

Fig. 1.12. One-isomorphic graphs: (a) G_1 and (b) G_2

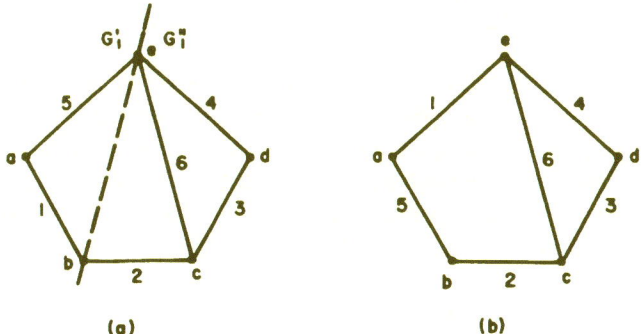

Fig. 1.13. Two-isomorphic graphs: (a) G_1 and (b) G_2

switching networks" since two contact-networks corresponding to 2-isomorphic graphs will have the identical switching behaviors.

Two-semi-isomorphic graphs: We are given a graph G_1 in Fig. 1.14a. Remove edges 3, 4, and 5 in G_1 and reconnect them between node-pairs (ac), (bd), and (cd). This results in G_2 as shown in Fig. 1.14b; G_1 and G_2 are 2-semi-isomorphic with each other.

From the definition of operation 1 and the example, it is noted that operation 1 is valid on G_1 only when G_1 is a separable connected graph or collection of disconnected subgraphs. However, a graph corresponding to an electrical network is in general not separable.

Therefore, we eliminate operation 1 from our further discussion. That is, we assume that a graph G_2 which is 2-semi-isomorphic to G_1 is obtained by operations 2 and/or 3.

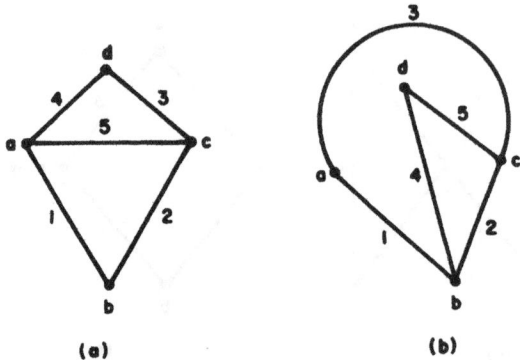

Fig. 1.14. Two-semi-isomorphic graphs: (a) G_1 and (b) G_2

Definition 21. In 2-semi-isomorphic graphs G_1 and G_2, edge e_i in G_1 and the same edge in G_2 together are called the "edge-pair e_i" of G_1 and G_2.

Definition 22. Let the terminal-nodes of the edges in an edge-pair of 2-semi-isomorphic graphs G_1 and G_2 be (i, j) and (p, q), respectively. If $i = p$ and $j = q$, and the orientations of both edges in the edge-pair are the same, then the edge-pair is said to be "ordinary"; otherwise, it is "active." Each edge of an ordinary and active edge-pair is called an "ordinary edge" and "active edge," respectively. The definition of the classification of edges in 2-semi-isomorphic graphs will become evident when their applications are illustrated.

Definition 23. A pair of trees of 2-semi-isomorphic graphs G_1 and G_2 that contain the same edges are called a "tree-pair" and the product of weights of the edges constituting a tree-pair is a "common tree-product"† of G_1 and G_2.

Definitions 21 through 23 are illustrated in the following example.

Example 5. Consider the 2-semi-isomorphic graphs G_1 and G_2 as shown in Fig. 1.15.

Edge-pairs w_1 and w_4 are the ordinary edge-pairs, but w_3, w_2, w_5, w_6, and w_7 are the active edge-pairs. If we choose a tree-pair of G_1 and G_2 as shown in Fig. 1.16, where node 2 is chosen as the reference node, then the principal and minor nodes of each edge in the trees are given in Table 2.

† Sometimes it is called a "complete tree-product."

Table 2. Principal and minor nodes for edges in the tree-pair of Fig. 1.16

	Tree of G_1				Tree of G_2			
Edges	w_1	w_2	w_4	w_6	w_1	w_2	w_4	w_6
Principal nodes	1	5	3	4	1	4	3	5
Minor nodes	2	1	2	5	2	1	2	1

Let us now introduce the orientation of edges in 2-semi-isomorphic graphs G_1 and G_2 of $(n+1)$ nodes and b edges, and let us choose a tree-pair t_1 and t_2 from G_1 and G_2. If we denote the incidence matrices

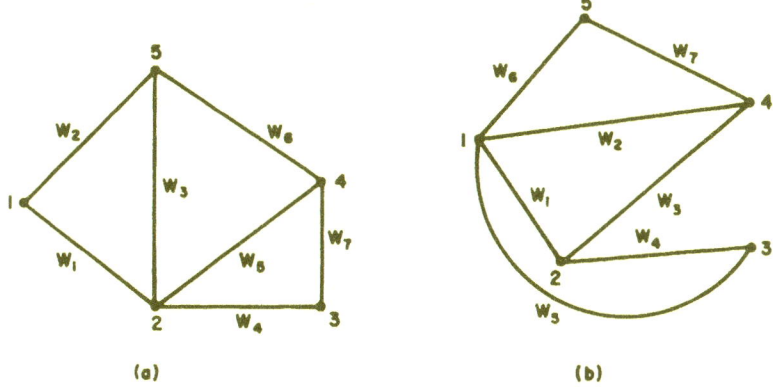

(a)   (b)

Fig. 1.15. Two-semi-isomorphic graphs: (a) G_1 and (b) G_2

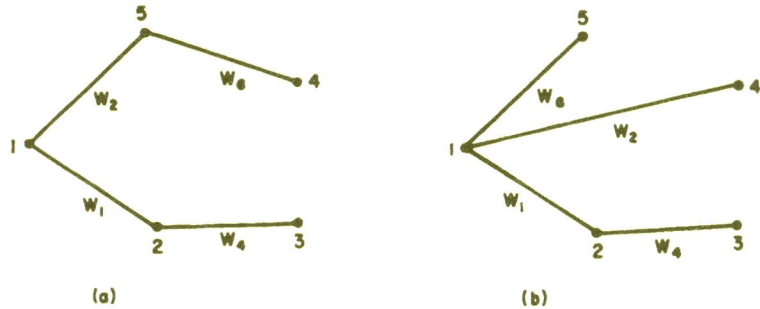

(a)   (b)

Fig. 1.16. A tree-pair of G_1 and G_2 of Fig. 1.15
A tree of (a) G_1 and (b) G_2.

of t_1 and t_2 by $\mathbf{A}_{t_1}$ and $\mathbf{A}_{t_2}$, respectively, then the sign of the common tree-product ϵ of the tree-pair is given by

$$\epsilon = |\mathbf{A}_{t_1}\mathbf{A}_{t_2}^t|. \tag{1.33}$$

This is obvious because of Property 6.

Property 18. The sign of the common tree-product of a tree-pair of 2-semi-isomorphic graphs is determined only by the active edge-pairs in the tree-pair.

Proof: Consider a tree-pair t_1 and t_2 of $(n+1)$ nodes, containing k active edge-pairs and $(n-k)$ ordinary edge-pairs. Then, arrange the columns of the incidence matrices of t_1 and t_2, $\mathbf{A}_{t_1}$ and $\mathbf{A}_{t_2}$, such that the first $(n-k)$ columns correspond to the ordinary edge-pairs and the last k columns to the active edge-pairs. Thus, we get

$$\mathbf{A}_{t_1} = \begin{array}{c} \text{ordinary edges} \quad \text{active edges} \\ \left[\begin{array}{ccc|ccc} a_{11} & \ldots & a_{1,n-k} & a_{1,n-k+1} & \ldots & a_{1n} \\ . & . & . & . & . & . \\ a_{nn} & \ldots & a_{n,n-k} & a_{n,n-k+1} & \ldots & a_{nn} \end{array} \right] \end{array} = [\mathbf{P}_{11} \quad \mathbf{P}_{12}] \tag{1.34a}$$

$$\mathbf{A}_{t_2} = \begin{array}{c} \text{ordinary edges} \quad \text{active edges} \\ \left[\begin{array}{ccc|ccc} b_{11} & \ldots & b_{1,n-k} & b_{1n-k+1} & \ldots & b_{1n} \\ . & . & . & . & . & . \\ b_{nn} & \ldots & b_{n,n-k} & b_{n,n-k+1} & \ldots & b_{nn} \end{array} \right] \end{array} = [\mathbf{Q}_{11} \quad \mathbf{Q}_{12}], \tag{1.34b}$$

where

$$\mathbf{P}_{11} = \begin{bmatrix} a_{11} & \ldots & a_{1,n-k} \\ . & . & . \\ a_{n1} & \ldots & a_{n,-kn} \end{bmatrix} \qquad \mathbf{P}_{12} = \begin{bmatrix} a_{1,n-k+1} & \ldots & a_{1n} \\ . & . & . \\ a_{n,n-k+1} & \ldots & a_{nn} \end{bmatrix} \tag{1.34c}$$

$$\mathbf{Q}_{11} = \begin{bmatrix} b_{11} & \ldots & b_{1,n-k} \\ . & . & . \\ b_{n1} & \ldots & b_{n,n-k} \end{bmatrix} \qquad \mathbf{Q}_{12} = \begin{bmatrix} b_{1,n-k+1} & \ldots & b_{1n} \\ . & . & . \\ b_{n,n\;k+1} & \ldots & b_{nn} \end{bmatrix} \tag{1.34d}$$

and since $\mathbf{P}_{11}$ and $\mathbf{Q}_{11}$ represent the incident relationship of the ordinary edge-pairs $\mathbf{P}_{11} = \mathbf{Q}_{11}$, i.e., $a_{ij} = b_{ij}$ for $i = 1, \ldots, n$ and $j = 1, \ldots, n-k$.

Now, assume that the edge 1 corresponding to the first column of $\mathbf{A}_{t_1}$ and $\mathbf{A}_{t_2}$ is incident at nodes r and s, i.e., a_{r1}, a_{s1}, b_{r1}, and b_{s1} are nonzero and $a_{rl} = b_{rl} = -a_{sl} = -b_{sl}$, and expand the determinants of $\mathbf{A}_{t_1}$ and $\mathbf{A}_{t_2}$ about the first column. If node r is the principal node of edge 1 of t_1, node s is the principal node of edge 1 of t_2 and $s \neq r+1$, then, interchange sth and $(r+1)$th rows in the matrices $\mathbf{A}_{t_1}$ and $\mathbf{A}_{t_2}$, respectively, before expanding the determinants of the matrices $\mathbf{A}_{t_1}$ and $\mathbf{A}_{t_2}$. If $s = r+1$, then no interchanges of rows are necessary. Thus we get, using Properties 3 and 8

$$|\mathbf{A}_{t_1}| = (-1)^{r+1+u} a_{r1}|M_{t_1}|_{r1}$$
$$|\mathbf{A}_{t_2}| = (-1)^{r+1+1+u} b_{s1}|M_{t_2}|_{s1}. \tag{1.35}$$

where u is the number of interchanges of rows (equal to 1 or 0), $|M_t|_{pq}$ is the minor of the determinant $|\mathbf{A}_t|$ of (p,q)-position, and $[M_t]_{pq}$ is the matrix corresponding to the minor. Since edge 1 is an ordinary edge, $a_{r1} = -b_{s1}$ for $r \neq s$, Eq. (1.35) is rewritten as

$$|\mathbf{A}_{t_1}| = (-1)^{r+1+u} a_{r1}|M_{t_1}|_{r1}$$
$$|\mathbf{A}_{t_2}| = (-1)^{r+1+u} a_{r1}|M_{t_2}|_{s1}. \tag{1.36}$$

The signs prefixing minors $|M_{t_1}|_{r1}$ and $|M_{t_2}|_{s1}$ are therefore the same.

When edge 1 has node r as its principal node both in t_1 and t_2, i.e., $a_{r1} = b_{r2}$, then the determinants of the incidence matrices are expanded about the element a_{r1} and we get

$$|\mathbf{A}_{t_1}| = (-1)^{r+1} a_{r1}|M_{t_1}|_{r1}$$
$$|\mathbf{A}_{t_2}| = (-1)^{r+1} b_{r1}|M_{t_2}|_{r1} = (-1)^{r+1} c_{r1}|M_{t_2}|_{r1}. \tag{1.37}$$

Thus the minors in Eq. (1.37) have the same sign prefixing them.

Next, expand the determinants of $[M_{t_1}]_{r1}$ and $[M_{t_2}]_{s1}$ (or $[M_{t_2}]_{r1}$) about the first column of the submatrices. If edge 2 of the tree-pair which corresponds to the first column of the submatrices has the same node as its principal node both in t_1 and t_2, the process described by Eq. (1.37) is repeated. If the principal node of edge 2 in t_1 is different from that of edge 2 in t_2, then the process described by Eq. (1.36) is repeated. Therefore, one continues to expand the determinants of the incidence matrices of t_1 and t_2 by the Laplace expansion as described by Eqs. (1.36) and (1.37) until the remaining minors of the determinants contain only columns corresponding to active edge-pairs in the tree-pairs. It is clear that the sign of the remainders are the same.

Furthermore, suppose we relabel row s as row (r, s) when a_{r1} is used as a complementary minor in expanding about a column, where s is the minor node of the edge corresponding to the column. Then the final resulting submatrix will be the incidence matrix of the reduced graph of the tree in the tree-pair. The sign of the tree-product then depends only on this final reduced tree. Hence, Property 18.

Definition 24. The "reduced" tree-pair of a tree-pair of 2-semi-isomorphic graphs is a pair of subgraphs derived from the tree-pair by removing all ordinary edge-pairs and identifying their terminal nodes in the pair.

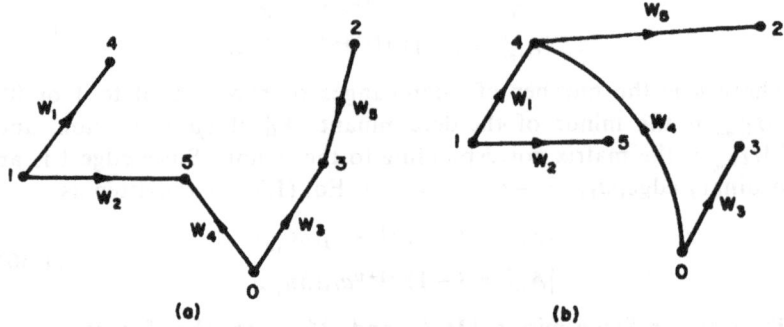

Fig. 1.17. A tree-pair t_1 and t_2

(a) A tree t_1; (b) t_2.

Example 6. Given: a tree-pair t_1 and t_2 as shown in Fig. 1.17, where node 0 is the reference node. The incidence matrices of t_1 and t_2, A_{t_1} and A_{t_2}, are found as:

$$A_{t_1} = \begin{array}{c} \\ 1 \\ 2 \\ 3 \\ 4 \\ 5 \end{array}\begin{array}{ccccc} 1 & 2 & 3 & 4 & 5 \\ \left[\begin{array}{ccccc} 1 & 1 & 0 & 0 & 0 \\ 0 & 0 & 0 & 0 & 1 \\ 0 & 0 & -1 & 0 & 1 \\ -1 & 0 & 0 & 0 & 0 \\ 0 & -1 & 0 & -1 & 0 \end{array}\right] \end{array} \qquad (1.38a)$$

$$A_{t_2} = \begin{array}{c} \\ 1 \\ 2 \\ 3 \\ 4 \\ 5 \end{array}\begin{array}{ccccc} 1 & 2 & 3 & 4 & 5 \\ \left[\begin{array}{ccccc} 1 & 1 & 0 & 0 & 0 \\ 0 & 0 & 0 & 0 & -1 \\ 0 & 0 & -1 & 0 & 0 \\ -1 & 0 & 0 & -1 & 1 \\ 0 & -1 & 0 & 0 & 0 \end{array}\right] \end{array}, \qquad (1.38b)$$

where the first three columns of $\mathbf{A}_{t_1}$ and $\mathbf{A}_{t_2}$ correspond to the ordinary edge-pairs 1, 2, and 3.

First, interchange rows 2 and 4 in the matrices, then expand the determinants. Thus, we get

$$
|\mathbf{A}_{t_1}| = (-1)\begin{array}{c}
\\
1 \\
4 \\
3 \\
2 \\
5
\end{array}
\begin{array}{ccccc}
w_1 & w_2 & w_3 & w_4 & w_5 \\
\left|\,1 & 1 & 0 & 0 & 0\,\right. \\
-1 & 0 & 0 & 0 & 0 \\
0 & 0 & -1 & 0 & -1 \\
0 & 0 & 0 & 0 & 1 \\
\left.0 & -1 & 0 & -1 & 0\,\right|
\end{array}
$$

$$
= (-1)\begin{array}{c}
\\
(1, 4) \\
3 \\
2 \\
5
\end{array}
\begin{array}{cccc}
w_2 & w_3 & w_4 & w_5 \\
\left|\,1 & 0 & 0 & 0\,\right. \\
0 & -1 & 0 & -1 \\
0 & 0 & 0 & 1 \\
\left.-1 & 0 & -1 & 0\,\right|
\end{array}
\qquad (1.39\text{a})
$$

$$
|\mathbf{A}_{t_2}| = (-1)\begin{array}{c}
\\
1 \\
4 \\
3 \\
2 \\
5
\end{array}
\begin{array}{ccccc}
w_1 & w_2 & w_3 & w_4 & w_5 \\
\left|\,1 & 1 & 0 & 0 & 0\,\right. \\
-1 & 0 & 0 & -1 & 1 \\
0 & 0 & -1 & 0 & 0 \\
0 & -1 & 0 & 0 & 0 \\
\left.0 & -1 & 0 & 0 & 0\,\right|
\end{array}
$$

$$
= (-1)\begin{array}{c}
\\
(1, 4) \\
3 \\
2 \\
5
\end{array}
\begin{array}{cccc}
w_2 & w_3 & w_4 & w_5 \\
\left|\,0 & 0 & -1 & 1\,\right. \\
0 & -1 & 0 & 0 \\
0 & 0 & 0 & -1 \\
\left.-1 & 0 & 0 & 0\,\right|
\end{array}
\qquad (1.39\text{b})
$$

Interchanging rows 5 and 3 we get

$$
|\mathbf{A}_{t_2}| = (-1)^2\begin{array}{c}
\\
(1, 4) \\
5 \\
2 \\
3
\end{array}
\begin{array}{cccc}
w_2 & w_3 & w_4 & w_5 \\
\left|\,1 & 0 & 0 & 0\,\right. \\
-1 & 0 & -1 & 0 \\
0 & 0 & 0 & 1 \\
\left.0 & -1 & 0 & -1\,\right|
\end{array}
$$

$$
= \quad
\begin{array}{c}
(1,\,4,\,5) \\
3 \\
2
\end{array}
\left|
\begin{array}{ccc}
w_3 & w_4 & w_5 \\
0 & -1 & 0 \\
0 & 0 & 1 \\
-1 & 0 & -1
\end{array}
\right|
\qquad (1.40a)
$$

$$
|\mathbf{A}_{t_1}| = (-1)^2 \quad
\begin{array}{c}
(1,\,4) \\
5 \\
2 \\
3
\end{array}
\left|
\begin{array}{cccc}
w_2 & w_3 & w_4 & w_5 \\
0 & 0 & -1 & 1 \\
-1 & 0 & 0 & 0 \\
0 & 0 & 0 & -1 \\
0 & -1 & 0 & 0
\end{array}
\right|
$$

$$
= \quad
\begin{array}{c}
(1,\,4,\,5) \\
3 \\
2
\end{array}
\left|
\begin{array}{ccc}
w_3 & w_4 & w_5 \\
0 & -1 & 1 \\
0 & 0 & -1 \\
-1 & 0 & 0
\end{array}
\right|.
\qquad (1.40b)
$$

Expanding the minors of Eqs. (1.40a) and (1.40b) about the element of (3, 1)-position, respectively, we have

$$
|\mathbf{A}_{t_1}| = (-1) \quad
\begin{array}{c}
(1,\,4,\,5) \\
3
\end{array}
\left|
\begin{array}{cc}
w_4 & w_5 \\
-1 & 0 \\
0 & 1
\end{array}
\right|
\qquad (1.41a)
$$

$$
|\mathbf{A}_{t_2}| = (-1) \quad
\begin{array}{c}
(1,\,4,\,5) \\
3
\end{array}
\left|
\begin{array}{cc}
w_4 & w_5 \\
-1 & 1 \\
0 & -1
\end{array}
\right|.
\qquad (1.41b)
$$

It is noted that the matrices corresponding to the resulting minors are the incidence matrices of the reduced trees shown in Fig. 1.18.

Definition 25. The sign of an active edge-pair is defined as $+1$ if both edges in the pair are directed away or toward their principal nodes and -1 otherwise. We therefore have Property 19.

Property 19. The sign of the common tree-product of a tree-pair, ϵ is given by†

$$
\epsilon = (-1)^\gamma \overset{k}{\Pi} \text{ [sign of active edge-pairs of the reduced tree-pair of}
$$
a tree-pair],

† Coates[18] first proposed the identical formula but used a different graphical approach. Mayeda[19] also proposed a formula; however, it is more laborious than the one derived here.

where γ is the number of interchanges of edges needed to give all active edge-pairs in the reduced tree-pair the same principal nodes, and k is the number of active edge-pairs in the reduced tree-pair.

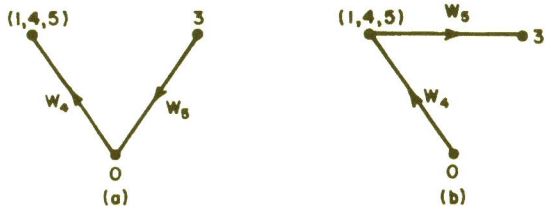

Fig. 1.18. Reduced tree-pair of the tree-pair of Fig. 1.17
Reduced tree of (a) t_1; (b) t_2.

Proof: By Property 18 we need only consider the reduced graphs of the trees in the tree-pair to determine the sign of the common tree-product. Let t'_1 and t'_2 denote the reduced trees and $\mathbf{A}_{t'_1} = [a'_{ij}]$, and $\mathbf{A}_{t'_2} = [b'_{ij}]$ their respective incidence matrices. Now interchange columns in $\mathbf{A}_{t'_1}$ until all edges in t'_1 and t'_2 have the same principal nodes, the resultant matrix is $\mathbf{A}'_{t'_1}$. By Property 3 this can always be done.

Suppose that in column 1 of $\mathbf{A}'_{t_1}$ and $\mathbf{A}_{t'_2}$ the respective nonzero entries corresponding to principal nodes are a'_{1s} and b'_{1s}. Then

$$|\mathbf{A}'_{t'_1}| = (-1)^{1+s}a'_{1s}|M_{t'_1}|_{1s}(-1)^\gamma$$
$$|\mathbf{A}_{t'_2}| = (-1)^{1+s}b'_{1s}|M_{t'_2}|_{1s}, \qquad (1.42)$$

where $|M_{t'}|_{ij}$ is the minor of the determinant of $\mathbf{A}_{t'}$, of (i,j)-position, and γ is the number of interchanges of columns in $\mathbf{A}_{t'_1}$ to give all edge-pairs in t'_1 and t'_2 the same principal nodes.

If the sign of edge-pair 1 is positive, then $b'_{1s} = a'_{1s}$ and

$$\epsilon = |\mathbf{A}'_{t'_1}||\mathbf{A}_{t'_2}| = [(-1)^{(1+s)}a'_{1s}]^2|M_{t'_1}||M_{t'_2}|(-1)^\gamma$$
$$= (-1)^\gamma|M_{t'_1}||M_{t'_2}|. \qquad (1.43)$$

If the sign of edge-pair 1 is negative, then $b'_{1s} = -a'_{1s}$ and

$$\epsilon = |\mathbf{A}'_{t'_1}||\mathbf{A}_{t'_2}| = [(-1)^{(1+s)}a'_{1s}]^2|M_{t'_1}||M_{t'_2}|(-1)^{\gamma+1}$$
$$= (-1)^{\gamma+1}|M_{t'_1}||M_{t'_2}|. \qquad (1.44)$$

We next expand minors $|M_{t'_1}|$ and $|M_{t'_2}|$ about edge 2. The process described by Eqs. (1.43) and (1.44) is thus repeated until we have expanded about all columns of $|\mathbf{A}'_{t'_1}|$ and $|\mathbf{A}_{t'_2}|$. The final result is $\epsilon = (-1)^\gamma(-1)^\beta$, where β is the number of active edge-pairs with a negative sign. This completes the proof.

We shall now briefly review the concept of duality and planarity of graphs. They will become essential for the discussion of the realizability of loop and cut-set matrices in connection with the synthesis of single-contact switching networks (see Part IV). However, we list a few definitions and properties on the subjects which will suffice for our purpose.

Let us consider a graph G of b edges and n nodes, consisting of P maximally connected subgraphs. Then, as defined previously, the rank of the graph is $(n-P)$ and the nullity of the graph is $(b-n+P)$. For $P = 1$, i.e., G being connected, the rank of $G = n-1$ and the nullity of $G = b-n+1$. Bearing this in mind, we define the following.

Definition 26. Let two connected graphs G and G' have one-to-one correspondence between their edges. We also let G_i and G'_i be the corresponding subgraphs of G and G', respectively. We now denote the complements of G_i and G'_i in G and G', respectively, by R_i and R'_i. Then, G' is a "dual" of G if, for every subgraph G_i in G,

$$\text{rank of } R_i = \text{nullity of } R'_i.$$

From Definition 26, the following property follows directly.

Property 20. If a graph G' is a dual of a graph G, then:

G is a dual of G';
Rank of G = nullity of G';
Rank of G' = nullity of G.[9,15]

We now need to define a so-called "planar graph" in order to explore further properties of duality.

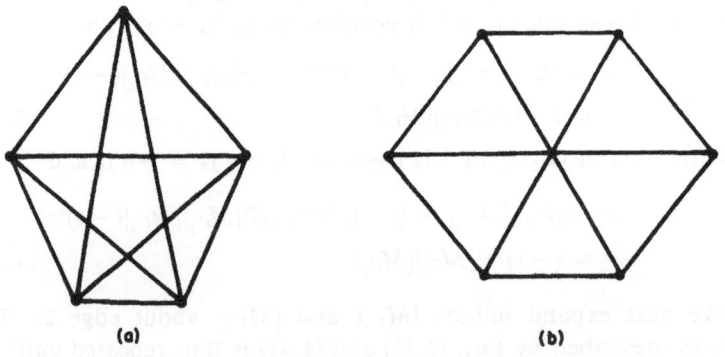

(a)   (b)

Fig. 1.19. Kuratowski's basic nonplanar graphs

Definition 27. If a graph is "planar" it can be mapped continuously upon a plane† or a sphere.

The following properties show some of the close relationships which exist between dual graphs and planar graphs.

Property 21. A planar graph G has the following properties:

G is planar if and only if it has a dual;

G is planar if and only if it does not contain either of Kuratowski's[21] graphs (see Fig. 1.19) or their subgraphs;

All cut-set edges of G correspond to loop-set edges, and conversely.

The proofs of the properties are very much involved and are therefore not presented here.‡

The concepts and definitions of linear graphs studied so far may not be sufficient to discuss the various problems which arise in the analysis and synthesis of electrical networks, switching networks, communication nets, and feedback control systems. We will therefore define some more as they become necessary.

† In other words, a graph is mapped upon a plane such that no edges of the graph cross with each other at the points other than the nodes of the graph.

‡ Readers should refer to References 7 (Chapter 3, Section 2, pp. 39–53), 9b, d, and 15 (pp. 1–26, 1–33, and 4–12) for the proofs of the property and more extensive study on dual graphs and planar graphs.

CHAPTER 2

Analysis of the Ordinary Two-Port Networks

WHEN AN ELECTRICAL NETWORK consists entirely of ordinary two-terminal elements such as resistors, inductors, and capacitors, then the network can be completely characterized by an abstract graph. However, each edge of the graph needs to be identified by a number or some other quantity. Since we are dealing with an electrical network, each element of the network is associated with a current, voltage, impedance, or admittance. Therefore, we shall use one of these quantities to identify or label each edge of a graph corresponding to an electrical network. A graph in which each edge is associated with a quantity, called the "edge-weight," is a "weighted graph."

We shall first study the application of graph theory to the analysis of the ordinary networks, e.g., RLC networks. Then, we will extend the method to networks containing dependent node-pairs in the next chapter.

The networks under consideration are linear with fixed parameters. We first formulate the Kirchhoff laws, which are the basis of the theory of electrical networks, in terms of topological expressions defined in the preceding chapter.

Kirchhoff's Current Law (KCL)

If the orientation of each edge of an ordinary network of b edges† and n nodes is identified with the orientation of the current flowing through the edge, then we have KCL equations for an ordinary network:

$$\mathbf{AI}_e = \mathbf{O} \tag{2.1a}$$

† An "edge" and a "two-terminal element" are synonymous terms when we speak of a RLC ordinary network.

and

$$\mathbf{I}_e = \begin{bmatrix} i_1(t) \\ \vdots \\ i_b(t) \end{bmatrix}, \tag{2.1b}$$

where $i_k(t)$ is the current of edge k, i.e., "edge-current k." In general, KCL equations may be written in terms of the cut-set matrix of the network, as:

$$\mathbf{C}\mathbf{I}_e = \mathbf{O}. \tag{2.2}$$

Kirchhoff's Voltage Law (KVL)

Let us denote the voltage across edge b, i.e., "edge-voltage b," by $v_b(t)$. The orientation of each loop is assigned arbitrarily, but the orientation of the edge is decided by the polarity of each edge-voltage. Then, for an ordinary network of b edges and $(b - n + 1)$ independent loops, we have KVL equations:

$$\mathbf{B}\mathbf{V}_e = \mathbf{O}, \tag{2.3a}$$

where $\mathbf{B}$ is a loop matrix of a set of $(b - n + 1)$ independent loops and

$$\mathbf{V}_e = \begin{bmatrix} v_1(t) \\ \vdots \\ v_b(t) \end{bmatrix}. \tag{2.3b}$$

Since the rank of both the incidence and cut-set matrices of a connected graph of n nodes is $(n - 1)$, that is, the number of independent KCL equations of an ordinary network of n nodes is $(n - 1)$, we shall rewrite Eqs. (2.1) and (2.2) in the form:

$$\mathbf{A}_1\mathbf{I}_e = \mathbf{O} \tag{2.4a}$$

$$\mathbf{C}_f\mathbf{I}_e = \mathbf{O}. \tag{2.4b}$$

In forming $\mathbf{A}_1$ from $\mathbf{A}$, one row corresponding to the reference node of the network is deleted from $\mathbf{A}$. If we arrange the columns of $\mathbf{A}_1$ such that the first $(b - n + 1)$ columns correspond to chords of a tree, then from Eqs. (2.4),

$$\mathbf{A}_1\mathbf{I}_e = \begin{bmatrix} \mathbf{A}_{11} & \mathbf{A}_{12} \end{bmatrix} \begin{bmatrix} \mathbf{I}_c \\ \mathbf{I}_b \end{bmatrix} = \mathbf{A}_{11}\mathbf{I}_c + \mathbf{A}_{12}\mathbf{I}_b = \mathbf{O} \tag{2.5a}$$

$$\mathbf{C}_f\mathbf{I}_e = \begin{bmatrix} \mathbf{C}_{f_{11}} & \mathbf{U}_1 \end{bmatrix} \begin{bmatrix} \mathbf{I}_c \\ \mathbf{I}_b \end{bmatrix} = \mathbf{C}_{f_{11}}\mathbf{I}_c + \mathbf{I}_b = \mathbf{O}, \tag{2.5b}$$

where I_b is a submatrix of I_e which corresponds to branch-currents and I_c for chord-currents. The columns of submatrix A_{12} and the unit matrix of order $(n-1)$, U_1, correspond to the branches of the tree, so A_{12} is nonsingular.

From Eqs. (2.5) we derive an expression with which one may be able to evaluate all branch currents for given values of chord-currents. That is,

$$I_b = -A_{12}^{-1}A_{11}I_c \tag{2.6a}$$

$$I_b = -C_{f_{1s}}I_c. \tag{2.6b}$$

We apply Properties 9 and 13 to Eq. (2.6a) and we obtain

$$I_b = B_{f_{1s}}{}^tI_c = (B_{11}^{-1}B_{12})^tI_c \tag{2.7a}$$

$$I_e = \begin{bmatrix} U_2 \\ B_{12}{}^t(B_{11}^{-1})^t \end{bmatrix} I_c = B_f{}^tI_c. \tag{2.7b}$$

We are now in a position to prove that a basic loop matrix of a connected graph is unimodular (see Property 4). On an ordinary network of b edges and n nodes let us choose two different trees t_r and t_k and let us also denote the basic loop matrices with respect to the chosen trees by B_{fr} and B_{fk}, respectively. If we arrange the columns of both matrices in the same order, then from Eqs. (2.7) we have

$$I_e = B_{fk}{}^tI_{kc} = B_{fr}{}^tI_{rc}, \tag{2.8}$$

where the subscripts k and r indicate trees t_k and t_r.

We now arrange the columns of both matrices B_{fr} and B_{fk} so that the first $(b-n+1)$ columns of both matrices correspond to the chords of tree t_k. Then we have

$$B_{fk} = [U_2 \quad B_{fk_{1s}}] \tag{2.9a}$$

$$B_{fr} = [B_{fr_{11}} \quad B_{fr_{1s}}], \tag{2.9b}$$

where B_{fr11} is nonsingular due to Property 11, and U_2 is a unit matrix of order $(b-n+1)$. If we arrange the columns of both matrices with respect to the set of chords of tree t_r, then

$$B'_{fk} = [B_{fk_{11}} \quad B'_{fk_{1s}}] \tag{2.10a}$$

$$B'_{fr} = [U_2 \quad B'_{fr_{1s}}], \tag{2.10b}$$

where $B'_{fr_{11}}$ is again nonsingular.

The substitution of Eqs. (2.9) and (2.10) into Eq. (2.8) will give us

$$\mathbf{I}_{kc} = \mathbf{B}_{fr_{11}}{}^t\mathbf{I}_{rc} \tag{2.11a}$$

$$\mathbf{I}_{rc} = (\mathbf{B}'_{fk_{11}})^t\mathbf{I}_{kc}. \tag{2.11b}$$

We therefore have

$$\mathbf{B}_{fr_{11}}{}^t = (\mathbf{B}'_{fk_{11}}{}^t)^{-1}. \tag{2.12}$$

Since each element of the submatrices of a basic loop matrix of a graph (or a network) must be 0, 1, or −1,† in order to have Eq. (2.12) valid,

$$\det \mathbf{B}_{fr_{11}} = \pm 1 \tag{2.13a}$$

$$\det \mathbf{B}'_{fk_{11}} = \pm 1. \tag{2.13b}$$

In other words, every nonsingular subdeterminant of order $(b - n + 1)$ of $\mathbf{B}_f$ is 1 or −1. A subdeterminant of order k, $k = 1, 2, \ldots$, of $\mathbf{B}_f$, will also be 0, 1, or −1 because if we cross out k rows of $\mathbf{B}_f$, i.e., opening the k chords corresponding to the k rows which are not taken into account, then the foregoing arguments must again be valid for the resulting graph. It is therefore clear that a basic loop matrix of a connected graph is unimodular because every subdeterminant of the matrix is 0, 1, or −1. This proves the third part of Property 4.

An expression similar to Eqs. (2.7) for chord-voltages in terms of branch-voltages can be derived directly from KVL equations. That is,

$$\mathbf{BV}_e = [\mathbf{B}_{11} \quad \mathbf{B}_{12}]\begin{bmatrix} \mathbf{V}_c \\ \mathbf{V}_b \end{bmatrix} = \mathbf{O}, \tag{2.14}$$

where again the columns of $\mathbf{B}_{11}$ include all chords of a tree and those of $\mathbf{B}_{12}$ correspond to branches of the tree, i.e., $\mathbf{B}_{11}$ is a nonsingular submatrix of order $(b - n + 1)$. Here $\mathbf{V}_c$ and $\mathbf{V}_b$ represent the chord-voltages and branch-voltages, respectively. From Eq. (2.14),

$$\mathbf{V}_c = -\mathbf{B}_{11}{}^{-1}\mathbf{B}_{12}\mathbf{V}_b. \tag{2.15}$$

Again using Properties 9 and 13 to Eq. (2.15), we have

$$\mathbf{V}_c = -\mathbf{B}_{f_{11}}\mathbf{V}_b = (\mathbf{A}_{12}{}^{-1}\mathbf{A}_{11})^t\mathbf{V}_b \tag{2.16a}$$

$$\mathbf{V}_e = \begin{bmatrix} \mathbf{A}_{11}{}^t(\mathbf{A}_{12}{}^{-1}) \\ \mathbf{U}_1 \end{bmatrix}\mathbf{V}_b. \tag{2.16b}$$

† As long as Eqs. (2.12) are valid and the elements of both matrices are real rational numbers, then Eqs. (2.13) follow.

Using Property 15 we have from Eq. (2.16b)

$$V_e = C_f{}^tV_b. \tag{2.17}$$

It is therefore clear from Eqs. (2.7) and (2.16) that a loop matrix of $(b-n+1)$ independent loops identifies all edge-currents of an ordinary network if chord-currents of a tree are given, and so do the incidence and basic cut-set matrices identify all edge-voltages if branch voltages of a tree are known.

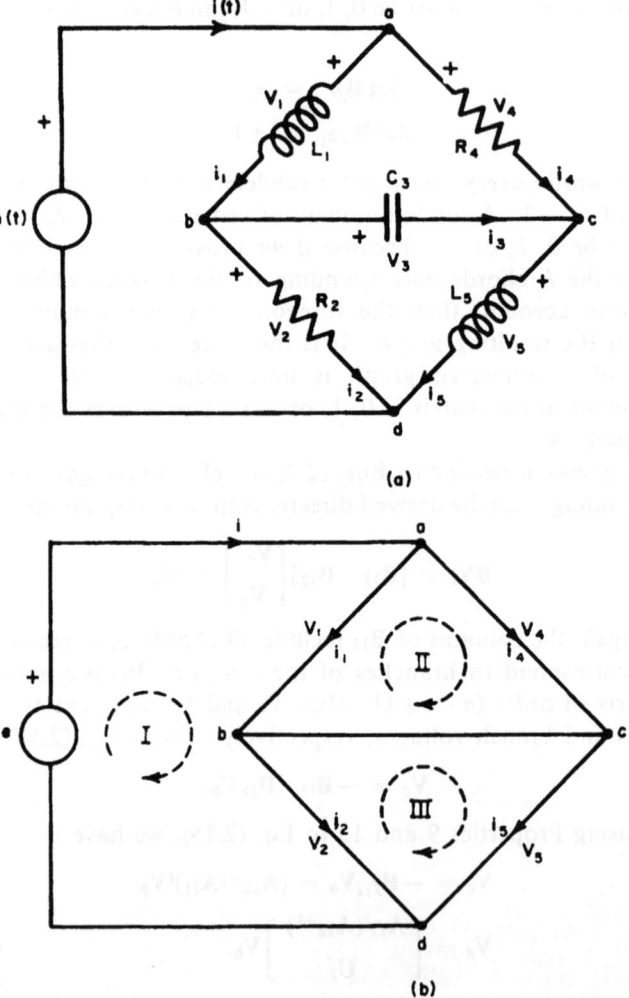

(a)

(b)

Fig. 2.1. (a) An ordinary electrical network and (b) the corresponding graph

As an illustration, let us consider an ordinary network shown in Fig. 2.1a and the corresponding graph with each edge-weight in terms of edge-current or edge-voltage in Fig. 2.1b.

In order to formulate KCL equations of the network from its graphical representation, one may open the driving-point terminals; node a and node d, the reference node of the network. Thus we have

$$\mathbf{A}_1\mathbf{I}_e = \begin{bmatrix} 1 & 0 & 0 & 1 & 0 \\ -1 & 1 & 1 & 0 & 0 \\ 0 & 0 & -1 & -1 & 1 \end{bmatrix} \begin{bmatrix} i_1 \\ i_2 \\ i_3 \\ i_4 \\ i_5 \end{bmatrix} = \begin{bmatrix} i \\ 0 \\ 0 \end{bmatrix}, \qquad (2.18a)$$

where

$$\mathbf{A}_1 = \begin{array}{c} \\ a \\ b \\ c \end{array} \begin{array}{ccccc} i_1 & i_2 & i_3 & i_4 & i_5 \\ \begin{bmatrix} 1 & 0 & 0 & 1 & 0 \\ -1 & 1 & 1 & 0 & 0 \\ 0 & 0 & -1 & -1 & 1 \end{bmatrix} \end{array}. \qquad (2.18b)$$

For setting up KVL equations, the driving-point terminals are identified such that the first loop is formed by edges 1 and 2. We therefore have for the orientation of each loop as indicated in the figure,

$$\mathbf{B}\mathbf{V}_e = \begin{bmatrix} 1 & 1 & 0 & 0 & 0 \\ -1 & 0 & -1 & 1 & 0 \\ 0 & -1 & 1 & 0 & 1 \end{bmatrix} \begin{bmatrix} v_1 \\ v_2 \\ v_3 \\ v_4 \\ v_5 \end{bmatrix} = \begin{bmatrix} e \\ 0 \\ 0 \end{bmatrix} \qquad (2.19a)$$

and

$$\mathbf{B} = \begin{array}{c} \\ \text{loop I} \\ \text{loop II} \\ \text{loop III} \end{array} \begin{array}{ccccc} v_1 & v_2 & v_3 & v_4 & v_5 \\ \begin{bmatrix} 1 & 1 & 0 & 0 & 0 \\ -1 & 0 & -1 & 1 & 0 \\ 0 & -1 & 1 & 0 & 1 \end{bmatrix} \end{array}. \qquad (2.19b)$$

Now, let us consider the so-called "node-datum voltage" or "node-pair voltage" for each node with respect to the reference node of the network. We shall denote the voltage which appears at node a and is measured with respect to node d by v_{ad}. Then, we shall study the following expression,

$$\mathbf{A}_1^t \mathbf{V}_n = \begin{bmatrix} 1 & -1 & 0 \\ 0 & 1 & 0 \\ 0 & 1 & -1 \\ 1 & 0 & -1 \\ 0 & 0 & 1 \end{bmatrix} \begin{bmatrix} v_{ad} \\ v_{bd} \\ v_{cd} \end{bmatrix} = \begin{bmatrix} v_{ad} - v_{bd} \\ v_{bd} \\ v_{bd} - v_{cd} \\ v_{ad} - v_{cd} \\ v_{cd} \end{bmatrix}. \qquad (2.20)$$

The right-hand expression of Eq. (2.20) then will be observed as nothing but the edge-voltages of the network. That is,

$$
\begin{aligned}
v_{ad} - v_{bd} &= v_1 \\
v_{bd} &= v_2 \\
v_{bd} - v_{cd} &= v_3 \\
v_{ad} - v_{cd} &= v_4 \\
v_{cd} &= v_5 .
\end{aligned}
\qquad (2.21)
$$

It is therefore clear that in general

$$\mathbf{A}_1^t \mathbf{V}_n = \mathbf{V}_e. \qquad (2.22)$$

This expression is called the "node transformation" because the node-datum voltage for each node is found in terms of edge-currents.

One may deduce a similar relationship between loop-currents and edge-currents. That is, if we consider the expression

$$\mathbf{B}^t \mathbf{I}_l = \begin{bmatrix} 1 & -1 & 0 \\ 1 & 0 & -1 \\ 0 & -1 & 1 \\ 0 & 1 & 0 \\ 0 & 0 & 1 \end{bmatrix} \begin{bmatrix} i_\mathrm{I} \\ i_\mathrm{II} \\ i_\mathrm{III} \end{bmatrix} = \begin{bmatrix} i_\mathrm{I} - i_\mathrm{II} \\ i_\mathrm{I} - i_\mathrm{III} \\ i_\mathrm{III} - i_\mathrm{II} \\ i_\mathrm{II} \\ i_\mathrm{III} \end{bmatrix}, \qquad (2.23)$$

then we can identify

$$
\begin{aligned}
i_\mathrm{I} - i_\mathrm{II} &= i_1 \\
i_\mathrm{I} - i_\mathrm{III} &= i_2 \qquad\qquad i_\mathrm{II} = i_4. \\
i_\mathrm{III} - i_\mathrm{II} &= i_3. \qquad\qquad i_\mathrm{III} = i_5
\end{aligned}
\qquad (2.24)
$$

Thus, we find that one may write in general

$$\mathbf{B}^t\mathbf{I}_l = \mathbf{I}_e, \tag{2.25}$$

which is called the "loop transformation." This formula together with the node transformation describes the structure of a network, i.e., the topology of a network.

One more set of expressions which characterize the behavior of an electrical network is the so-called "Ohm's law." That is,

$$\mathbf{Z}_e\mathbf{I}_e(s) = \mathbf{V}_e(s) \tag{2.26a}$$

$$\mathbf{Y}_e\mathbf{V}_e(s) = \mathbf{I}_e(s), \tag{2.26a'}$$

where $\mathbf{Z}_e$ and $\mathbf{Y}_e$ are "edge impedance matrix" and "edge admittance matrix" such that

$$\mathbf{Z}_e = \begin{bmatrix} z_1 & & & 0 \\ & z_2 & & \\ & & \cdot & \\ & & & \cdot \\ 0 & & & z_b \end{bmatrix} \quad \mathbf{Y}_e = \begin{bmatrix} y_1 & & & 0 \\ & y_2 & & \\ & & \cdot & \\ & & & \cdot \\ 0 & & & y_b \end{bmatrix} \tag{2.26b}$$

and z_k and y_k are impedance and admittance of edge k, respectively. Here $\mathbf{I}_e(s)$ and $\mathbf{V}_e(s)$ are column matrices such that

$$\mathbf{I}_e(s) = \begin{bmatrix} I_1(s) \\ I_2(s) \\ \cdot \\ \cdot \\ \cdot \\ I_b(s) \end{bmatrix} \quad \mathbf{V}_e(s) = \begin{bmatrix} V_1(s) \\ V_2(s) \\ \cdot \\ \cdot \\ \cdot \\ V_b(s) \end{bmatrix} \tag{2.26c}$$

in which $I_k(s)$ and $V_k(s)$ are the Laplace transformation of edge-current $i_k(t)$ and edge-voltage $v_k(t)$, respectively.

We have thus derived the following set of formulas:

Kirchhoff's law

$$\mathbf{A}_1\mathbf{I}_e(s) = \mathbf{O} \tag{2.27a}$$

$$\mathbf{B}_f\mathbf{V}_e(s) = \mathbf{O} \tag{2.27b}$$

Ohm's law

$$Y_e V_e(s) = I_e(s) \tag{2.27c}$$

$$Z_e I_e(s) = V_e(s) \tag{2.27d}$$

topological relationships

$$V_e(s) = A_1{}^t V_n(s) \tag{2.27e}$$

$$I_e(s) = B_f{}^t I_l(s). \tag{2.27f}$$

In Eqs. (2.27), if we substitute (c) into (a), then, using (e), we get

$$(A_1 Y_e A_1{}^t) V_n(s) = 0.\dagger \tag{2.28a}$$

From (b), (d), and (f), we obtain

$$(B_f Z_e B_f{}^t) I_l(s) = 0. \tag{2.28b}$$

In general, Eqs. (2.28) are written for an ordinary network containing independent ideal current and voltage sources as:

$$(A_1 Y_e A_1{}^t) V_n(s) = J(s) \tag{2.29a}$$

$$(B_f Z_e B_f{}^t) I_l(s) = E(s), \tag{2.29b}$$

where

$$J(s) = \begin{bmatrix} J_1 \\ J_2 \\ \cdot \\ \cdot \\ \cdot \\ J_n \end{bmatrix} \qquad E(s=) \begin{bmatrix} E_1 \\ E_2 \\ \cdot \\ \cdot \\ \cdot \\ E_l \end{bmatrix} \tag{2.29c}$$

in which J_k and E_m are Laplace transforms of $j_k(t)$ and $e_m(t)$, and $j_k(t)$ is the sum of independent current sources connected at node k and $e_m(t)$ the sum of independent voltage sources contained in loop m. One must note here that if a network under consideration contains both current and voltage sources, then before forming the incidence matrix of the network all the voltage sources should be converted into the corresponding current sources and then all the current sources are open-circuited. For the formulation of a loop matrix all the current sources should be converted into voltage sources, and then all the voltage sources in the network are short-circuited.

† The formation of the node admittance matrix in terms of cut-set matrices will be considered in Part III.

Then, the right-hand expression of Eq. (2.28a) or Eq. (2.29a) is the so-called "node admittance matrix" and its determinant is the "node admittance determinant" or simply "node determinant," which is very familiar to us in the analysis of electrical networks. Similarly, the right-hand of Eq. (2.28b) or Eq. (2.29b) is the "loop impedance matrix" and its determinant is the "loop impedance determinant" or simply "loop determinant" of a network.†

We therefore have, if we denote the node and loop determinants of a network by Δ_n and Δ_l, respectively, from Eqs. (2.29) and (2.29) as:

$$\Delta_n = |\mathbf{A}_1 \mathbf{Y}_e \mathbf{A}_1{}^t| \qquad (2.30a)$$

$$\Delta_l = |\mathbf{B}_f \mathbf{Z}_e \mathbf{B}_f{}^t|. \qquad (2.30b)$$

Now, applying our familiar Binet–Cauchy theorem in the evaluation of each determinant of Eqs. (2.30) it follows that

Δ_n = sum of products of all pairs of corresponding
major determinants of $(\mathbf{A}_1 \mathbf{Y}_e)$ and $\mathbf{A}_1{}^t$ (2.31)

Δ_l = sum of products of all pairs of corresponding
major determinants of $(\mathbf{B}_f \mathbf{Z}_e)$ and $\mathbf{B}_f{}^t$.   (2.31b)

As one may already have recognized in the expression of Eq. (2.31a), if the network under consideration consists entirely of one-ohm resistors then $\mathbf{Y}_e$ becomes a unit matrix and then each determinant is equal to the "total number of trees" (see Property 7). However, this time we have the effect of diagonal matrix to consider. Because of diagonal matrix $\mathbf{Y}_e$ the evaluation of the node determinant will give us the sum of products of admittance of branches of a tree for all possible trees of a network. So, if we call the product of the branch admittances of a tree a "tree admittance product" or simply a "tree-product," then

Δ_n = sum of tree admittance products for all
possible trees of an ordinary network N.‡   (2.32)

Similarly one can deduce from Eq. (2.31b), using Property 10, that

---

† Since the incidence matrix is a unimodular matrix, Eq. (2.28a) or Eq. (2.29a) is called a "unimodular congruence." By the same token, if Eq. (2.28b) or Eq. (2.29b) is written in terms of a basic loop matrix, i.e. $(\mathbf{B}_f \mathbf{Z}_e \mathbf{B}_f{}^t)$, then it is called a "unimodular congruence" because a basic loop matrix is a unimodular matrix. This will be thoroughly discussed in Part III.

‡ This formula was originally proposed by Maxwell[22] and is therefore called the "Maxwell formula."

$$\Delta_l = \text{sum of chord impedance products for all}$$
$$\text{possible trees of an ordinary network } N. \quad (2.33)$$

Although we have thus derived two topological formulas for the analysis of a network the first is preferred to the second. There are a number of reasons to support this practice. We shall, however, see one reason which should be self-evident. That is, it is necessary to find all possible trees of a network to use either formula, but the second formula requires more work to identify each set of chords of a tree even after each tree has been found. There has actually been a considerable amount of work[23-27] done on digital computer analysis of an ordinary network based on the formula of Eq. (2.32).

We now extend the formula of Eq. (2.32) in order to complete the topological analysis method of a network. A cofactor of the node-determinant of an ordinary network Δ_n, $\Delta_{n_{ij}}$, can be deduced to be†

$$\Delta_{n_{ij}} = |A_{1_{-i}} Y_e A_{1_{-j}}{}^t| = \text{sum of tree-products for all possible}$$
trees which are common to both subnetworks $N^{(ir)}$ and $N^{(jr)}$;
$$(2.34)$$

$A_{1_{-i}}$ is the submatrix of A_1 with ith row deleted and this corresponds to the subnetwork of $N^{(ir)}$ of network N with node i and reference node r identified. The reference node is the node which was deleted in forming A_1.

Let us define[33c]:

T = sum of all tree (complete tree) admittance products in a network. A tree admittance product is the product of admittance of all branches of a tree. The tree admittance product of a node is defined to be unity.

$T^{(abc\cdots)}$ = sum of all tree admittance products in the network derived from the original one with a number of nodes, say nodes a, b, and c, . . . , made coincident. Assume $a \neq b \neq c \neq \ldots$; otherwise $T^{(abc\cdots)} = 0$. For example, $T^{(aab\cdots)} = 0$.

$T^{(a'b'c'\cdots)(a'b'c'\cdots)}$ = sum of all tree admittance products in the network derived from the original one with a number of groups of nodes, say groups of nodes $(abc \ldots)$ and $(a'b'c' \ldots)$ made coincident, respectively. Assume $a \neq b \neq c \neq \ldots$, and $a' \neq b' \neq c' \neq \ldots$; otherwise the tree-product vanishes. For example, $T^{(aac\cdots)(a'b'c'\cdots)} = 0$.

$T^{(abc\cdots)} \cap {(a'b'c'\cdots)}$ = sum of all tree admittance products common to both $T^{(abc\cdots)}$ and $T^{(a'b'c'\cdots)}$.

These definitions are illustrated in the following example.

† See Reference 6 for the rigorous proof of Eq. (2.34).

Example 7. Let a network be given in Fig. 2.2a and some of the subnetworks of the network shown in Figs. 2b, 2c, and 2d.

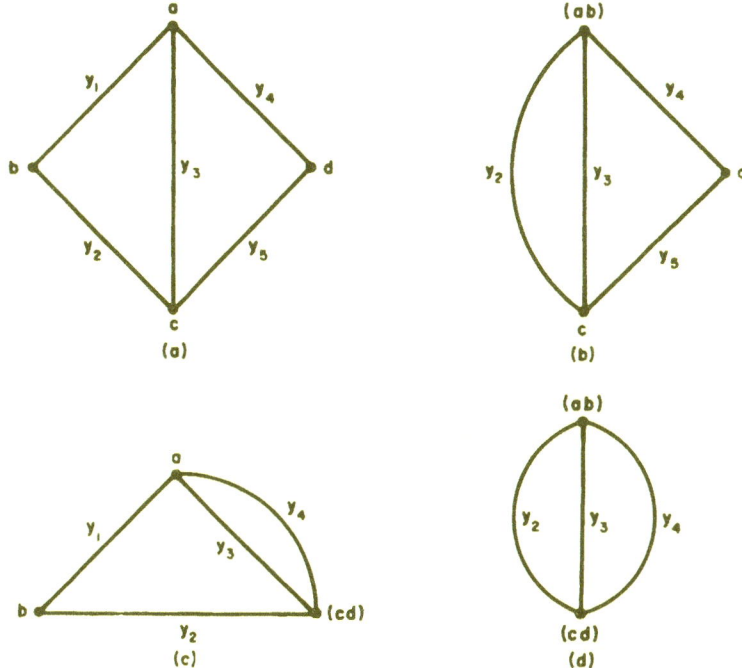

Fig. 2.2. *Various subgraphs derived from a graph G*
(a) A graph G; (b) nodes a and b made coincident; (c) nodes c and d made coincident; (d) two groups of nodes, (ab) and (cd), made coincident.

Then, T is the sum of all tree admittance products of the network of Fig. 2.2a and is given by

$$T = y_1y_2y_4 + y_1y_2y_5 + y_1y_4y_5 + y_1y_3y_4 + y_1y_3y_5 + y_2y_3y_4 + y_2y_3y_5 + y_2y_4y_5;$$

$$(2.35)$$

$T^{(ab)}$ is the sum of all tree admittance products of the subnetwork of Fig. 2b, and

$$T^{(ab)} = y_2y_4 + y_2y_5 + y_3y_4 + y_3y_5 + y_4y_5; \qquad (2.36)$$

$T^{(cd)}$ is the sum of all tree admittance products of the subnetwork of Fig. 2c, and

$$T^{(cd)} = y_1y_2 + y_1y_3 + y_1y_4 + y_2y_3 + y_2y_4; \qquad (2.37)$$

$T^{(ab)\cap(cd)}$ is the sum of tree admittance products common to both $T^{(ab)}$ and $T^{(cd)}$, and

$$T^{(ab)\cap(cd)} = y_2 y_4; \tag{2.38}$$

$T^{(ab)(cd)}$ is the sum of all tree admittance products of the subnetwork of Fig. 2d, and

$$T^{(ab)(cd)} = y_2 + y_3 + y_4. \tag{2.39}$$

Finally,

$$T^{(abcd)} = 1 \text{ by definition.} \tag{2.40}$$

Again, let

Δ_n = the node determinant of an ordinary network;
$\Delta_{n_{ij}}$ = the cofactor of the node determinant for the (i, j)-position;
$\Delta_{n_{ij,kl}}$ = the algebraic complement of the second-order minor of the node determinant. $\tag{2.41}$

Then, from Eqs. (2.34) and (2.32),

$$\Delta_n = T$$
$$\Delta_{n_{ij}} = T^{(ri)} \qquad \text{for } i = j \tag{2.42}$$
$$\Delta_{n_{ij}} = T^{(ri)\cap(rj)} \qquad \text{for } i \neq j,$$

where the rth node is the reference node of the network. It can also be verified that

$$\Delta_{n_{ij,kl}} = T^{(rik)} \qquad \text{for } i = j \text{ and } k = l$$
$$\Delta_{n_{ij,kl}} = T^{(rik)\cap(ril)} \qquad \text{for } i = j \text{ and } k \neq l$$
$$\Delta_{n_{ij,kl}} = T^{(rik\cap(rjk)} \qquad \text{for } i \neq j \text{ and } k = l \tag{2.43}$$
$$\Delta_{n_{ij,kl}} = T^{(rik)\cap(rjl)} \qquad \text{for } i \neq j \text{ and } k \neq l.$$

Using the notations defined and Eqs. (2.42) and (2.43) the following identities are obtained:

1. $T^{(ri)} = T^{(ri)\cap(rj)} + T^{(ri)\cap(ij)}$
2. $T^{(ij)} = T^{(ri)\cap(ij)} + T^{(rj)\cap(ij)}$
3. $T^{(rij)} = T^{(rij)\cap(rik)} + T^{(rij)\cap(ijk)} + T^{(rij)\cap(rjk)}$
4. $T^{(ijk)} = T^{(rij)\cap(ijk)} + T^{(rjk)\cap(ijk)} + T^{(rik)\cap(ijk)}$
5. $T^{(ij)}T^{(ik)} - [T^{(ij)\cap(ik)}]^2 = TT^{(ijk)}$ $\tag{2.44}$
6. $T^{(ijk)}T^{(ijl)} - [T^{(ijk)\cap(ijl)}]^2 = T^{(ij)}T^{(ijkl)}$
7. $T^{(ij)} + T^{(ik)} - 2T^{(ij)\cap(ik)} = T^{(jk)}$
8. $T^{(ijk)} + T^{(ijl)} - 2T^{(ijk)\cap(ijl)} = T^{(ij)(kl)}.$

Proof for identity 1: Let

$$
\Delta_n = \begin{vmatrix}
y_{11} & \cdots & -y_{1h} & -y_{1i} & -y_{1j} & \cdots & -y_{n} \\
\cdot & \cdot & \cdot & \cdot & \cdot & \cdot & \cdot \\
-y_{hi} & \cdot & \cdot & \cdot & \cdot & \cdot & -y_{hn} \\
-y_{i1} & \cdot & \cdot & \cdot & \cdot & \cdot & -y_{in} \\
-y_{j1} & \cdot & \cdot & \cdot & \cdot & \cdot & -y_{jn} \\
\cdot & \cdot & \cdot & \cdot & \cdot & \cdot & \cdot \\
-y_{n1} & \cdot & \cdot & \cdot & \cdot & \cdot & y_{nn}
\end{vmatrix}. \tag{2.45}
$$

Now, consider a cofactor of the determinant for the (i,j)-position, and in the cofactor add all the columns for each row at jth column. Thus, we have

$$
\Delta_{n_{ij}} = \begin{vmatrix}
y_{11} & \cdots & -y_{1h} & (y_{1r}+y_{1i}) & \cdots & -y_{1n} \\
\cdot & \cdot & \cdot & \cdot & \cdot & \cdot \\
-y_{h1} & \cdots & y_{hh} & (y_{hr}+y_{hi}) & \cdots & -y_{hn} \\
-y_{j1} & \cdots & -y_{jh} & (y_{jr}+y_{ji}) & \cdots & -y_{jn} \\
\cdot & \cdot & \cdot & \cdot & \cdot & \cdot \\
-y_{n1} & \cdots & -y_{nh} & (y_{nr}+y_{ni}) & \cdots & y_{nn}
\end{vmatrix}
$$

$$
= \begin{vmatrix}
y_{11} & \cdots & -y_{1h} & y_{1i} & \cdots & -y_{1n} \\
\cdot & \cdot & \cdot & \cdot & \cdot & \cdot \\
-y_{h1} & \cdots & y_{hh} & y_{hi} & \cdots & -y_{hn} \\
-y_{j1} & \cdots & -y_{jh} & y_{ji} & \cdots & -y_{jn} \\
\cdot & \cdot & \cdot & \cdot & \cdot & \cdot \\
-y_{n1} & \cdots & -y_{nh} & y_{ni} & \cdots & y_{nn}
\end{vmatrix}
+ \begin{vmatrix}
y_{11} & \cdots & -y_{1h} & y_{1r} & \cdots & -y_{1n} \\
\cdot & \cdot & \cdot & \cdot & \cdot & \cdot \\
-y_{h1} & \cdots & y_{hh} & y_{hr} & \cdots & -y_{hn} \\
-y_{j1} & \cdots & -y_{jh} & y_{jr} & \cdots & -y_{jn} \\
\cdot & \cdot & \cdot & \cdot & \cdot & \cdot \\
-y_{n1} & \cdots & -y_{nh} & y_{nr} & \cdots & y_{nn}
\end{vmatrix}
$$

$$
= \Delta_{n_{ij}} + \Delta'_{n_{rj}}, \tag{2.46}
$$

where Δ'_n is the node-determinant of the same network with the ith node as its reference node. Therefore, from Eqs. (2.42) and (2.43) one has

$$
\Delta_{n_{ij}} = T^{(ir)\cap(jr)},
$$
$$
\Delta'_{n_{rj}} = T^{(ir)\cap(ij)}. \tag{2.47}
$$

The rest of the identities can be proved by a process similar to that used for the proof of identity 1.[33c] Some of the identities can be derived in terms of two-tree admittance products.[6]

The various network functions of an ordinary network are expressible in terms of the topological identities just derived. For a two-port network with four terminals as shown in Fig. 2.3, the open-

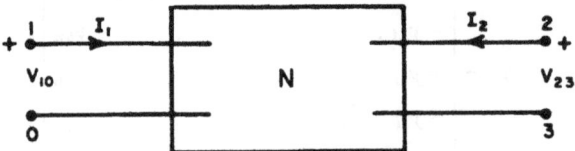

Fig. 2.3. A four-terminal network

circuit impedance parameters of the network are easily verified as

$$
\begin{bmatrix} z_{11} & z_{12} \\ z_{21} & z_{22} \end{bmatrix} = \frac{1}{\Delta_n} \begin{bmatrix} \Delta_{n_{11}} & \Delta_{n_{12}} - \Delta_{n_{13}} \\ \Delta_{n_{12}} - \Delta_{n_{13}} & \Delta_{n_{22}} + \Delta_{n_{33}} - 2\Delta_{n_{23}} \end{bmatrix}.
$$

$$
= \frac{1}{T} \begin{bmatrix} T^{(10)} & T^{(10)\cap(20)} - T^{(10)\cap(30)} \\ T^{(10)\cap(20)} - T^{(10)\cap(30)} & T^{(23)} \end{bmatrix}. \quad (2.48)
$$

The short-circuit admittance parameters are also directly derived from Eq. (2.48) and are given by

$$
\begin{bmatrix} y_{11} & y_{12} \\ y_{21} & y_{22} \end{bmatrix} = \frac{1}{T^{(10)(23)}} \begin{bmatrix} T^{(23)} & T^{(10)\cap(30)} - T^{(10)\cap(20)} \\ T^{(10)\cap(30)} - T^{(10)\cap(20)} & T^{(10)} \end{bmatrix}. \quad (2.49)
$$

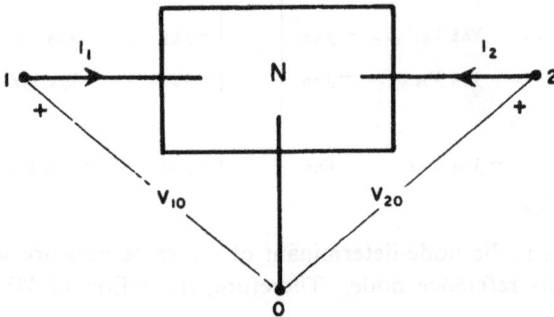

Fig. 2.4. A three-terminal network

For the three-terminal network of Fig. 2.4, the formulas of Eqs. (2.48) and (2.49) are reduced to

$$
\begin{bmatrix} z_{11} & z_{12} \\ z_{21} & z_{22} \end{bmatrix} = \frac{1}{T} \begin{bmatrix} T^{(10)} & T^{(10)\cap(20)} \\ T^{(10)\cap(20)} & T^{(20)} \end{bmatrix} \quad (2.50)
$$

$$\begin{bmatrix} y_{11} & y_{12} \\ y_{21} & y_{22} \end{bmatrix} = \frac{1}{T^{(012)}} \begin{bmatrix} T^{(20)} & -T^{(10)\cap(20)} \\ -T^{(10)\cap(20)} & T^{(10)} \end{bmatrix}. \quad (2.51)$$

Example 8. Given a network in Fig. 2.5, find the z parameters of the network. The node determinant of the network Δ_n is found to be

$$\Delta_n = T = y_1 y_2 (y_4 + y_5) + y_1 y_5 (y_3 + y_4) + y_2 y_4 (y_3 + y_5) + y_1 y_3 y_4 + y_2 y_3 y_5. \quad (2.52)$$

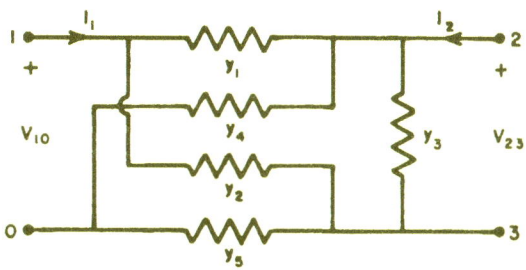

Fig. 2.5. *A four-terminal network used for Example 8*

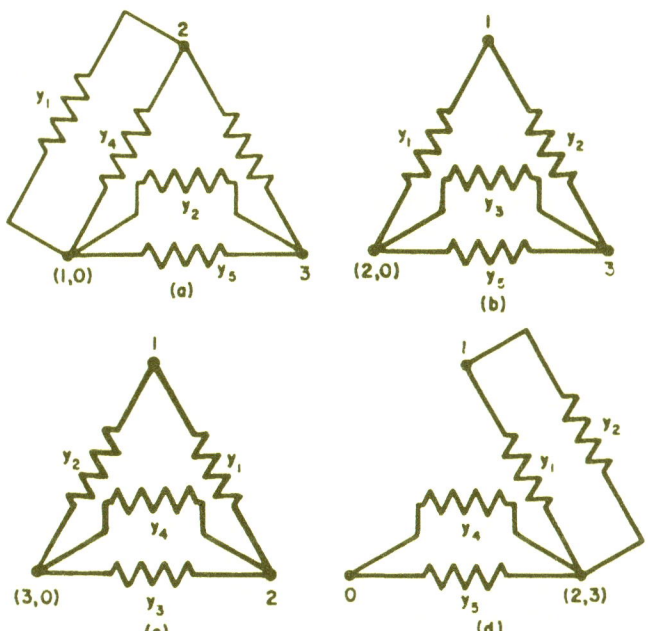

Fig. 2.6. *Subnetworks necessary to evaluate z parameters of the network of Fig. 2.5*

(a) $N^{(10)}$; (b) $N^{(20)}$; (c) $N^{(30)}$; (d) $N^{(23)}$.

The various cofactors necessary to evaluate the z parameters are found respectively from Figs. 2.6a, 2.6b, 2.6c, and 2.6d as:

$$T^{(10)} = (y_1+y_4)(y_2+y_3+y_5)+y_3(y_2+y_5) \qquad (2.53\text{a})$$

$$T^{(20)} = (y_3+y_5)(y_1+y_2)+y_1y_2 \qquad (2.53\text{b})$$

$$T^{(30)} = (y_3+y_4)(y_1+y_2)+y_1y_2 \qquad (2.53\text{c})$$

$$T^{(23)} = (y_4+y_5)(y_1+y_2) \qquad (2.53\text{d})$$

$$T^{(10)\cap(20)} = y_1(y_2+y_3+y_5)+y_2y_3 \qquad (2.53\text{e})$$

$$T^{(10)\cap(30)} = y_1y_2+y_1y_3+y_2y_3+y_2y_4. \qquad (2.53\text{f})$$

We have, therefore,

$$z_{11} = (y_1+y_4)(y_2+y_3+y_5)+y_3(y_2+y_5)/T$$

$$z_{12} = z_{21} = T^{(10)\cap(20)} - T^{(10)\cap(30)}/\,T = (y_1y_5-y_2y_4)/T \qquad (2.54)$$

$$z_{22} = T^{(23)} = (y_4+y_5)(y_1+y_2)/T.$$

Let us consider the cofactor $\Delta_{n_{11}}$ of the node determinant Δ_n of the network. Then,

$$\Delta_{n_{11}} = T^{(10)} = (y_1+y_4)(y_2+y_3+y_5)+y_3(y_2+y_5). \qquad (2.55)$$

If we define a "2-tree admittance product" (or simply a 2-tree product) as the product of the admittances of the branches which constitute a 2-tree, with our convention that the tree-product of a single node is unity, then each tree-product term in Eq. (2.55) corresponds to a 2-tree of the network as shown in Fig. 2.7.

Since all the tree-products which appear in Eq. (2.55) correspond to all the 2-tree products of network N, one can express the cofactor $\Delta_{n_{11}}$ in terms of 2-tree products of N instead of in terms of tree-products of the subnetworks. Each of the 2-trees shown in Fig. 2.7 has two separated pieces such that one part contains node 1 and the other node 0; the reference node. This may be clearly understood for the following reasons.

The product terms in Eq. (2.55) are the trees of $N^{(10)}$ which is the subnetwork of N with nodes 1 and 0 identified. Therefore, the number of nodes of the subnetwork is one less than that of the original one.

The edges connected between nodes 1 and 0 would not appear in the subnetwork. If we choose a 2-tree of the original network such that one part of the 2-tree contains node 1 and the other node 0, then the edge(s) connected between nodes 1 and 0 will never be counted.

In general, a tree of a network N (or a connected graph) of $(n+1)$ nodes contains n branches. But a 2-tree of the same network has

$(n-1)$ branches while the number of branches of a tree of subnetwork $N^{(ij)}$ of N is also $(n-1)$. Thus, denoting the sum of 2-tree products of a network such that one part of each of the 2-trees contains node i and the other node j by $T_{i,j}$, we have the following identity:

$$T^{(ij)} = 0 \qquad \text{for } i = j$$
$$T^{(ij)} = T_{j,i} \qquad \text{for } i \neq j. \tag{2.56}$$

Extending the concept of a 2-tree product defined in Eq. (2.56) one obtains

$$T^{(ij) \cap (kj)} = T_{ik,j} \qquad \text{for } i \neq j \neq k. \tag{2.57}$$

The substitution of Eqs. (2.56) and (2.57) into Eq. (2.44) yields the following identities in terms of 2-tree products of a network:

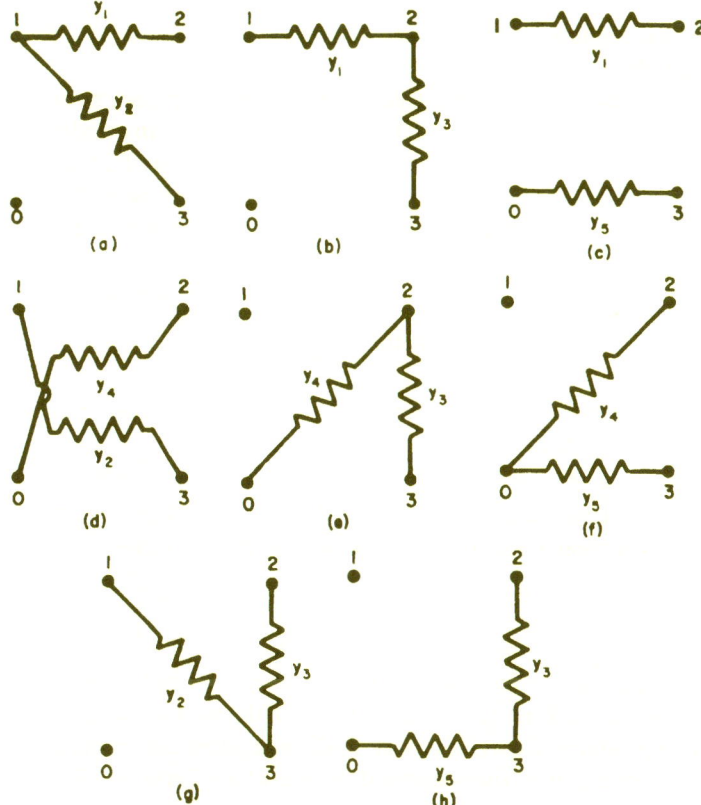

Fig. 2.7. *Two-tree products which appear in Eq.* (2.55)
(a) y_1y_2; (b) y_1y_3; (c) y_1y_5; (d) y_2y_4; (e) y_3y_4; (f) y_4y_5; (g) y_2y_3; (h) y_3y_5.

$$T_{i,j} = T_{ik,r} + T_{i,kr} \qquad \text{for } i \neq k \neq j = r$$
$$T_{i,j} = T_{ir,j} + T_{i,rj} \qquad \text{for } i \neq j \neq r \qquad (2.58)$$
$$T_{i,j} = 0 \qquad \text{for } i = j.$$

By studying Eq. (2.58), one can extend the formulas into the following new identities:

$$T_{ij,k} = T_{ijl,r} + T_{ij,r} \qquad \text{for } i \neq j \neq l \neq k = r$$
$$T_{ij,k} = T_{ijr,k} + T_{ij,kr} \qquad \text{for } i \neq j \neq k \neq r. \qquad (2.59)$$

If we use Eq. (2.59) in order to evaluate the quantity $\Delta_{n_{12}} - \Delta_{n_{13}}$, we obtain

$$\Delta_{n_{12}} - \Delta_{n_{13}} = T_{12,0} - T_{13,0} = (T_{123,0} + T_{12,30}) - (T_{123,0} + T_{13,20})$$
$$= T_{12,30} - T_{13,20}. \qquad (2.60)$$

The expression of Eq. (2.60) was intuitively derived by Percival[28] using the following scheme:

$\Delta_{n_{12}} - \Delta_{n_{13}} =$ (sum of 2-tree products of a two-port network such that one part of each 2-tree contains nodes 1 and 2 and the other nodes 3 and 0) − (sum of 2-tree products of the same network such that one part of each 2-tree contains nodes 2 and 3 and the other nodes 2 and 0)

= (2-tree products of) $\begin{smallmatrix}1\\0\end{smallmatrix}\left(\begin{smallmatrix}\circ\!-\!-\!-\!\circ\\\circ\!-\!-\!-\!\circ\end{smallmatrix}\right)\begin{smallmatrix}2\\3\end{smallmatrix}$ minus

(2-tree products of) $\begin{smallmatrix}1\\0\end{smallmatrix}\left(\begin{smallmatrix}\circ\;\;\circ\\\circ\;\;\circ\end{smallmatrix}\right)\begin{smallmatrix}2\\3\end{smallmatrix}$.

Therefore, the z parameters of a four-terminal network of Fig. 2.3 are found in terms of 2-tree products as[†]:

$$\begin{bmatrix} z_{11} & z_{12} \\ z_{21} & z_{22} \end{bmatrix} = \frac{1}{T} \begin{bmatrix} T_{1,0} & T_{12,30} - T_{13,20} \\ T_{12,30} - T_{13,20} & T_{2,3} \end{bmatrix}, \qquad (2.61)$$

and for a three-terminal network of Fig. 2.4 we have

$$\begin{bmatrix} z_{11} & z_{12} \\ z_{21} & z_{22} \end{bmatrix} = \frac{1}{T} \begin{bmatrix} T_{1,0} & T_{12,0} \\ T_{12,0} & T_{2,0} \end{bmatrix}. \qquad (2.62)$$

Comparing the formulas of the z parameters in terms of common tree-products, Eqs. (2.48) and (2.50), with the formulas in terms of 2-tree products, Eqs. (2.61) and (2.62), one notes that in the evaluation of z_{12} or z_{21} of a four-terminal network the formula of Eq. (2.48) may involve some terms which will be canceled with each other

† See Reference 6 for the various formulas in terms of 2-tree and 3-tree products.

because there may be some of the tree-products common to both $T^{(10)\cap(20)}$ and $T^{(10)\cap(30)}$. But by using Eqs. (2.61) it is clear that there will be no canceling term(s) in the computation of z_{12}. For this reason, Eq. (2.61) is called the "minimum effort formula" by some authors.[7] Of course, for a three-terminal two-port network there would be no difference between Eqs. (2.50) and (2.62). In general, the set of formulas (2.48) and (2.50) are more suitable to digital computer analysis than the other set of formulas.

Using the topological formulas, we can easily find a number of necessary conditions for the realizability of the voltage-ratio transfer function of an ordinary two-port network. Let us denote the voltage-ratio transfer function by $G(s)$. Then, $G(s)$ may be written, in general, in the following form:[33c]

$$G(s) = \frac{P(s)}{Q(s)} = \frac{a_m s^m + a_{m-1} s^{m-1} + \ldots + a_0}{b_n s^n + b_{n-1} s^{n-1} + \ldots + b_0}, \qquad (2.63)$$

where $P(s)$ and $Q(s)$ have no common factor(s), that is, all common factors have been canceled out.

One also finds $G(s)$ in terms of tree-products or 2-tree products as:

$$G(s) = \frac{z_{12}}{z_{11}} = \frac{T^{(10)\cap(20)}}{T^{(10)}} = \frac{T_{12,0}}{T_{1,0}}$$

for a three-terminal network,

$$G(s) = \frac{z_{12}}{z_{11}} = \frac{T^{(10)\cap(20)} - T^{(10)\cap(30)}}{T^{(10)}} \qquad \frac{T_{12,30} - T_{13,20}}{T_{1,0}} \qquad (2.64)$$

for a four-terminal network.

From Eq. (2.64) it follows that every tree-product or 2-tree product which appears in the numerator of $G(s)$ will be included in the denominator but not every term in the denominator will appear in the numerator. Therefore, one can conclude that[†]: (1) $n \geqslant m$, that is, no pole is at infinity; (2) $b_i \geqslant |a_i|$ for all values of i; (3) no pole is at the origin, that is, $b_n \neq 0$. Readers should refer to the work of Hakimi and Mayeda[14b] for a more thorough discussion of this subject including two element-kind networks.

The topological formulas derived thus far are certainly useful in network analysis since there may be no canceling terms involved in the evaluation of the various network functions. Furthermore, the

---

[†] The same set of conditions was found by Fialkow and Gerst (see Reference 29) and also by Weinberg (see Reference 30).

formulas may be more suitable for programming into a digital computer. However, they become no longer very efficient if a network under consideration contains a large number of elements, because the number of trees or 2-trees of the network increases rapidly as the number of elements of a network increases. In such cases, it may be desirable to decompose a network into a number of subnetworks, and compute pertinent tree-products or 2-tree products of each subnetwork. Then, by putting the results together, one may evaluate the original network function. Based on this motivation, several decomposition techniques have been proposed.[31,33b,c]

At present, although a few attempts have been made,[32,33a,c] there exists no system for the application of topological relationships to network synthesis.

CHAPTER 3

Linear Active Multipoles

THE TOPOLOGICAL METHOD is not directly applicable to a network containing linear active devices, such as vacuum tubes and transistors, or mutually coupled passive devices, such as transformers and gyrators. The difficulty lies in the graphical representation of mutually dependent node-pairs due to a dependent source or the mutual coupling effect imbedded in a device. In order to characterize a dependent node-pair in a network by graphical representation, Percival[28c] recently introduced artificial two-terminal elements, so-called "current and voltage elements," and hence made it possible for the theory of linear graphs to be extended to the analysis of active as well as mutually coupled networks.

GRAPHICAL CHARACTERIZATION OF MUTUALLY DEPENDENT NODE-PAIRS[18, 19, 28c, 33f, 33e]

Let us consider node-pairs (pq) and (mn) in the linear network N shown in Fig. 3.1. If the current $i_{pq}(t)$ flowing from node p to node q

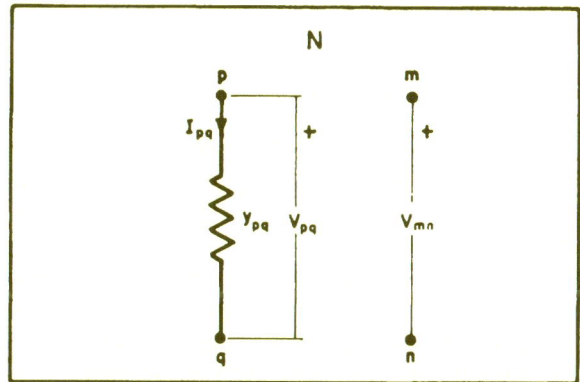

Fig. 3.1. A linear network with mutually dependent node-pairs

depends not only on the voltage-difference between nodes p and q, $v_{pq}(t)$, but also on the voltage difference between nodes m and n, $v_{mn}(t)$, then, since the network is linear, one may write

$$y_{pq}V_{pq} + y_{pq,mn}V_{mn} = I_{pq}, \qquad (3.1)$$

where the y's are proportional factors and $I = \mathscr{L}i(t)$, $V = \mathscr{L}v(t)$.

In Eq. (3.1), since y_{pq} relates the voltage and current for the same node-pair, it is called the "self-admittance" or simply "admittance" of the element (or elements) connected between nodes p and q. On the other hand, $y_{pq,mn}$ relates the voltage and current for different node-pairs, and hence is called the "mutual admittance."

If the network N consists entirely of passive devices, then Eq. (3.1) is divided into two isolated parts,

$$y_{pq}V_{pq} = I_{pq} \qquad (3.2a)$$

or

$$y_{pq,mn}V_{mn} = I_{pq}. \qquad (3.2b)$$

It is clear, of course, that if N contains only the ordinary two-terminal devices (R, L, and C elements), then $y_{pq} = y_{qp}$ for all node-pairs in the network, which is said to be "passive and reciprocal."[†] However, if transformers or gyrators are contained in N, then for some node-pairs in the network $y_{pq,mn} \neq y_{mn,pq}$, and N is called "passive and nonreciprocal." Therefore, Eq. (3.1) holds for some node-pairs in N if it contains a linear active device or devices.

In Eq. (3.1), I_{pq}, which depends on both V_{pq} and V_{mn}, can be decomposed into two components: one due to V_{pq} denoted by I'_{pq} and the other, due to V_{mn}, denoted by I''_{pq}. Thus

$$y_{pq}V_{pq} = I'_{pq} \qquad (3.3a)$$

$$y_{pq,mn}V_{mn} = I''_{pq} \qquad (3.3b)$$

$$I'_{pq} + I''_{pq} = I_{pq}. \qquad (3.3c)$$

The voltage–current relationship of Eq. (3.3a) represents an ordinary two-terminal element. Equation (3.3b) shows the dependent relationship of node-pairs (pq) and (mn). A voltage-difference between nodes m and n, with m positive, causes a current flowing from nodes p to q.

In order to apply the concept of graphs, it is therefore desirable to introduce artificial two-terminal elements into node-pairs (mn) and (pq) to characterize their dependent relationship. The element inserted between nodes m and n, called a "voltage element," is to sense

† Note here that we are not concerned with a negative resistor or a negative element, and we imply dependent sources by the word "active."

a voltage-difference in the node-pair. The element introduced in node-pair (*pq*), called a "current element," produces a current of magnitude I'' flowing from node *p* to node *q* due to the voltage-difference *V* in node-pair (*mn*). The voltage and current elements are thus related by mutual admittance $y_{pq,mn}$. Note that voltage and current elements always occur in a pair but between different node-pairs. If they occur in the same node-pair, that is, $p = m$ and $q = n$, the mutual admittance $y_{pq,mn}$ is reduced to the self-admittance y_{pq}. The graphical representation of voltage and current elements for dependent node-pairs (*pq*) and (*mn*) is shown in Fig. 3.2.

Since the current and voltage elements are two-terminal elements they can be represented by the edges of a linear graph as "current

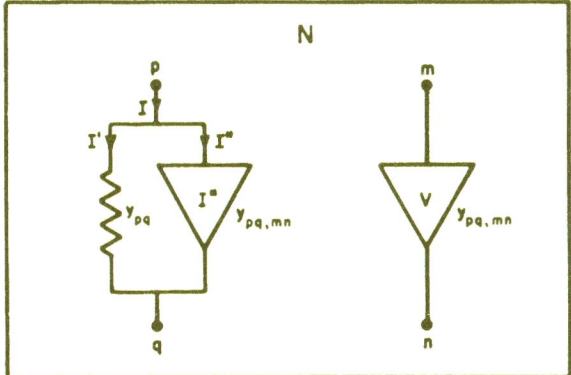

Fig. 3.2. Use of voltage and current elements for dependent node-pairs

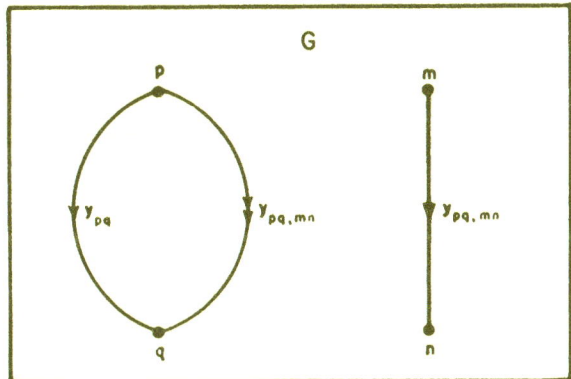

Fig. 3.3. Graphical representation of dependent node-pairs

and voltage edges," respectively. The weight of the edges is the mutual admittance. The orientation of a current edge is determined by the direction of the current-flow in the corresponding current element, the orientation of a voltage edge by the polarity of the corresponding voltage element. The graphical representation of the network N of Fig. 3.2 is shown in Fig. 3.3, where G denotes the graph corresponding to the network N. The current edge is indicated by double arrows, and the voltage edge by a triangular-shaped arrow.

The following examples illustrate the concept and use of current and voltage elements in characterizing a network containing linear active devices as well as mutually coupled multiterminal passive elements.

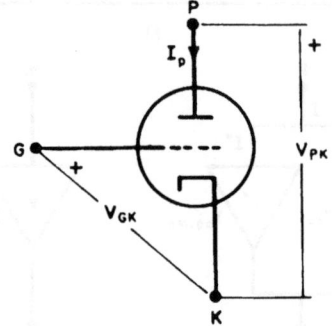

Fig. 3.4. A triode

Example 9. A triode of Fig. 3.4 is characterized as a linear active device by

$$I_P = g_m V_{GK} + g_p V_{PK}, \tag{3.4}$$

where g_m is the transconductance and g_p the plate-conductance. An equivalent circuit of the triode characterized by Eq. (3.4) in terms of voltage and current elements is known as a "mathematical equivalent circuit," and is shown in Fig. 3.5a. Figure 3.5b shows the conventional equivalent circuit of a triode for comparison. The graphical representation of Fig. 3.5a is given in Fig. 3.6.

A two-channel gyrator, shown in Fig. 3.7, may be characterized by the following set of equations:

$$\begin{aligned} I_1 &= \alpha_1 V_2 \\ I_2 &= -\alpha_2 V_1, \end{aligned} \tag{3.5}$$

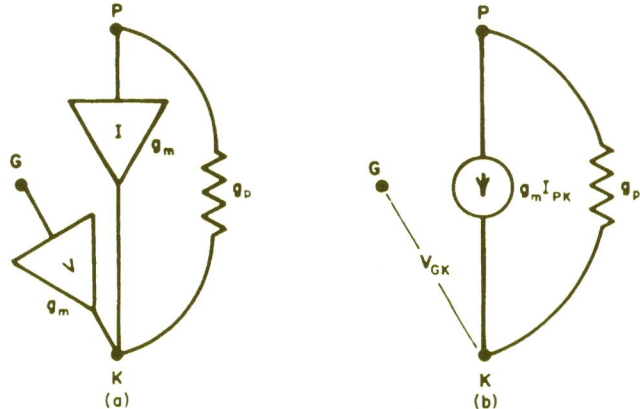

Fig. 3.5. Equivalent circuits of a triode
(a) Mathematical equivalent circuit; (b) conventional equivalent circuit.

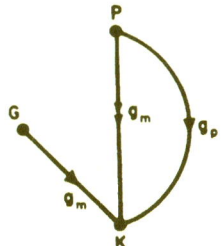

Fig. 3.6. Graphical representation of a triode

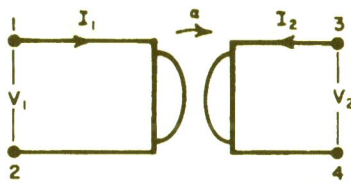

Fig. 3.7. A gyrator

where $\alpha_1 = \alpha_2 = \alpha$ is the gyrator admittance. The equivalent circuit of the gyrator represented by the terminal characterization of Eq. (3.5) is shown in Fig. 3.8a and its graphical representation in Fig. 3.8b.

Now, a graph representing a device or a network can be decomposed into two subgraphs. One is the "current graph" containing the edges corresponding to the ordinary elements and current elements. The other is the "voltage graph," with the edges corresponding to the ordinary elements and voltage elements. One may immediately see that if a network consists entirely of ordinary elements, the current and voltage graphs of the network are identical;

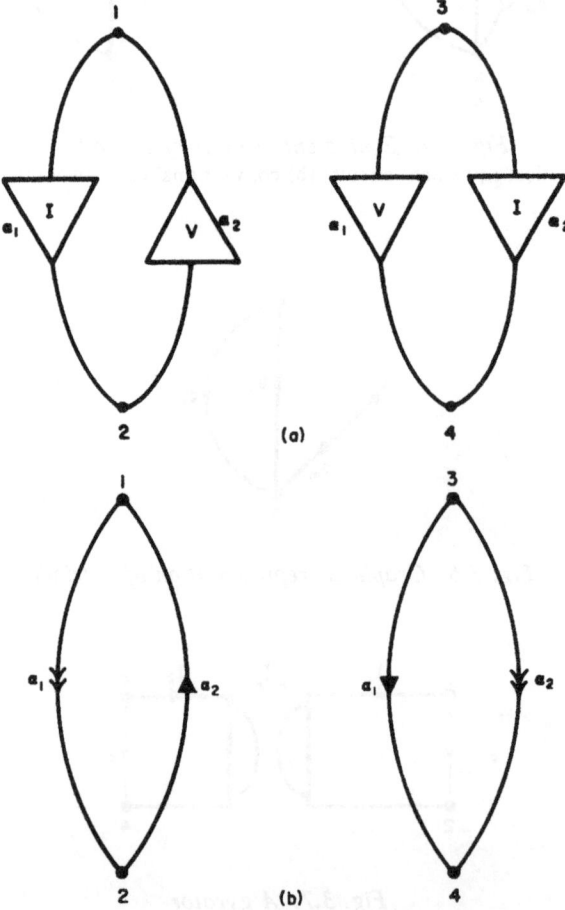

Fig. 3.8. Graphical characterization of a gyrator, where $\alpha_1 = \alpha_2 = \alpha$ (a) Mathematical equivalent circuit; (b) graphical representation.

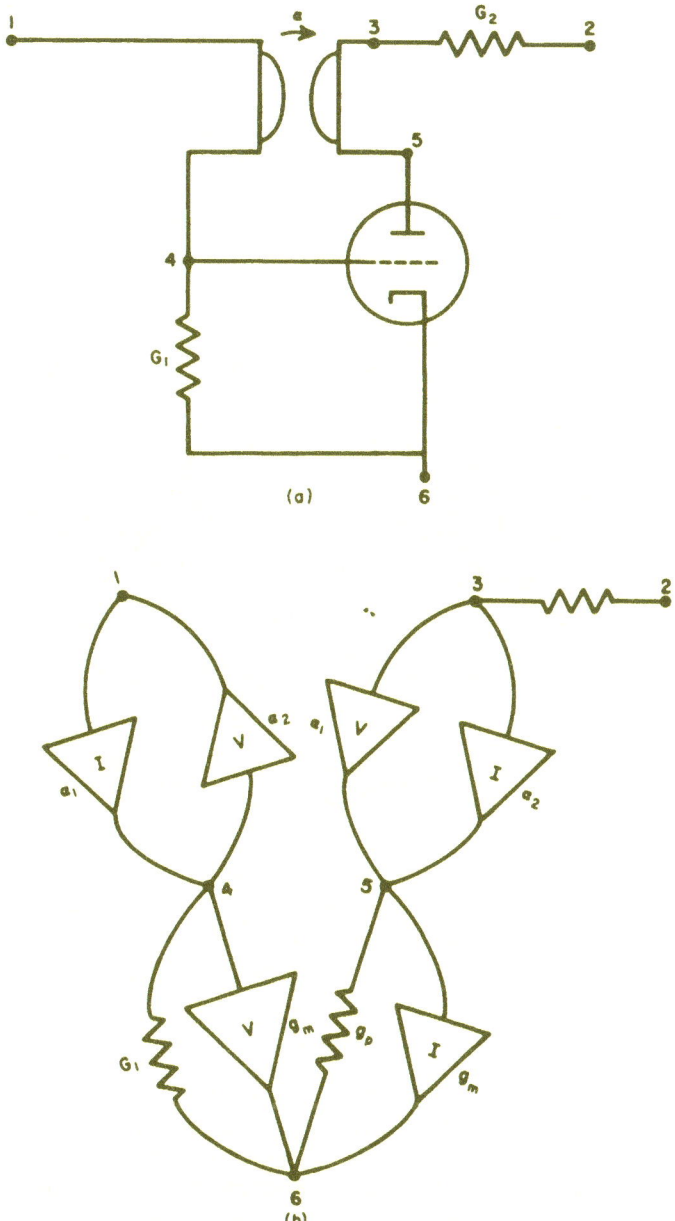

Fig. 3.9. A network and its equivalent circuit, where $\alpha_1 = \alpha_2 = \alpha$
(a) A network with dependent node-pairs; (b) mathematical equivalent circuit.

i.e., a graph corresponding to an ordinary network is the current graph of the network and at the same time the voltage graph. The orientation of each edge of a graph of an ordinary network is arbitrarily assigned. When a network contains active devices or mutually coupled devices, the current and voltage graphs of the network may not be identical. This is illustrated in Example 10.

Example 10. Given the network of Fig. 3.9a; then, the mathematical equivalent circuit of the network is shown in Fig. 3.9b. The current and voltage graphs of the network are shown in Figs. 3.10a

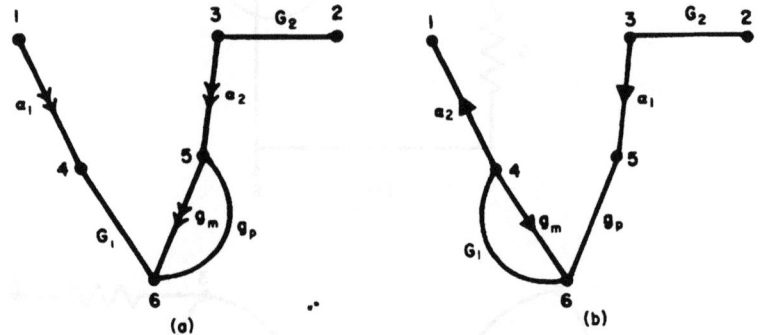

(a)   (b)

Fig. 3.10. (a) Current and (b) voltage graphs of the network of Fig. 3.9

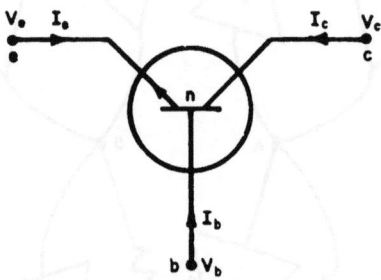

Fig. 3.11. A three-terminal transistor

and 3.10b, respectively, where the orientation of the ordinary edges is not indicated since it can be arbitrarily assigned.

Next, let us consider the three-terminal transistor shown in Fig. 3.11, where e, c, and b represent the emitter, collector, and base terminals, respectively. It may be characterized as a linear active device as follows:

$$I_e = g_e(V_e - V_n) \tag{3.6a}$$

$$I_c = -ag_e(V_e - V_n) + g_c(V_c - V_n) \tag{3.6b}$$

$$I_b = g_b(V_b - V_n) \tag{3.6c}$$

$$I_e + I_c + I_b = 0, \tag{3.6d}$$

where g_e is the emitter conductance, g_c is the collector conductance, g_b is the base conductance, and a the current amplification factor. If one characterizes the first term of the right-hand expression of Eq. (3.6b) by current and voltage elements with mutual admittance ag_e, then the mathematical equivalent circuit of a three-terminal

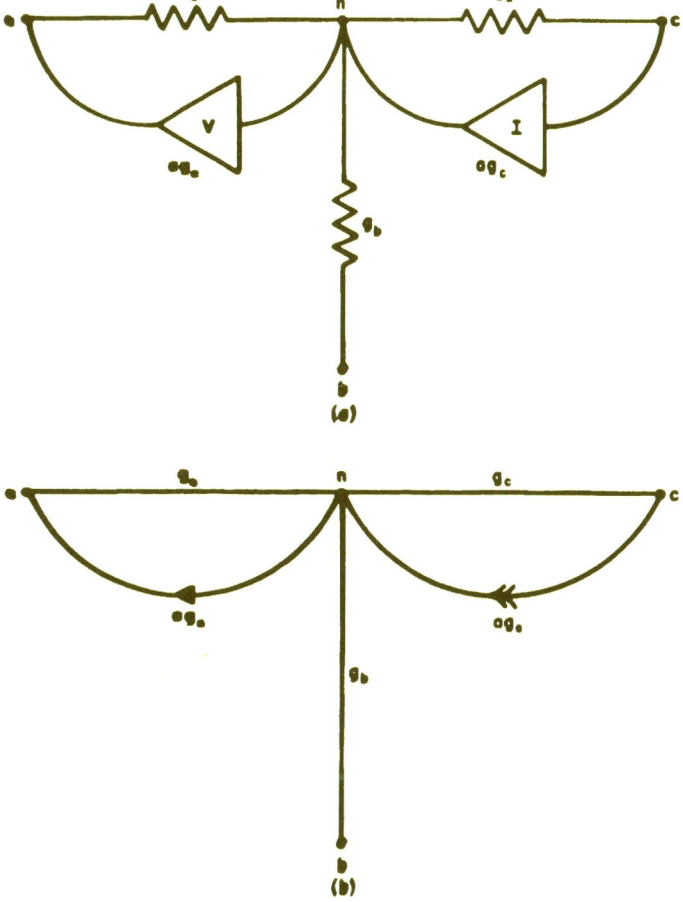

Fig. 3.12. *Graphical characterization of a three-terminal transistor* (a) Mathematical equivalent circuit; (b) graphical representation.

transistor and its graphical representation are as given in Fig. 3.12.

By this time it should be clear that the current and voltage graphs of a network with dependent node-pairs are *2-semi-isomorphic* with each other. This is so because they always contain the same number of nodes and edges since all current and voltage edges occur in pairs with the identical edge-weight. We therefore will use the properties of 2-semi-isomorphic graphs which were developed in Chapter I.

TOPOLOGICAL FORMULAS OF NODE-DETERMINANTS OF ACTIVE NETWORKS

With the aid of the graphical representation of a linear network containing dependent node-pairs in terms of current and voltage graphs, one can derive a set of topological formulas for the analysis of the network. Furthermore, making use of the properties of 2-semi-isomorphic graphs, we shall deduce a formula for the computation of the node determinant of the network which would be similar to the one which we found in the previous chapter for the ordinary networks.

Let us consider the mathematical equivalent circuit of a linear network N and denote its current and voltage graphs by G_I and G_V, respectively. We also assume that each graph has b edges and n nodes. We shall now define:

A_{1_V}, A_{1_I} as the incidence matrix of G_I and G_V, respectively, with the reference node being deleted;

B_V, B_I as the loop matrix of $(b-n+1)$ independent loops in G_I and G_V, respectively;

V_e, I_e as the column matrix of all edge-voltages and edge-currents, respectively;

V_n as the column matrix of node voltages with respect to the reference node;

I_l as the column matrix of loop currents for the $(b-n+1)$ independent loops;

Z_e, Y_e as the diagonal matrix of impedance and admittance of each edge and for all the edges in G_I and G_V, respectively;

E, J as the column matrix of the independent voltage sources and current sources connected at the terminal nodes, respectively.

It is assumed that all initial conditions are equal to zero. Then, we have

Kirchhoff's laws

$$\mathbf{A}_{1_t}\mathbf{I}_e = \mathbf{J} \tag{3.7a}$$

$$\mathbf{B}_V\mathbf{V}_e = \mathbf{E}. \tag{3.7b}$$

Node and loop transformations

$$\mathbf{V}_e = \mathbf{A}_{1_{v}}{}^t\mathbf{V}_n \tag{3.8a}$$

$$\mathbf{I}_e = \mathbf{B}_l{}^t\mathbf{I}_l \tag{3.8b}$$

Ohm's laws

$$\mathbf{I}_e = \mathbf{Y}_e\mathbf{V}_e \tag{3.9a}$$

$$\mathbf{V}_e = \mathbf{Z}_e\mathbf{I}_e. \tag{3.9b}$$

The substitution of Eq. (3.8a) into Eq. (3.9a) will yield

$$\mathbf{I}_e = \mathbf{Y}_e\mathbf{A}_{1_v}{}^t\mathbf{V}_n. \tag{3.10}$$

Substituting Eq. (3.10) again into Eq. (3.7a) we have

$$(\mathbf{A}_{1_t}\mathbf{Y}_e\mathbf{A}_{1_v}{}^t)\mathbf{V}_n = \mathbf{J}. \tag{3.11}$$

It is, therefore, clear that the node-determinant Δ_n of the network is[†]

$$\Delta_n = |\mathbf{A}_{1_t}\mathbf{Y}_e\mathbf{A}_{1_v}{}^t|. \tag{3.12}$$

A similar formula exists for the loop determinant Δ_l of the network using Eqs. (3.7b), (3.8b), and (3.9b), as,

$$\Delta_l = |\mathbf{B}_V\mathbf{Z}_e\mathbf{B}_l{}^t|. \tag{3.13}$$

Example 11. Given an ideal three-channel circulator as shown in Fig. 3.13, the circulator is characterized by

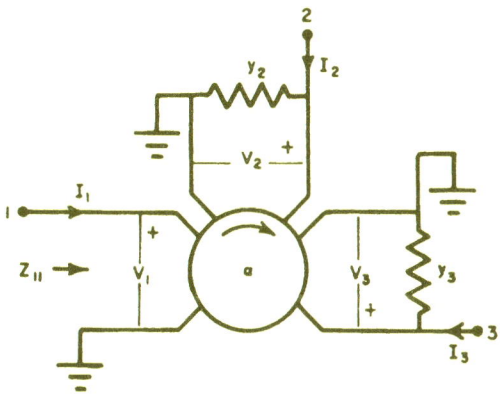

Fig. 3.13. A three-channel circulator

[†] This formula was first derived by Coates[18] and later also by Mayeda.[19]

$$I_1 = \alpha_1(V_3 - V_2)$$

$$I_2 = \alpha_2(V_1 - V_3) + y_2 V_2 \qquad (3.14)$$

$$I_3 = \alpha_3(V_2 - V_1) + y_3 V_3,$$

where $\alpha_1 = \alpha_2 = \alpha_3 = \alpha$ is the circulator admittance. The mathematical equivalent circuit and its current and voltage graphs are given in Fig. 3.14, where node n is the reference node and the orienta-

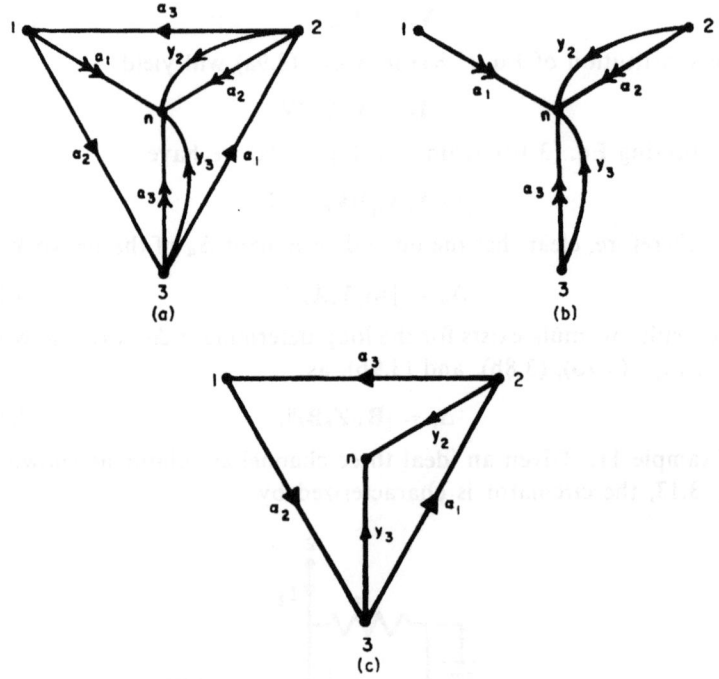

Fig. 3.14. *Graphical characterization cf the circulator of Fig. 3.13*
(a) Mathematical equivalent circuit; (b) current graph G_I; (c) voltage graph G_V.

tion of the ordinary edges is arbitrarily assigned.

The node-determinant of the circulator Δ_n is given as

$$\Delta_n = |A_{1_I} Y_e A_{1_V}{}^t| = \begin{vmatrix} 0 & -\alpha_1 & \alpha_1 \\ \alpha_2 & y_2 & -\alpha_2 \\ -\alpha_3 & \alpha_3 & y_3 \end{vmatrix} \qquad (3.15a)$$

and

$$
\mathbf{A_{1_I}} = \begin{array}{c} 1 \\ 2 \\ 3 \end{array} \begin{array}{ccccc} \alpha_1 & \alpha_2 & \alpha_3 & y_2 & y_3 \\ \left[\begin{array}{ccccc} 1 & 0 & 0 & 0 & 0 \\ 0 & 1 & 0 & 1 & 0 \\ 0 & 0 & 1 & 0 & 1 \end{array} \right] \end{array} \tag{3.15b}
$$

$$
\mathbf{A_{1_V}} = \begin{array}{c} 1 \\ 2 \\ 3 \end{array} \begin{array}{ccccc} \alpha_1 & \alpha_2 & \alpha_3 & y_2 & y_3 \\ \left[\begin{array}{ccccc} 0 & 1 & -1 & 0 & 0 \\ -1 & 0 & 1 & 1 & 0 \\ 1 & -1 & 0 & 0 & 1 \end{array} \right] \end{array} \tag{3.15c}
$$

$$
\mathbf{Y_e} = \left[\begin{array}{ccccc} \alpha_1 & 0 & 0 & 0 & 0 \\ 0 & \alpha_2 & 0 & 0 & 0 \\ 0 & 0 & \alpha_3 & 0 & 0 \\ 0 & 0 & 0 & y_2 & 0 \\ 0 & 0 & 0 & 0 & y_3 \end{array} \right]. \tag{3.15d}
$$

By the use of the Binet–Cauchy theorem the node-determinant given in Eq. (3.12) is rewritten as:

Δ_n = sum of the products of all possible pairs of the corresponding majors of $(\mathbf{A_{1_I}}\mathbf{Y_e})$ and $\mathbf{A_{1_V}}{}^t$. (3.16)

If we apply Properties 6 and 11 of a connected graph we should be able to deduce a topological formula similar to the one which we obtained in Eq. (2.32) for the ordinary networks. It should be noted, however, that two matrices $\mathbf{A_{1_I}}$ and $\mathbf{A_{1_V}}{}^t$ may not be identical with each other if the network under consideration contains dependent node-pairs. In other words, the sign associated with each tree-product which is common to both G_I and G_V may not be the same. Thus, in general,

$$
\Delta_n = \sum_i \epsilon_i \times (\text{common tree-product } t_i \text{ of current graph } G_I \text{ and voltage graph } G_V \text{ of network } N), \tag{3.17}
$$

where a common tree-product is a tree-product which is common to both current and voltage graphs and ϵ_i is the sign of common tree-product t_i taking the value of 1 or -1. The summation is for all possible common tree-products.

When a network contains only the ordinary edges, the current and voltage graphs of the network are identical, i.e., $A_I = A_V$. Therefore, the sign of a tree-product is always positive. However, when a network includes dependent node-pairs, its current and voltage graphs are different because of active edges in the graphs. Hence, a tree-product which is common to both current and voltage graphs may *not* always be positive. However, since the current and voltage graphs of a network are 2-semi-isomorphic with each other, we can use directly the formula proposed in Property 19 for the determination of the sign for each common tree. That is, each pair of current and voltage edges which will occur in the current and voltage graphs of a network is an active edge-pair as defined in the first chapter. An edge corresponding to an ordinary network element, such as a resistor, inductor, or capacitor, is an ordinary edge and the direction of the edge can be assigned arbitrarily. As we have proved previously, the sign of a common tree-product of the current and voltage graphs will be determined only by the sign of active edge-pairs included in the corresponding tree-pair of the two graphs. In other words, the ordinary edges contained in each tree of the current and voltage graphs should be reduced to zero, i.e., two nodes of an ordinary edge are identified with each other and we call the resulting subgraph of the tree the "reduced tree" of a tree. We therefore rewrite the formula for the determination of the sign ϵ_i for a common tree-product t_i of current and voltage graphs G_I and G_V as:

$$\epsilon_i = (-1)^{\gamma} \overset{k}{\Pi} \text{ (sign of active edge-pairs of the reduced tree-pair}$$

$$\text{of a tree-pair } t_i \text{ of } G_I \text{ and } G_V), \tag{3.18}$$

where γ is the number of interchanges of edges needed to give all active edge-pairs in the reduced tree-pair the same principal nodes, k is the number of active edge-pairs in the reduced tree-pair, and the sign of an active edge-pair is defined in Definition 24.

Example 12. Given: a network containing a tube and a gyrator as shown in Fig. 3.15a. The mathematical equivalent circuit of the network in terms of current and voltage elements is then found as shown in Fig. 3.15b. Then, the current graph G_I of the equivalent circuit, which contains the current elements and the ordinary elements, is given in Fig. 3.16a, and its voltage graph G_V with the voltage elements and the ordinary elements is shown in Fig. 3.16b. The orientations of the ordinary edges, y_1, y_2, y_3, and g_p are assigned arbitrarily.

By inspection of Fig. 3.16 it is clear that there exist four tree-pairs of the two graphs as shown in Fig. 3.17.

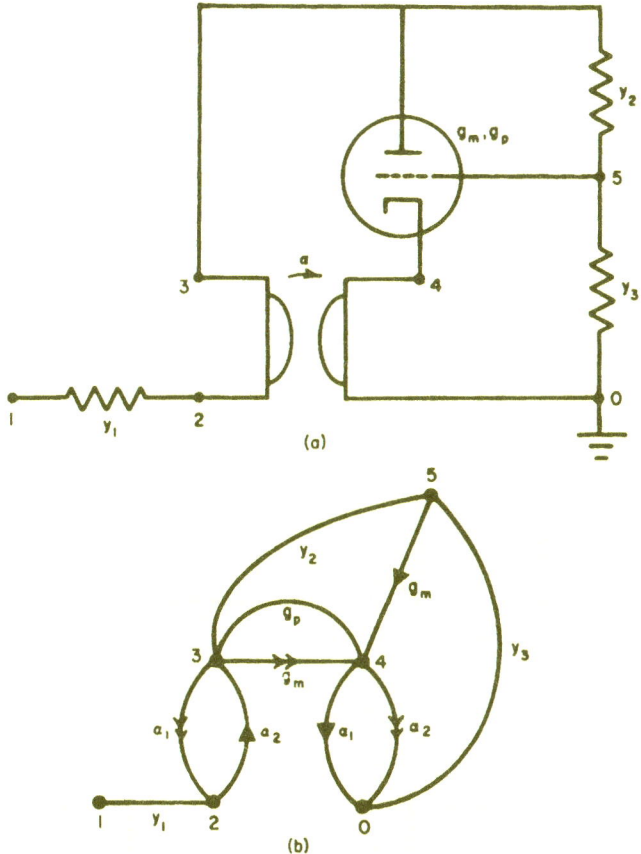

Fig. 3.15. A network and its equivalent circuit, where $\alpha_1 = \alpha_2 = \alpha$
(a) An active network; (b) mathematical equivalent circuit.

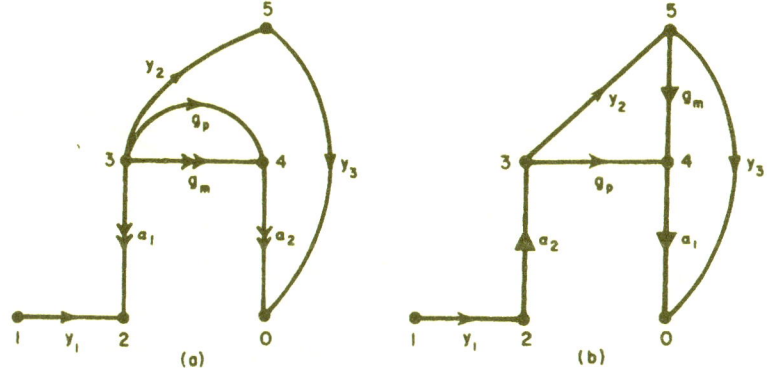

Fig. 3.16. (a) Current (G_I) and (b) voltage (G_V) graphs of the network
of Fig. 3.15

Tree-pair 1:

Tree-pair 2:

Tree-pair 3:

Tree-pair 4:

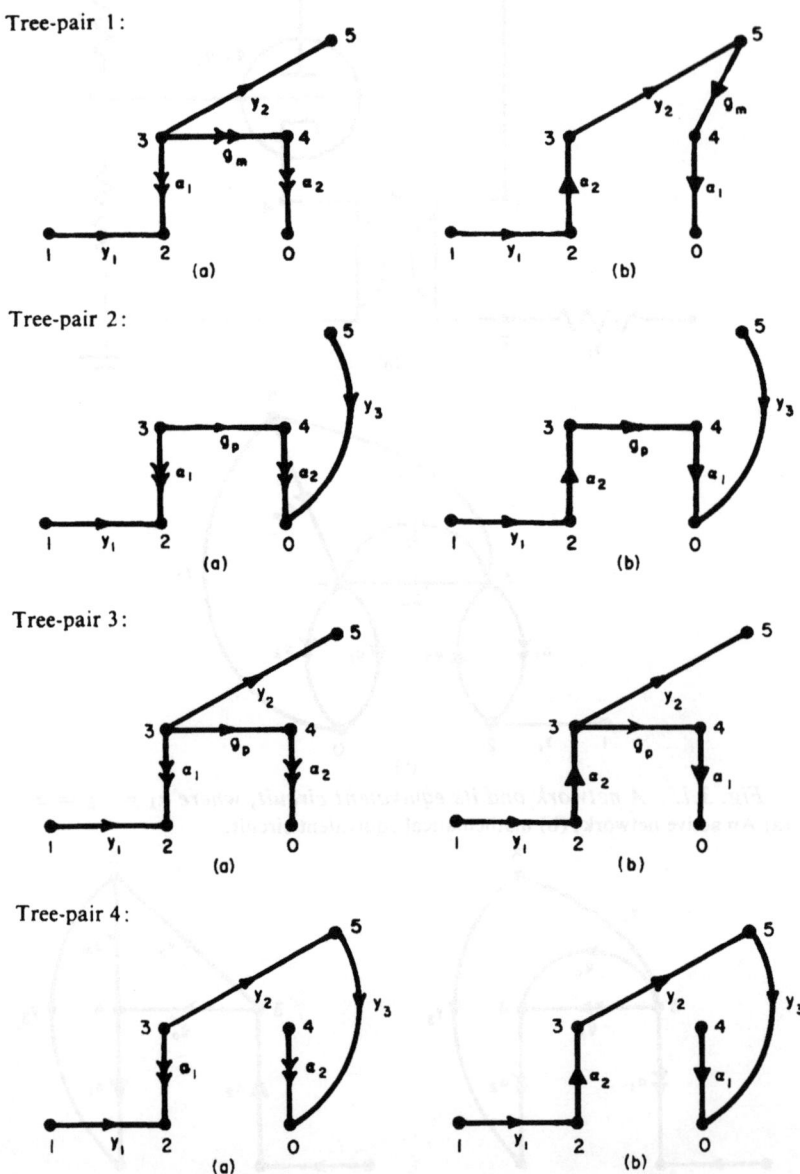

Fig. 3.17. Tree-pairs of G_I and G_V of Fig. 3.16

Tree-pair 1: (a) tree 1 of G_I; (b) tree 1 of G_V. Tree-pair 2: (a) tree 2 of G_I; (b) tree 2 of G_V. Tree-pair 3: (a) tree 3 of G_I; (b) tree 3 of G_V. Tree-pair 4: (a) tree 4 of G_I; (b) tree 4 of G_V.

In order to find the sign of each common tree-product of the tree-pairs by the formula of Eq. (3.18), the ordinary edge-pairs in the tree-pairs are reduced and the reduced tree-pairs are shown in Fig. 3.18.

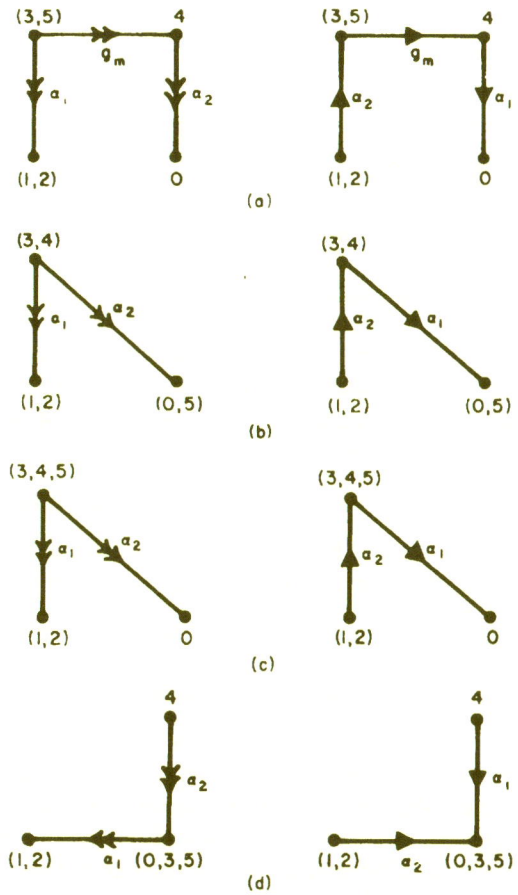

Fig. 3.18. Reduced tree-pairs of the tree-pairs of Fig. 3.17
Of tree-pair (a) 1; (b) 2; (c) 3; (d) 4.

For the reduced tree-pair 1, the number of active edge-pairs is three. However, edge-pair g_m becomes an ordinary edge-pair in the reduced tree. Therefore, the reduced tree-pair 1 of Fig. 3.18 is again reduced as shown in Fig. 3.19.

In Fig. 3.19, the signs of edge-pairs α_1 and α_2 are found to be -1 and $+1$, respectively.

The number of interchanges of the nodes or edges needed to give edge-pairs α_1 and α_2 the same principal nodes is one. Therefore, from Eq. (3.18) we have

(the sign of the common tree-product of tree-pair 1) = (-1)(sign of edge-pair α_1)(sign of edge-pair α_2) = $(-1)(1)(-1) = 1$.

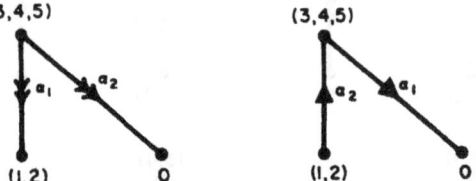

Fig. 3.19. Reduced tree-pair of Fig. 3.18 is further reduced

Similarly, one gets the sign of the common tree-product of tree-pair 2 = $+1$, the sign of the common tree-product of tree-pair 3 = $+1$, and the sign of the common tree-product of tree-pair 4 = $+1$. Thus, the node-determinant of the network Δ_n is found by

$$\Delta_n = y_1 y_2 \alpha_1 \alpha_2 g_m + y_1 y_3 g_p \alpha_1 \alpha_2 + y_1 y_2 g_p \alpha_1 \alpha_2 + y_1 y_2 y_3 \alpha_1 \alpha_2. \quad (3.19)$$

Extending the formula of Eq. (3.18) into a cofactor of (p, q)-position, $\Delta_{n_{pq}}$, of the node-determinant, Δ_n, of a network, one obtains

$$\Delta_{n_{pq}} = \sum_i \epsilon'_i \times \text{ a common tree-product } u_i \text{ of current graph } G_I{}^{(pr)}$$

and the corresponding voltage graph $G_V{}^{(qr)}$, $\qquad\qquad (3.20)$

where $G_I{}^{(pr)}$ is the subgraph of the current graph G_I of the network derived from G with node p and the reference node r made coincident, and $G_V{}^{(qr)}$ is the subgraph of the corresponding voltage graph G_V with node p and the reference node made coincident. The sign for a common tree u_i, ϵ'_i, is determined by the rules stated in Eq. (3.18) for the corresponding current and voltage trees in $G_I{}^{(pr)}$ and $G_V{}^{(qr)}$.

Example 13. Find the driving-point impedance Z_{11} at channel 1 of the circulator shown in Fig. 3.13 for $y_2 = y_3 = \alpha$. Then, the node-determinant Δ_n is found by Eq. (3.15a) as:

$$\Delta_n = \alpha_1 \alpha_3 y_2 + \alpha_1 \alpha_2 y_3 = 2\alpha^3. \quad (3.21)$$

In order to evaluate the cofactor of the node-determinant $\Delta_{n_{11}}$ by the topological method proposed, we must first obtain the subgraphs of current and voltage graphs of the circulator, $G_I{}^{(1n)}$ and

$G_V^{(1n)}$. They are shown in Fig. 3.20. Inspecting the current and voltage subgraphs of Fig. 3.20, there exist two complete trees u_1 and u_2 as shown in Fig. 3.21.

Since u_2 consists of the ordinary edges, the sign of tree products is positive. For u_1 active edge-pairs α_2 and α_3 (shown in Fig. 3.22) will have signs of -1 and $+1$, respectively. If we interchange the

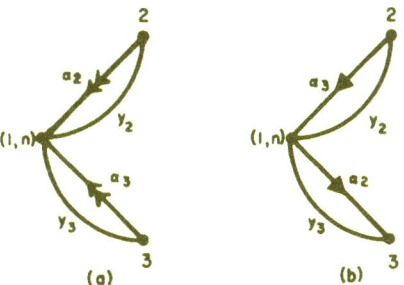

Fig. 3.20. Subgraphs of the graphs of Fig. 3.14: (a) $G_I^{(1n)}$, (b) $G_V^{(1n}$

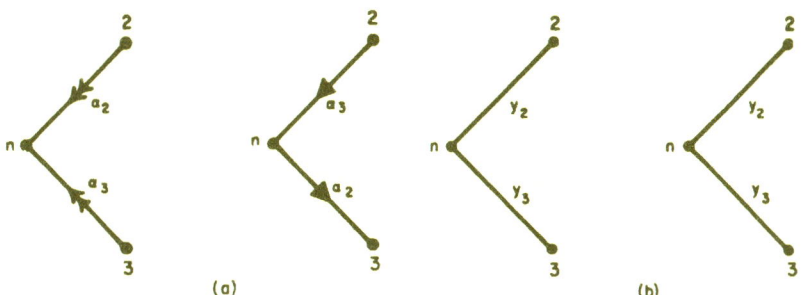

Fig. 3.21. Tree-pairs of the subgraphs of Fig. 3.20
Tree-pair (a) u_1; (b) u_2.

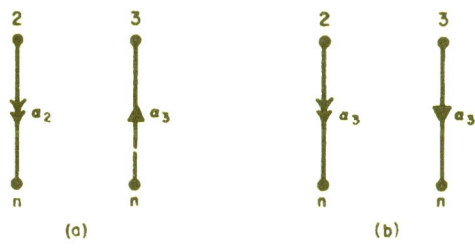

Fig. 3.22. Active edge-pairs in tree-pair u_1 of Fig. 3.21
Edge-pair (a) α_2; (b) α_3.

principal nodes of the edges in the pair α_3 only once, they will be the same as the principal nodes of the edges in the pair α_2. Therefore, the sign associated with the common tree-product of Fig. 3.22a, ϵ'_i, is found to be

$$\epsilon'_i = (-1)^1(-1)(1) = 1. \tag{3.22a}$$

Hence,

$$\Delta_{n_{11}} = \alpha_1 y_2 + y_2 y_3 = 2\alpha^2 \tag{3.22b}$$

and

$$Z_{11} = \Delta_{n_{11}}/\Delta_n = 1/\alpha. \tag{3.22c}$$

TOPOLOGICAL FORMULAS OF NETWORK FUNCTIONS AND TRANSFORMATIONS

Let us denote the graph corresponding to the mathematical equivalent circuit of network N by G, and the current and voltage graphs of the equivalent circuit by G_I and G_V, respectively. Now, let us define [33e,t]:

$G^{(ij)}$	$\triangleq$ the subgraph of G with nodes i and j made coincident;
$G^{\langle ij \rangle}$	$\triangleq$ the subgraph of G with all the nodes except nodes i and j made coincident;
T_ϵ	$\triangleq$ sum of all tree-products in G;
$T_I \bigcap T_V$	$\triangleq$ sum of all tree-products which are common to both G_I and G_V with the proper sign for each tree-product defined by Eq. (3.18);
$T_I^{(ij)} \bigcap^\epsilon T_V^{(mn)}$	$\triangleq$ sum of all tree-products which are common to both $G_I^{(ij)}$ and $G_V^{(mn)}$ with the proper sign for each tree-product;
$T_I^{\langle ij \rangle} \bigcap^\epsilon T_V^{\langle mn \rangle}$	$\triangleq$ sum of all tree-products which are common to both $G_I^{\langle ij \rangle}$ and $G_V^{\langle mn \rangle}$ with proper sign for each tree-product.

Then, from Eq. (3.17) and (3.20),

$$\Delta_n = T_I \bigcap^\epsilon T_V. \tag{3.23a}$$

$$\Delta_{n_{ij}} = T_I^{(ir)} \bigcap^\epsilon T_V^{(jr)}, \tag{3.23b}$$

where the rth node is the reference node of N.

If N consists entirely of the ordinary elements, then Eqs. (3.23) may be reduced to

$$\Delta_n = T$$
$$\Delta_{n_{pq}} = T^{(pr)\cap(qr)} \qquad (3.24)$$

as we found in the previous chapter.

Example 14. Given: the tube circuit shown in Fig. 3.23a. Find the transfer function V_2/V_1.

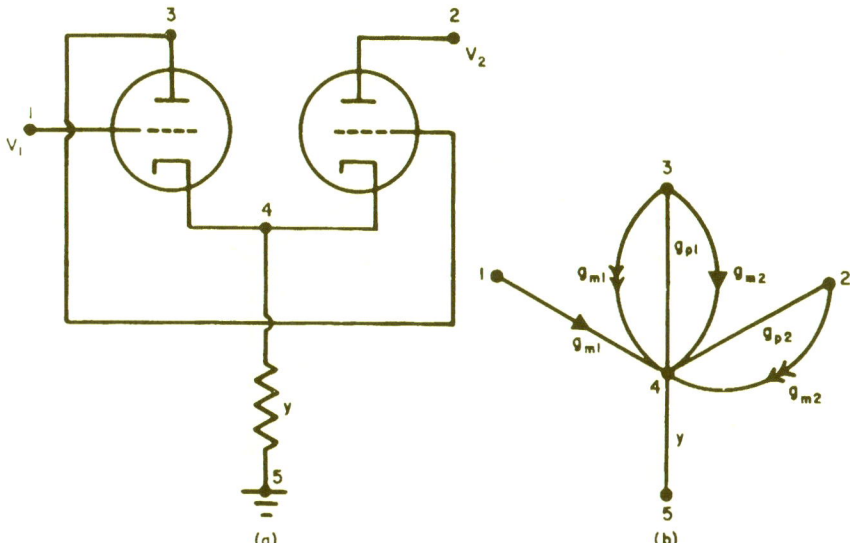

Fig. 3.23. (a) Tube circuit and (b) its mathematical equivalent circuit G

We first find the mathematical equivalent circuit of the network as shown in Fig. 3.24 in terms of current and voltage elements, where g_{p_1} and g_{p_2} are the plate conductance and g_{m_1} and g_{m_2} are the transconductances of the first and second tubes, respectively.

If we denote the node-determinant of the network by Δ_n, then the ratio of V_2 and V_1 may be expressed as:

$$V_2/V_1 = \Delta_{n_{12}}/\Delta_{n_{11}} = [T_I^{(15)} \overset{\epsilon}{\cap} T_V^{(25)}] / [T_I^{(15)} \overset{\epsilon}{\cap} T_V^{(15)}]. \qquad (3.25)$$

In order to evaluate Eq. (3.25) it is necessary to derive the subgraphs of the current and voltage graphs $G_I^{(15)}$, $G_V^{(15)}$, and $G_V^{(25)}$. They are shown in Fig. 3.24. From Fig. 3.24 and the rule of sign-determination for each tree-product stated, one obtains

$$T_I^{(15)} \overset{\epsilon}{\bigcap} T_V^{(15)} = yg_{p_1}g_{p_2}$$ (3.26a)

$$T_I^{(15)} \overset{\epsilon}{\bigcap} T_V^{(25)} = yg_{m_1}g_{m_2}.$$ (3.26b)

Therefore, $$V_2/V_1 = g_{m_1}g_{m_2}/g_{p_1}g_{p_2}.$$ (3.27)

When a network contains multiterminal devices such as vacuum tubes, transistors, transformers, gyrators, circulators, etc., it is often necessary to suppress a number of internal nodes of the network in order to characterize it by a set of the terminal equations. This is the "node condensation" technique. In other words, a "wye-delta" transformation or its successive applications on a network may be needed.

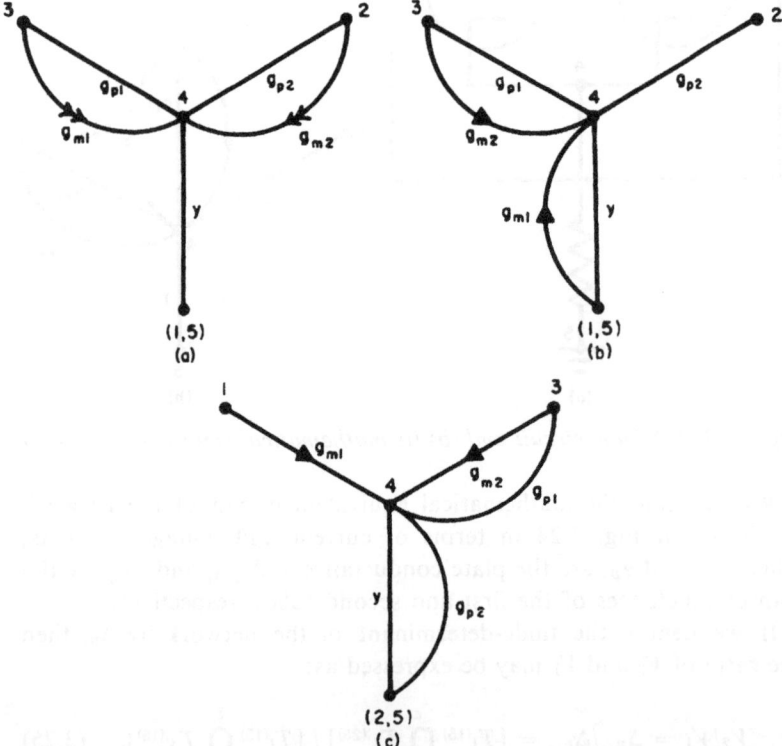

Fig. 3.24. *Subgraphs of the current and voltage graphs of the network of Fig. 3.23*

(a) $G_I^{(1,5)}$; (b) $G_V^{(1,5)}$; (c) $G_V^{(2,5)}$.

Let us consider the mathematical equivalent circuit of a network, and assume that the equivalent circuit contains n nodes. We now find the short-circuit admittances between every pair of the nodes. Thus, we obtain the so-called "indefinite admittance coefficient matrix,"[34,35] $\mathscr{Y} = [y_{ij}]$ for $i, j = 1, 2, \ldots, n$, where $y_{ij} = I_i/V_j$, provided that all other node-voltages, V's, are reduced to zero.

Now, if we suppress node n, the admittance coefficients of the resultant network can be obtained by applying the pivotal condensation to the admittance matrix about the element in (n, n)-position. We therefore have for the admittance matrix of the resultant network, $\mathscr{Y}^* = [y_{ij}^*]$,

$$\mathscr{Y}^* = \frac{1}{y_{nn}} \begin{bmatrix} \begin{vmatrix} y_{11} & y_{1n} \\ y_{n1} & y_{nn} \end{vmatrix} & \cdots & \begin{vmatrix} y_{1,n-1} & y_{1n} \\ y_{n,n-1} & y_{nn} \end{vmatrix} \\ \cdot \quad \cdot \quad \cdot \quad \cdot \quad \cdot \quad \cdot \quad \cdot \quad \cdot \\ \begin{vmatrix} y_{n-1,1} & y_{n-1,n} \\ y_{n1} & y_{nn} \end{vmatrix} & \cdots & \begin{vmatrix} y_{n-1,n-1} & y_{n-1,n} \\ y_{n,n-1} & y_{nn} \end{vmatrix} \end{bmatrix}. \tag{3.28}$$

Then, y_{pq}^* is found to be

$$y_{pq}^* = \begin{vmatrix} y_{pq} & y_{pn} \\ y_{nq} & y_{nn} \end{vmatrix}$$

$$= |A_{I\langle pn\rangle} Y_e A^t_{V\langle n\rangle}| = (-1)^\alpha [T_I^{\langle pn\rangle} \overset{\epsilon}{\bigcap} T_V^{\langle qn\rangle}], \tag{3.29}$$

where $A_{\langle ij\rangle}$ is the submatrix of A with all but the ith and jth rows deleted, and $\alpha = 0$ for $p = q$ and $\alpha = 1$ for $p \neq q$.

Thus, Eq. (3.28) may be rewritten in topological form as

$$\mathscr{Y}^* = \frac{1}{T_I^{\langle n\rangle} \overset{\epsilon}{\bigcap} T_V^{\langle n\rangle}}$$

$$\begin{bmatrix} (T_I^{\langle 1n\rangle} \overset{\epsilon}{\bigcap} T_V^{\langle 1n\rangle}) \ldots -(T_I^{\langle 1n\rangle} \overset{\epsilon}{\bigcap} T_V^{\langle n-1,n\rangle}) \\ \cdot \quad \cdot \quad \cdot \quad \cdot \quad \cdot \quad \cdot \quad \cdot \quad \cdot \quad \cdot \quad \cdot \\ -(T_I^{\langle n-1,n\rangle} \overset{\epsilon}{\bigcap} T_V^{\langle 1n\rangle}) \ldots (T_I^{\langle n-1,n\rangle} \overset{\epsilon}{\bigcap} T_V^{\langle n-1,n\rangle}) \end{bmatrix}. \tag{3.30}$$

It should be clear that if a network under consideration contains only the ordinary elements, Eq. (3.30) is reduced to

$$\mathscr{Y}^* = \frac{1}{T^{\langle n\rangle}} \begin{bmatrix} T^{\langle 1n\rangle} & \ldots & -T^{\langle 1n\rangle \cap \langle n-1,n\rangle} \\ \cdot \quad \cdot \quad \cdot \quad \cdot \quad \cdot \quad \cdot \quad \cdot \quad \cdot \quad \cdot \\ -T^{\langle n-1,n\rangle \cap \langle 1n\rangle} & \ldots & T^{\langle n-1,n\rangle} \end{bmatrix}. \tag{3.31}$$

Example 15. The wye-delta transformation of tube circuits will be investigated in terms of the topological formulas proposed in Eq. (3.29). Let us consider the so-called "common-grid transformation"[33e, 36] which is shown in Fig. 3.25.

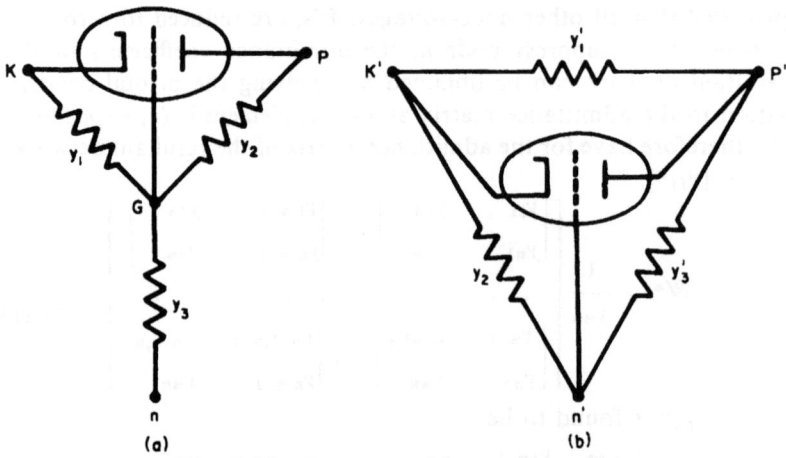

(a)   (b)

Fig. 3.25. Common grid transformation
(a) Wye-connected circuit N; (b) delta-connected circuit N^*.

First the mathematical equivalent circuits of both networks are found. This is shown in Fig. 3.26, where g_m, g'_m are the transconductances and g_p and g'_p are the plate-conductances of the tubes of Figs. 3.25a and 3.25b, respectively.

The indefinite admittance matrices of networks N and N^*, $\mathscr{Y}$ and $\mathscr{Y}^*$, are found by

$$\mathscr{Y} = \begin{array}{c} \\ P \\ K \\ n \\ G \end{array} \overset{\begin{array}{cccc} P & K & n & G \end{array}}{\begin{bmatrix} g_p+y_2 & -g_p-g_m & 0 & -y_2+g_m \\ -g_p & g_p+y_1+g_m & 0 & -y_1-g_m \\ 0 & 0 & y_3 & -y_3 \\ -y_2 & -y_1 & -y_3 & y_1+y_2+y_3 \end{bmatrix}} \quad (3.32a)$$

$$\mathscr{Y}^* = \begin{array}{c} \\ P' \\ K' \\ n' \end{array} \overset{\begin{array}{ccc} P' & K' & n' \end{array}}{\begin{bmatrix} g'_p+y'_1+y'_3 & -g'_p-y'_1-g'_m & -y'_3+g'_m \\ -g'_p-y'_1 & g'_p+y'_1+y'_2+g'_m & -y'_2-g'_m \\ -y'_3 & -y'_2 & y'_2+y'_3 \end{bmatrix}} . \quad (3.32b)$$

By applying the formula of Eq. (3.30), $\mathscr{Y}^*$ is found in terms of current and voltage graphs and their subgraphs of N to be

$$\mathbf{Y}^* = \frac{1}{y_1+y_2+y_3}\begin{bmatrix} T_I^{(Kn)} \overset{e}{\cap} T_V^{(Kn)} & -[T_I^{(Kn)} \overset{e}{\cap} T_V^{(Pn)}] & -[T_I^{(Kn)} \overset{e}{\cap} T_V^{(PK)}] \\[4pt] -[T_I^{(Pn)} \overset{e}{\cap} T_V^{(Kn)}] & T_I^{(Pn)} \overset{e}{\cap} T_V^{(Pn)} & -[T^{(Pn)} \overset{e}{\cap} T_V^{(PK)}] \\[4pt] -[T_I^{(Pn)} \overset{e}{\cap} T_V^{(Kn)}] & -[T_I^{(PK)} \overset{e}{\cap} T_V^{(Pn)}] & T^{(PK)} \overset{e}{\cap} T_V^{(PK)} \end{bmatrix}$$

$$= \frac{1}{y_1+y_2+y_3}\begin{bmatrix} g_p(y_1+y_2+y_3)+y_2(y_1+y_3+g_m) & -g_p(y_1+y_2+y_3)-y_2(y_1+g_m)-y_3 g_m & -y_3(y_2-g_m) \\[4pt] -g_p(y_1+y_2+y_3)-y_2(y_1+g_m) & g_p(y_1+y_2+y_3)+y_1(y_2+y_3)+g_m(y_2+y_3) & -y_3(y_1+g_m) \\[4pt] -y_2 y_3 & -y_1 y_3 & y_3(y_1+y_2) \end{bmatrix}. \quad (3.33)$$

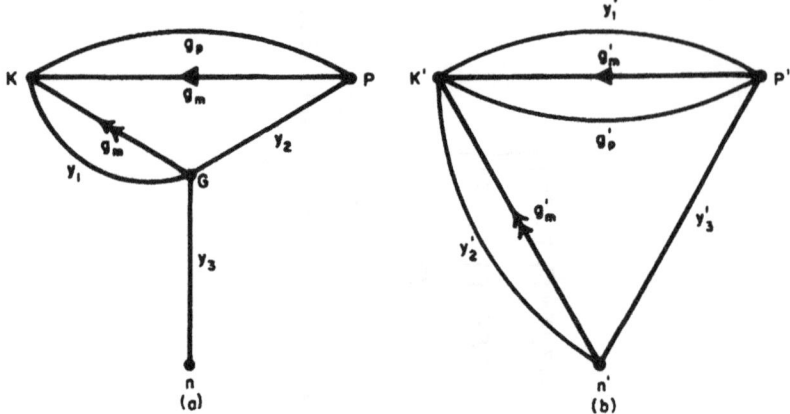

Fig. 3.26. Graphs (a) of N and (b) of N* of the circuits of Fig. 3.25

From Eqs. (3.32a) and (3.33) one finds that

$$y'_1 = y_2(y_1 + g_m)/S$$
$$g'_p = g_p(y_1 + y_2 + y_3)/S$$
$$y'_2 = y_1 y_3/S \qquad\qquad (3.34)$$
$$g'_m = g_m y_3/S$$
$$y'_3 = y_2 y_3/S,$$

where $S = y_1 + y_2 + y_3$.

Although our discussions have been so far restricted only to topological analysis, the concepts of the characterization of a dependent node-pair by the artificial current and voltage elements have great potential utility in inventing a systematic synthesis method of a network, particularly for active as well as mutually coupled networks.

Let us explore a little more a possible application of the concept of the mathematical equivalent circuit to a synthesis problem. Let us assume that $\mathbf{Y}_n = [y_{ij}]$ is the node admittance matrix of a network containing mutually dependent node-pairs. Then, each element of $\mathbf{Y}_n$, y_{ij}, is determined by†

$y_{ij} = (-1)^\alpha \Sigma$ admittances of the ordinary edges connected between nodes i and $j + \Sigma$ sign × (mutual admittance of a corresponding active edge-pair such that the current edge terminates at node i and the voltage edge terminates at node j), (3.35)

---

† This formulation was proposed by Percival.[28b]

where $\alpha = 0$ for $i = j$ and include all the ordinary edges incident at node i and $\alpha = 1$ for $i \neq j$. The sign of the second term is positive if both active edges in a pair are directed away or toward nodes i and j, respectively, and negative otherwise.

For the realization of an asymmetric matrix, $\mathscr{Q} = [q_{ij}]$ as a node admittance matrix of a network, Percival's rule, Eq. (3.25), is very useful. That is, we shall first decompose a given matrix $\mathscr{Q}$ into the sum of two submatrices $\mathscr{Q}_1$ and $\mathscr{Q}_2$ such that

$$\mathscr{Q} = \mathscr{Q}_1 + \mathscr{Q}_2, \tag{3.36}$$

where $\mathscr{Q}_1$ is a symmetric matrix. Then, the symmetric matrix $\mathscr{Q}_1$ may be always realizable in terms of the ordinary R,L,C elements. Since $\mathscr{Q}_2$ is asymmetric, the subnetwork corresponding to the matrix should contain dependent node-pairs. We therefore apply Percival's rule in order to determine which node-pairs are dependent with each other, and then insert the current and voltage elements between the proper dependent node-pairs.

After we have completed the insertion process of the artificial elements (if it is possible), both subnetworks realized by $\mathscr{Q}_1$ and $\mathscr{Q}_2$ should be superimposed. This resulting network is the mathematical equivalent circuit of a network corresponding to $\mathscr{Q}$. The realization will be completed if the equivalent circuit is converted into a network with the physical devices which are available in our engineering practice of today. There exist, of course, a number of problems in systematizing this possible realization technique. These are left to readers as research problems.

REFERENCES

1. L. Euler, "Solutis Problematis ad Geometriam Situs Pertinantis," in *Commentarii Academiae Scientiarum Imperialis Petropolitanae*, Vol. 8, pp. 128–40, 1736.

2. F. Reza, "Some topological considerations in network theory," *IRE Trans. PGCT-5* (1958), 30–42.

3. G. Kirchhoff, "Uber die Auflosung der Gleichungen, auf welche man bei der Untersuchungen der Linearen Verteilung Galvanisher Strome Gefuhrt Wird," *Poggendorf Ann. Physik*, 72 (1847), 497–508. English translation by J. B. O'Toole. Also published in *IRE Trans. PGCT-5* (1958), 4–7.

4. O. Veblen, *Analysis Situs* (Cambridge Colloquium Publications). New York, American Mathematical Society, 1931.

5. D. König, *Theorie der Endlichen und Unendlichen Graphen*. New York, Chelsea Pub. Co., 1950.

6. W. Mayeda and S. Seshu, "Topological formulas for network functions," Bull. No. 446, Univ. of Illinois, Urbana, Illinois, Engineering Experiment Station, 1957.

7. S. Seshu and M. B. Reed, *Linear Graphs and Electrical Networks.* Reading, Mass., Addison-Wesley, 1961.

8. M. B. Reed, *Foundation for Electrical Network Theory.* Englewood Cliffs, New Jersey, Prentice-Hall, 1961.

9a. H. Whitney, "Congruent graphs and connectivity of graphs," *Am. J. Math.*, 54 (1932), 150–68.

9b. H. Whitney, "Non-separable and planar graphs," *Trans. Am. Math. Soc.*, 34 (1932), 339–62.

9c. H. Whitney, "On the classification of graphs," *Am. J. Math.*, 55 (1933), 236–44.

9d. H. Whitney, "Planar graphs," *Fundamental Math.*, 21 (1933), 73–84.

9e. H. Whitney, "2-Isomorphic graphs," *Am. J. Math.*, 55 (1933), 245–54.

9f. H. Whitney, "On the abstract properties of linear dependence," *Am. J. Math.*, 57 (1935), 509–33.

10. I. Cederbaum, "Matrices all of whose elements and subdeterminants are 1, −1, or 0," *J. Math. and Phys.*, 36 (1958), 351–61.

11. H. M. Trent, "Note on the enumeration and listing of all possible trees in a connected linear graph," *Proc. Natl. Acad. Sci. U.S.*, 40 (1954), 1004–07.

12a. S. Okada, "Algebraic and Topological Foundations of Network Synthesis," in *Proceedings of a Symposium on Modern Network Synthesis*, Vol. 4, pp. 283–322. Brooklyn Polytechnic Institute, New York, 1955.

12b. S. Okada, "On node and mesh determinants," *Proc. IRE*, 43 (1955), 1527.

13. A. C. Aitken, *Determinants and Matrices.* Edinburgh, Scotland, Oliver and Boyd Ltd., 1956.

14a. S. L. Hakimi, "On the realizability of a set of trees," *IRE Trans.* CT-8 (1961), 11–17.

14b. S. L. Hakimi and W. Mayeda, "On coefficients of polynomials in network functions," *IRE Trans.* CT-7 (1960), 40–44.

15. R. L. Gould, "Application of graph theory of the synthesis of contact networks," doctoral dissertation, Harvard University, Cambridge, Mass. 1957. Also published as Rept. No. BL-18, Computational Laboratory, Harvard University, 1957.

16a. O. Wing, "Reliability study of communication systems," Tech. Rept. T-32/B, AF Contract 18 (600)-677, Columbia University, New York, 1958. Also published in *Proceedings of Midwest Symposium on Circuit Theory*, pp. k1–k5. East Lansing, Michigan, Michigan State University, 1959.

16b. O. Wing and W. H. Kim, "The path matrix and its realizability," *IRE Trans.* CT-6 (1959), 267–72.

16c. O. Wing and W. H. Kim, "The path matrix and switching functions," *J. Franklin Inst.* 268 (1959), 251–69.

17. R. L. Ashenhurst, "A uniqueness theorem for abstract two-terminal switching networks," Progress Rept. No. BL-10, Harvard Computational Laboratory, Cambridge, Mass., 1954.

18. C. L. Coates, "General topological formulas for linear network functions," General Electric Research Laboratory, Schenectady, New York, Rept. No. 57-RL-1746, August 1957. Also published in *IRE Trans. CT*-5 (1959), 30–42, March, 1958.

19. W. Mayeda, "Topological formulas for active networks," Interim Tech. Rept. No. 8, U.S. Army Contract No. Da-11-022-ORD-1983, Univ. of Illinois, Urbana, Illinois, January, 1958. Also published in *Proceedings of National Electronics Conference*, edited by M. H. Crothery, Vol. 15, pp. 1–13. Chicago, The Conference, 1958.

20a. I. T. Frisch and W. H. Kim, "Properties of 2-semi-isomorphic graphs and their applications: active network analysis," *J. Math. Phys.*, 2 (1961), 627–35.

20b. I. T. Frisch and W. H. Kim, "Realization of communication nets with maximum information flow," in *Proceedings of Seventh National Communication Symposium Record*, pp. 254–65, 1961.

21. C. Kuratowski, "Sur le problème des courbes gauches en topologie," *Fundamental Math.*, 15 (1930), 271–83.

22a. J. C. Maxwell, *Electricity and Magnetism.* Oxford, England, Clarendon Press, 1892, 3rd ed., Vol. 1.

22b. Y. H. Ku, "Resume of Maxwell's and Kirchhoff's Rules," *J. Franklin Inst.*, 253 (1952), 211–24.

23. W. Mayeda and M. E. Van Valkenburg, "Network analysis and synthesis by digital computers," in *IRE Wescon Convention Record*, Part 2, pp. 137–44, 1957.

24. F. J. MacWilliams, "Topological network analysis as a computer program," *IRE Trans. PGCT*-5 (1958), 228–29.

25. E. W. Hobbs, "Topological network analysis as a computer program (discussion)," *IRE Trans. PGCT*-6 (1959), 135–36.

26. W. Mayeda, "Reducing computation time in the analysis of networks by digital computers," *IRE Trans. PGCT*-6 (1959), 136–37.

27. W. H. Kim, D. H. Younger, C. V. Freiman, and W. Mayeda, "On Iterative Factorization in Network Analysis by Digital Computer," in *Proceedings of the Eastern Joint Computer Conference*, pp. 241–53, 1960.

28a. W. S. Percival, "Solution of passive electrical networks by means of mathematical trees," *Proc. Inst. Elec. Engrs. (London)*, 100 (1953), 143–50.

28b. W. S. Percival, "Improved matrix and determinant methods of solving networks," *Proc. Inst. Elec. Engrs. (London)*, 101 (1954), 258–65.

28c. W. S. Percival, "Graphs of active networks," *Proc. Inst. Elec. Engrs. (London)*, 102 (1955), 270–78.

29a. A. D. Fialkow and I. Gerst, "The transfer function of general two-terminal-pair RC networks," *Quart. App. Math.*, 10 (1952), 113–27.

29b. A. D. Fialkow and I. Gerst, "The transfer function of networks without mutual reactance," *Quart. App. Math.*, 12 (1954), 117–31.

30. L. Weinberg, "Kirchhoff's third and fourth laws," *IRE Trans. PGCT*-5 (1958), 8–30.

31. H. Watanabe, "A method of tree expansion in network topology," *IRE Trans. CT*-8 (1961), 4–10.

32. S. Seshu, "Topological considerations in the design of driving point functions," *IRE Trans. CT*-2 (1955), 356–67.

33a. W. H. Kim, "A new method of synthesis of driving-point functions," Interim Tech. Rept. No. 1, U.S. Army Contract No. DA-11-022-ORD-1983, Univ. of Illinois, Urbana, Illinois, April, 1956.

33b. W. H. Kim, "On non-series-parallel realization of driving-point functions," in *IRE National Convention Record*, Part 2, pp. 76–81, 1958.

33c. W. H. Kim, "Topological evaluation of network functions," *J. Franklin Inst.*, 267 (1959), 283–93.

33d. W. H. Kim, "Role of Network Topology in Network Synthesis," in *Proceedings of Midwestern Symposium on Circuit Theory*, pp. 6.1–6.18. Ames, Iowa, Iowa State College, 1958.

33e. W. H. Kim, "Topological analysis of linear multipoles," *The Matrix and Tensor Quart. (London)*, 11 (1960), 33–41.

33f. W. H. Kim, "Application of graph theory to the analysis of active and mutually coupled networks," *J. Franklin Inst.*, 271 (1961), 200–21.

34a. L. A. Zadeh, "A note on the analysis of vacuum tube and transistor circuits," *IRE Proc.*, 41 (1953), 989–92.

34b. L. A. Zadeh, "Multipole analysis of active networks," *IRE Trans.* CT-4 (1957), 97–105.

35a. J. Shekel, "Matrix representation of transistor circuits," *IRE Proc.*, 40 (1952), 1493–97.

35b. J. Shekel, "The gyrator as a 3-terminal element," *IRE Proc.*, 41 (1953), 1014–16.

36. H. Hsu, "On transformation of linear active networks with applications at ultra high frequencies," *IRE Proc.*, 41 (1953), 59–67.

PART II

Flow-Graph Techniques for the Solution

of Linear and Sampled-Data Systems

Introduction

IN THIS PART we are concerned with the study of linear feedback control systems. Mason[1,2] has demonstrated a topological method for the analysis of such systems. A so-called "signal-flow graph" is constructed from the given system; the graph contains all the information necessary to predict the behavior of the system. Mason has shown that the transmission from input to output depends on the topological properties of the graph, specifically on the forward paths and feedback loops. Mason's "general gain formula" greatly simplifies the analysis of feedback control systems; by means of the formula, the system output may be computed by inspection of the signal-flow graph. His derivation of the formula was based on the theory of feedback systems, in particular on the concept of loop gain and return difference developed by Bode.[3] It would be desirable to find out in what way the topological structure of the signal-flow graph is responsible for the validity of the general gain formula. For this reason, a study of abstract properties of the signal-flow graph is necessary. Recently Ash[4] proposed a rigorous proof of Mason's formula.

Mason's result does not directly apply to an important class of control systems, namely, to sampled-data systems. It is natural to inquire whether the flow-graph approach can be modified and extended so as to apply to such systems. Also of interest would be a general gain formula for sampled-data systems, analogous to Mason's formula for linear continuous systems. With the aid of such a formula, the response of a sampled-data system could be computed directly from the original system without the necessity of solving any equations. With this motivation, Lendaris and Jury,[5] and Ash, Kim, and Kranc[6] have independently developed the topological analysis method for sampled-data systems as an extension of Mason's work.

The approach and materials which will be used for the proof of Mason's formula are based on the work of Ash.[4] In the second chapter the signal-flow-graph approach is modified and extended in order to analyze sampled-data systems. Then, we further extend the flow-graph techniques to analyze multirate sampled-data systems, i.e., systems containing switches which operate in synchronism but at different sampling rates.

CHAPTER 1

Topological Properties
of Signal-Flow Graphs

WE SHALL APPLY the concepts of linear graph theory to the signal-flow graph in order to derive its topological properties. We first define three matrices, called the "exit," "entrance," and "edge-weight" matrices, respectively, which completely characterize the graph. We then examine the properties of these matrices and establish a one-to-one correspondence between nonsingular sub-matrices and nontouching feedback loops of the graph. Finally, we shall give a graph-theoretic proof of Mason's general gain formula.

DEFINITION AND CHARACTERISTIC MATRICES OF SIGNAL-FLOW GRAPHS

The signal-flow graph is a convenient pictorial representation of a set of independent simultaneous linear equations. For convenience in establishing the results of this chapter, we shall assume that the equations are of the following form:

$$
\begin{aligned}
x_1 + a_{12}x_2 + \ldots + a_{1n}x_n &= y_1 \\
a_{21}x_1 + x_2 + \ldots + a_{2n}x_n &= y_2 \\
&\vdots \\
a_{n1}x_1 + a_{n2}x_2 + \ldots + x_n &= y_n.
\end{aligned}
\tag{1.1}
$$

Any independent set of n linear equations in n unknowns can be manipulated into the above form. In matrix form Eq. (1.1) becomes

$$
\mathscr{A}X = Y, \tag{1.2}
$$

where

$$\mathcal{A} = \begin{bmatrix} 1 & a_{12} \ldots a_{1n} \\ a_{21} & 1 \ldots a_{2n} \\ \cdot & \cdot \cdot \cdot \cdot \cdot \cdot \\ a_{n1} & a_{n2} \ldots 1 \end{bmatrix} \qquad (1.3a)$$

and

$$\mathbf{X} = \begin{bmatrix} x_1 \\ x_2 \\ \cdot \\ \cdot \\ \cdot \\ x_n \end{bmatrix} \qquad \mathbf{Y} = \begin{bmatrix} y_1 \\ y_2 \\ \cdot \\ \cdot \\ \cdot \\ y_n \end{bmatrix}. \qquad (1.3b)$$

Definition 1. The "signal-flow graph" associated with Eq. (1.1) is an oriented graph with a node corresponding to each unknown x_i $(i = 1, 2, \ldots, n)$ and a node corresponding to each nonzero constant y_j. We shall refer to the node corresponding to x_i as "node i" and the node (if any) corresponding to y_j as "node y_j." If $i \neq j$ and $a_{ij} \neq 0$, there is a directed edge with weight $(-a_{ij})$ incident on nodes i and j. The direction of the edge is toward node i. If $y_j \neq 0$, there is a directed edge with weight 1 incident on nodes j and y_j, with the direction of the edge toward node j.

For example, consider the following set of equations:

$$\begin{aligned} x_1 - f x_2 - e x_3 &= y_1 \\ -a x_1 + x_2 - d x_4 &= 0 \\ -b x_2 + x_3 - g x_4 &= y_3 \\ -c x_3 + x_4 &= 0. \end{aligned} \qquad (1.4)$$

The signal-flow graph corresponding to Eq. (1.4) is shown in Fig. 1.1.

Definition 2. A "proper edge" of a signal-flow graph is an edge not incident on any of the nodes y_j for $j = 1, 2, \ldots, n$ and a "proper node" is a node corresponding to each unknown x_i for $i = 1, 2, \ldots, n$. For the remainder of this chapter, the term "edge" will mean "proper edge" and the term "node" will mean "proper node" unless otherwise specified. In other words, we shall restrict our attention to the graph obtained from the original signal-flow graph by removing nodes y_j $(j = 1, 2, \ldots, n)$ and the edges incident on these nodes.

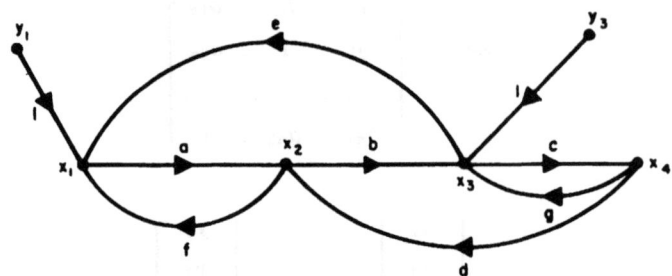

Fig. 1.1. Signal-flow graph corresponding to Eq. (1.4)

Consider a signal-flow graph with b edges, denoted by $b_1, b_2, \ldots, b_b$ and n nodes, denoted by $1, 2, \ldots, n$.

Definition 3. The "exit matrix," denoted by $\mathcal{M}$ of the signal-flow graph, is a matrix whose columns correspond to the edges of the graph and whose n rows correspond to the nodes of the graph. The element m_{ij} of $\mathcal{M}$ is defined by:

$m_{ij} = 1$,   if edge b_j is incident at node i, with the direction of b_j
            away from node i;
$m_{ij} = 0$,   otherwise.

Definition 4. The "entrance matrix," denoted by $\mathcal{D}$, of the signal-flow graph is a matrix whose b columns correspond to the edges of the graph and whose n rows correspond to the nodes of the graph. The element d_{ij} of $\mathcal{D}$ is determined by:

$d_{ij} = 1$,   if edge b_j is incident at node i, with the direction of b_j
            toward node i;
$d_{ij} = 0$,   otherwise.

For the graph of Fig. 1.1, we find that

$$\mathcal{M} = \begin{array}{c} \\ 1 \\ 2 \\ 3 \\ 4 \end{array} \begin{array}{cccccccc} a & b & c & d & e & f & g & \\ \left[\begin{array}{cccccccc} 1 & 0 & 0 & 0 & 0 & 0 & 0 & 0 \\ 0 & 1 & 0 & 0 & 0 & 1 & 0 \\ 0 & 0 & 1 & 0 & 1 & 0 & 0 \\ 0 & 0 & 0 & 1 & 0 & 0 & 1 \end{array}\right] \end{array} \quad (1.5)$$

$$\mathcal{D} = \begin{matrix} & a & b & c & d & e & f & g \\ 1 & 0 & 0 & 0 & 0 & 1 & 1 & 0 \\ 2 & 1 & 0 & 0 & 1 & 0 & 0 & 0 \\ 3 & 0 & 1 & 0 & 0 & 0 & 0 & 1 \\ 4 & 0 & 0 & 1 & 0 & 0 & 0 & 0 \end{matrix}. \tag{1.6}$$

The rows and columns of $\mathcal{M}$ and $\mathcal{D}$ are always assumed to be ordered in the same way. Notice that $(\mathcal{M} - \mathcal{D})$ is the incidence matrix of linear graph theory, which we defined in Part I.

Definition 5. The "edge-weight matrix," denoted by **W**, of a signal-flow graph with b edges is a diagonal matrix of order b whose rows and columns correspond to the edges of the graph. The element w_{ij} of **W** is defined by:

$$w_{ij} = 0, \quad \text{if } i \neq j;$$
$$w_{ii} = \text{the weight of edge } b_i.$$

The ordering of the columns of **W** is the same as the ordering of the columns of $\mathcal{M}$ and $\mathcal{D}$. For the graph of Fig. 1.1. the edge-weight matrix of the graph is found to be

$$\mathbf{W} = \begin{bmatrix} a & 0 & 0 & 0 & 0 & 0 & 0 \\ 0 & b & 0 & 0 & 0 & 0 & 0 \\ 0 & 0 & c & 0 & 0 & 0 & 0 \\ 0 & 0 & 0 & d & 0 & 0 & 0 \\ 0 & 0 & 0 & 0 & e & 0 & 0 \\ 0 & 0 & 0 & 0 & 0 & f & 0 \\ 0 & 0 & 0 & 0 & 0 & 0 & g \end{bmatrix}. \tag{1.7}$$

The three matrices $\mathcal{M}$, $\mathcal{D}$, and **W** completely characterize a signal-flow graph. The basis for this statement is the following theorem.

Theorem 1. Suppose that a signal-flow graph with matrices $\mathcal{M}$, $\mathcal{D}$, and **W** describes the set of equations $\mathcal{A}X = Y$, where $\mathcal{A}$ is the coefficient matrix of the form Eq. (1.3a). Then,

$$\mathbf{U} - \mathcal{D}\mathbf{W}\mathcal{M}^t = \mathcal{A},$$

where $\mathcal{M}^t$ is the transpose of $\mathcal{M}$ and **U** is a unit matrix whose order is equal to that of $\mathcal{A}$.

Proof: Let q_{ij} be the element in row i and column j of the matrix $(\mathbf{U} - \mathcal{D}\mathbf{W}\mathcal{M}^t)$. Then, if $i \neq j$, it is clear that

$$q_{ij} = -\sum_{k=1}^{b}\sum_{r=1}^{b} d_{ik}w_{kr}m_{jr}. \tag{1.8}$$

Since **W** is a diagonal matrix, the expression for q_{ij} reduces to

$$q_{ij} = -\sum_{k=1}^{b} d_{ik}w_{kk}m_{jk}. \tag{1.9}$$

It follows from the definition of the three matrices that

$$d_{ik}w_{kk}m_{jk} = w_{kk} = \text{the weight of edge } b_k, \quad \text{if } b_k \text{ leaves}$$
$$\text{node } j \text{ and enters node } i; \tag{1.10a}$$
$$d_{ik}w_{kk}m_{jk} = 0, \quad \text{otherwise.} \tag{1.10b}$$

By construction of the graph, there cannot be more than one edge between any pair of nodes. Hence, if an edge b_k leaves node j and enters node i, then $-q_{ij}$ is the weight of b_k. If there is no such edge, $q_{ij} = 0$. Therefore, $q_{ij} = a_{ij}$.

If $i = j$, we have

$$q_{ij} = q_{ii} = 1 - \sum_{k=1}^{b} d_{ik}w_{kk}m_{ik}. \tag{1.11}$$

By construction of the graph, no edge may enter and leave the same node. Hence, $d_{ik}w_{kk}m_{ik} = 0$ for $k = 1, 2, \ldots, b$. Thus, $q_{ii} = a_{ii} = 1$. This completes the proof.

Definition 6. An "exit-edge" of a node is an edge leaving that node. An "entrance-edge" of a node is an edge entering that node.

Definition 7. A "feedback loop" of a signal-flow graph is a single oriented loop, such that in the loop, each node has one exit-edge and one entrance-edge. A number of feedback loops are "nontouching" or "node-disjoint" if no two of the loops have a node in common. The "weight of a feedback loop" is the product of the weights of the edges in the loop.

In the graph of Fig. 1.1, there are four feedback loops with weights (af), (bcd), (abe), and (cg), respectively. Loops (af) and (cg) are the only nontouching feedback loops of the graph. To establish the desired result, we need a preliminary lemma.

Lemma 1. Let a square submatrix S of $\mathcal{M}$ (or $\mathcal{D}$) be defined by edges $b_1, b_2, \ldots, b_k$ and nodes $1, 2, \ldots, k$. Then S is nonsingular if and only if each node of the set $(1, 2, \ldots, k)$ has exactly one exit (or entrance) edge from the set $(b_1, b_2, \ldots, b_k)$.

Proof: We consider the matrix $\mathcal{M}$. The proof for $\mathcal{D}$ is identical. First suppose that each node of the set $(1, 2, \ldots, k)$ has exactly one exit-edge from the set $(b_1, b_2, \ldots, b_k)$. Then there is exactly one non-zero element s_1 in row 1 of S. There is exactly one nonzero element s_2 in row 2 of S; s_2 cannot be in the same column as s_1, for $\mathcal{M}$ (and therefore S) can have no more than one nonzero element per column. Continuing in this way. there is exactly one nonzero element s_k in row k of S, and s_k is not in the same column as any of the elements $s_1, s_2, \ldots, s_{k-1}$. Hence the determinant of S is $(\pm s_1, s_2, \ldots, s_k) = \pm 1$, and S is nonsingular.

Now suppose S is nonsingular. If node i $(1 \leqslant i \leqslant k)$, has no exit-edges from the set $(b_1, b_2, \ldots, b_k)$, then row i of S consists entirely of zeros, contradicting the hypothesis of nonsingularity. If node i has two exit-edges b_r and b_n $(1 \leqslant r, n \leqslant k)$, then columns r and n of S are identical, again contradicting the nonsingularity of S. Hence, the lemma.

The proof of Lemma 1 shows that all nonsingular submatrices of $\mathcal{M}$ and $\mathcal{D}$ have determinant ± 1. i.e., they are unimodular matrices.

Theorem 2. Corresponding square submatrices S_1 and S_2 of $\mathcal{M}$ and $\mathcal{D}$, respectively (submatrices defined by the same rows and columns) are *both* nonsingular if and only if the edges corresponding to the columns of S_1 and S_2 form a feedback loop or a set of non-touching feedback loops.

Proof: For convenience, assume that the edges corresponding to the columns of S_1 and S_2 are ordered $b_1, b_2, \ldots, b_k$ and that the nodes corresponding to the rows of S_1 and S_2 are ordered $1, 2, \ldots, k$. By Lemma 1, S_1 and S_2 are both nonsingular if and only if each node of the set $(1, 2, \ldots, k)$ has exactly one exit-edge and one entrance-edge from the set $(b_1, b_2, \ldots, b_k)$, However, a subgraph in which each node has exactly one exit-edge and exactly one entrance-edge is a feedback loop if the subgraph is connected, or a set of nontouching feedback loops if the subgraph is not connected. Hence, the theorem.

In the graph of Fig. 1.1, consider the feedback loop formed by edges b, c, and d. From Eqs. (1.5) and (1.6), the corresponding nonsingular submatrices of $\mathcal{M}$ and $\mathcal{D}$ are

$$
S_1 = \begin{matrix} 2 \\ 3 \\ 4 \end{matrix} \begin{matrix} b & c & d \\ \begin{bmatrix} 1 & 0 & 0 \\ 0 & 1 & 0 \\ 0 & 0 & 1 \end{bmatrix} \end{matrix} \qquad S_2 = \begin{matrix} 2 \\ 3 \\ 4 \end{matrix} \begin{matrix} b & c & d \\ \begin{bmatrix} 0 & 0 & 1 \\ 1 & 0 & 0 \\ 0 & 1 & 1 \end{bmatrix} \end{matrix}. \qquad (1.12)
$$

MASON'S GENERAL GAIN FORMULA

The solution to Eq. (1.1) may be obtained directly from the associated signal-flow graph by use of Mason's general gain formula. In many situations, including the analysis of linear feedback control systems, the use of Mason's formula is far more efficient than the standard algebraic methods of solution. In a later section we shall give a proof of Mason's formula based on linear graph theory. We first establish two preliminary lemmas.

Lemma 2. Let S_1 and S_2 be corresponding nonsingular submatrices of $\mathcal{M}$ and $\mathcal{D}$, respectively. Let $L_1, L_2, \ldots, L_r$ be the (nontouching) feedback loops formed by the edges corresponding to the columns of S_1 and S_2. Let N be the number of loops in the set $(L_1, L_2, \ldots, L_r)$ which have an even number of edges. Then S_1 and S_2 have the same determinant if and only if N is even.

Proof: Consider a feedback loop L_i for $i = 1, 2, \ldots, r$. Suppose that the edges of L_i are ordered $b_{i1}, b_{i2}, \ldots, b_{in_i}$ and that the nodes of L_i are ordered so that the edge b_{ij} leaves node i_j and enters node i_{j+1} $(j = 1, 2, \ldots, n_i)$. Let S_{1i} be the submatrix of S_1 formed by columns $b_{i1}, b_{i2}, \ldots, b_{in_i}$ and rows $i_1, i_2, \ldots, i_{n_i}$. Let S_{2i} be the corresponding submatrix of S_2. Then we have

$$
S_{1i} = \begin{matrix} & \begin{matrix} b_{i1} & b_{i2} \ldots b_{in_i} \end{matrix} \\ \begin{matrix} i_1 \\ i_2 \\ \\ \\ \\ i_{n_i} \end{matrix} & \begin{bmatrix} 1 & 0 \ldots 0 \\ 0 & 1 \ldots 0 \\ \cdot & \cdot \quad \cdot \\ \cdot & \cdot \quad \cdot \\ \cdot & \cdot \quad \cdot \\ 0 & 0 \ldots 1 \end{bmatrix} \end{matrix}
\qquad
S_{2i} = \begin{matrix} & \begin{matrix} b_{i1} & b_{i2} \ldots b_{in_i} \end{matrix} \\ \begin{matrix} i_1 \\ i_2 \\ i_3 \\ \\ \\ \\ i_{n_i} \end{matrix} & \begin{bmatrix} 0 & 0 \ldots 1 \\ 1 & 0 \ldots 0 \\ 0 & 1 \ldots 0 \\ \cdot & \cdot \quad \cdot \\ \cdot & \cdot \quad \cdot \\ \cdot & \cdot \quad \cdot \\ 0 & 0 \ldots 0 \end{bmatrix} \end{matrix} ; \quad (1.13)
$$

S_{1i} and S_{2i} are, of course, nonsingular. Then S_{2i} may be made to coincide with S_{1i} by $(n_i - 1)$ interchanges of adjacent rows. Thus $|S_{1i}| = |S_{2i}|$ if and only if n_i is odd, and $|S_{2i}| = -|S_{2i}|$ if and only if n_i is even.

Now let the loops $L_1, L_2, \ldots, L_r$ be ordered so that loops $L_1, L_2, \ldots, L_N$, $N \leqslant r$, have an even number of edges, and the remaining loops have an odd number of edges. We partition the matrices S_1 and S_2 as follows:

$$\mathbf{S}_1 = \begin{bmatrix} \mathbf{S}_{11} & & & & & & \\ & \mathbf{S}_{12} & & & & 0 & \\ & & \cdot & & & & \\ & & & \cdot & & & \\ & & & & \mathbf{S}_{1N} & & \\ & & & & \cdot & & \\ & & & & & \cdot & \\ & & & & & & \mathbf{S}_{1,N+1} \\ & & & & & & \cdot \\ & & & & & & \cdot \\ & 0 & & & & & \cdot \\ & & & & & & \mathbf{S}_{1r} \end{bmatrix} \qquad (1.14)$$

$$\mathbf{S}_2 = \begin{bmatrix} \mathbf{S}_{21} & & & & & 0 \\ \cdot & & & & & \\ \cdot & & & & & \\ \cdot & & & & & \\ & \mathbf{S}_{22} & & & & \\ & \cdot & & & & \\ & & \cdot & & & \\ & & \cdot & & & \\ & & & \mathbf{S}_{2N} & & \\ & & & \cdot & & \\ & & & & \cdot & \\ & & & & \mathbf{S}_{2,N+1} & \\ & & & & \cdot & \\ & & & & & \cdot \\ & & & & & \cdot \\ 0 & & & & & \mathbf{S}_{2r} \end{bmatrix} \qquad (1.15)$$

It follows from the above partitioning that

$$|S_1| = |S_{11}||S_{12}| \cdots |S_{1N}||S_{1,N+1}| \cdots |S_{1r}| \qquad (1.16a)$$

$$|S_2| = |S_{21}||S_{22}| \cdots |S_{2N}||S_{2,N+1}| \cdots |S_{2r}|. \qquad (1.16b)$$

Since loops $L_1, \ldots, L_N$ have an even number of edges and loops $L_{n+1}, \ldots, L_r$ have an odd number of edges,

$$|S_{1i}| = -|S_{2i}| \qquad (i \leqslant N) \qquad (1.17a)$$

$$|S_{1i}| = |S_{2i}| \qquad (i > N). \qquad (1.17b)$$

Hence, $|S_1| = (-1)^N |S_2|$, and the lemma follows.

As an illustration of Lemma 2, consider the graph of Fig. 1.1. The Loops $L_1 = (af)$ and $L_2 = (cg)$ are nontouching. From Eqs. (1.5) and (1.6), the submatrices of $\mathcal{M}$ and $\mathcal{D}$ corresponding to the set (L_1, L_2) are

$$S_1 = \begin{array}{c} \\ 1 \\ 2 \\ 3 \\ 4 \end{array} \begin{array}{cccc} a & c & f & g \\ \left[\begin{array}{cccc} 1 & 0 & 0 & 0 \\ 0 & 0 & 1 & 0 \\ 0 & 1 & 0 & 0 \\ 0 & 0 & 0 & 1 \end{array}\right] \end{array} \qquad S_2 = \begin{array}{c} \\ 1 \\ 2 \\ 3 \\ 4 \end{array} \begin{array}{cccc} a & c & f & g \\ \left[\begin{array}{cccc} 0 & 0 & 1 & 0 \\ 1 & 0 & 0 & 0 \\ 0 & 0 & 0 & 1 \\ 0 & 1 & 0 & 0 \end{array}\right] \end{array}. \qquad (1.18)$$

And, we find that $|S_1| = |S_2| = -1$, as we would expect from Lemma 2 since both loops have an even number of edges.

Lemma 3. Let S_1 and S_2 be corresponding nonsingular submatrices of $\mathcal{M}$ and $\mathcal{D}$, respectively. Let edges $b_1, b_2, \ldots, b_k$ correspond to the columns of S_1 and S_2. Let $L_1, L_2, \ldots, L_r$ be the (nontouching) feedback loops formed by the edges $b_1, b_2, \ldots, b_k$. Then S_1 and S_2 have the same determinant if and only if k and r are either both even or both odd.

Proof: Let N be the number of loops in the set $(L_1, L_2, \ldots, L_r)$ which have an even number of edges. We may then construct Table 1. The table is to be interpreted in the following way. Line 1

Table 1. Proof for Lemma 3

	N	$r-N$	r	k
1	even	even	even	even
2	even	odd	odd	odd
3	odd	even	odd	even
4	odd	odd	even	odd

states that if the numbers N and $r - N$ are both even then the number of loops r and the number of edges k must both be even. Lines 2, 3, and 4 have a similar interpretation. By Lemma 2, S_1 and S_2 have the same determinant if and only if N is even. Lemma 3 then follows from consideration of the table.

We are now ready to establish the first half of Mason's general gain formula.

Theorem 3. (Due to Mason.) Suppose a signal-flow graph G characterizes a set of equations $\mathscr{A}X = Y$ where $\mathscr{A}$ is the coefficient matrix of the form of Eq. (1.3a). Let the feedback loops of G be $L_1, L_2, \ldots, L_r$. Let W_i be the weight of L_i ($i = 1, \ldots, r$). Then the determinant of $\mathscr{A}$, Δ, is given by

$$\Delta = |\mathscr{A}| = 1 - \sum_{i=1}^{r} W_i + \sum_{\substack{i,j \\ L_i, L_j \text{ nontouching}}} W_i W_j -$$

$$- \sum_{\substack{i,j,k \\ L_i, L_j, L_k \text{ nontouching}}} W_i W_j W_k + \ldots + (-1)^\alpha \sum_{\substack{i,j,k,\ldots m \\ L_i, L_j, L_k, \ldots, L_m \text{ nontouching}}} W_i W_j W_k \ldots W_m, \quad (1.19)$$

where α, the number of elements in the set $(i, j, k, \ldots, m)$, is the maximum number of nontouching feedback loops of G. In words, Theorem 3 states that

$$\Delta = \det \mathscr{A} = 1 - (\text{sum of the weight-products of all feedback loops}) + (\text{sum of the products of weight-products of all nontouching feedback loops taken two at a time}) - (\text{sum of the products of weight-products of all nontouching feedback loops taken three at a time}) + \ldots + (-1)^\alpha (\text{the product of the weight-products of all nontouching feedback loops}), \quad (1.20)$$

where α is the maximum number of nontouching feedback loops.

For the graph of Fig. 1.1, $\alpha = 2$ and we have

$$\Delta = 1 - af - cg - abe - bcd + afcg. \quad (1.21)$$

Proof of Theorem 3: (Due to Ash.) The signal-flow graph G is characterized by matrices $\mathscr{M}$, $\mathscr{D}$, and W. Let $R = \mathscr{D}W\mathscr{M}^t$. Then $U - R = \mathscr{A}$ by Theorem 1. We may expand the determinant of $(U - R)$ by diagonal elements. Thus, we obtain

$$|\mathscr{A}| = |U - R| = 1 + \sum_{j=1}^{n} (-1)^j M_j, \quad (1.22)$$

where n is the order of $\mathcal{A}$ and M_j is the sum of all principal minors of $\mathbf{R}$ of order j. The elements along the main diagonal of $\mathbf{R}$ are zero; hence $M_1 = 0$. Now consider a typical principal minor $r_{11}r_{22} \ldots r_{kk}$, formed by rows 1, 2, $\ldots$, k and columns 1, 2, $\ldots$, k of $\mathbf{R}$. To evaluate the minor, we delete rows $k+1$, $k+2$, $\ldots$, n from $\mathbf{R}$ and also delete like-numbered columns. Equivalently, we have

$$|r_{11}r_{22}\ldots r_{kk}| = |\mathcal{D}_o\mathbf{W}\mathcal{M}_o{}^t|, \tag{1.23}$$

where $\mathcal{D}_o$ is formed from $\mathcal{D}$ by deleting rows $k+1$, $\ldots$, n and $\mathcal{M}_o$ is formed from $\mathcal{M}$ in the same way.

Since the minor $|r_{11}r_{22}\ldots r_{kk}|$ is the determinant of the product of the two matrices $(\mathcal{D}_o\mathbf{W})$ and $\mathcal{M}_o{}^t$, the Binet–Cauchy theorem applies. The matrix $(\mathcal{D}_o\mathbf{W})$ is formed from $\mathcal{D}_o$ by multiplying each column of $\mathcal{D}_o$ by the weight of the edge corresponding to that column. Hence, a square submatrix of $(\mathcal{D}_o\mathbf{W})$ is nonsingular if and only if the corresponding submatrix of $\mathcal{D}_o$ is nonsingular. By Theorem 2, a typical term in the expansion of $|\mathcal{D}_o\mathbf{W}\mathcal{M}_o{}^t|$ is $\pm w_1w_2 \ldots w_k$, where w_i is the weight of edge b_i, and edges $b_1, \ldots, b_k$ form a feedback loop or a set of non-touching feedback loops $L_1, \ldots, L_r$ incident on nodes 1, 2, $\ldots$, k. Since all possible combinations of nodes appear in Eq. (1.22), we have established the formula, Eq. (1.19) except for considerations of sign.

To complete the proof, consider Table 2. The table shows that the

Table 2. Sign associated with a set of nontouching feedback loops

k	r	Sign of $w_1 \ldots w_k$ in the expansion of $|r_{11}\ldots r_{kk}|$ (by Lemma 3)	Sign of $|r_{11}\ldots r_{kk}|$ in Eq. (1.22)	Sign of $w_1 \ldots w_k$ in the expansion of Δ
1	even	even	+	+	+
2	even	odd	−	+	−
3	odd	even	−	−	+
4	odd	odd	+	−	−

sign associated with a set of nontouching feedback loops is positive if and only if the number of loops in the set is even, exactly as stated in the formula Eq. (1.19). Thus, the proof is complete.

Definition 8. A "direct path" or "forward path" from node i to node j of a signal-flow graph is a connected, ordered sequence of edges whose initial node (first node relative to the ordering) is i and whose final node (last node relative to the ordering) is j. Node i has one exit-edge but no entrance-edges; node j has one entrance-edge but no exit-edges, and all other nodes have exactly one exit-edge and exactly one entrance-edge. The "forward path product" or the "weight of a forward path" is the product of the weights of the edges constituting a forward path.

Theorem 4. (Due to Mason.) Suppose a signal-flow graph G characterizes a set of equations $\mathscr{A}X = Y$, where $\mathscr{A}$ is the coefficient matrix of the form of Eq. (1.3a). Let $\mathscr{A}_{ij}$ be the cofactor of the element a_{ij} of $|\mathscr{A}|$, Then if $i \neq j$, $\mathscr{A}_{ij}$ is given by

$$\mathscr{A}_{ij} = \sum_k W(P_k)\Delta_k, \qquad (1.24)$$

where $W(P_k)$ is the weight of the forward path P_k from node i to node j, and Δ_k is the formula, Eq. (1.19), evaluated for the graph obtained from G by deleting all the nodes of P_k and all edges incident at these nodes. The summation is over all forward paths from node i to node j. The cofactor $\mathscr{A}_{ii}$ is given again by Eq. (1.19) evaluated for the graph obtained from G by deleting node i and all edges incident at node i.

As an illustration, for the graph of Fig. 1.1,

$$\mathscr{A}_{14} = abc$$
$$\mathscr{A}_{11} = 1 - bcd - cg$$
$$\mathscr{A}_{12} = a(1 - cg) \qquad (1.25)$$
$$\mathscr{A}_{33} = 1 - af$$
$$\mathscr{A}_{32} = ea + cd.$$

Proof of Theorem 4: Without loss of generality let us consider the cofactor $\mathscr{A}_{n1}$ for $n \neq 1$. Then $\mathscr{A}_{n1}$ may be written as follows:

$$\mathscr{A}_{n1} = (-1)^{n+1} \begin{vmatrix} a_{12} & a_{13} \dots a_{1n} \\ \rule{0pt}{0pt} & a_{2n} \\ Q & \vdots \\ & a_{n-1,n} \end{vmatrix}, \qquad (1.26)$$

where

$$
Q = \begin{array}{ccccc}
 & 2 & 3 & & n-1 \\
1 & 1 & a_{23} & \cdots & a_{2,n-1} \\
 & a_{32} & 1 & \cdots & a_{3,n-1} \\
 & \cdot & \cdot & & \cdot \\
 & \cdot & \cdot & & \cdot \\
 & \cdot & \cdot & & \cdot \\
 & a_{n-1,2} & a_{n-1,3} & \cdots & 1
\end{array}
$$

(1.27)

Let us now expand $\mathscr{A}_{n1}$ by the Cauchy expansion of a bordered determinant. We obtain

$$
\mathscr{A}_{n1} = (-1)^{n+1}[(-1)^n a_{1n}Q + (-1)^{n-1} \sum_{i,j=2}^{n-1} a_{in}a_{ij}Q_{ij}], \quad (1.28)
$$

where Q_{ij} is the cofactor of the element in row i and column j of Q for $i,j = 2, 3, \ldots, n-1$.

Now let $t_{ji} = -a_{ij}$. Then t_{ji} is the weight of the edge b_{ji} leaving node j and entering node i of G. With the substitution $t_{j1} = -a_{ij}$, Eq. (1.28) becomes

$$
\mathscr{A}_{n1} = t_{n1}Q + \sum_{i,j=2}^{n-1} t_{ni}t_{j1}Q_{ij}. \quad (1.29)
$$

The subgraph G' corresponding to Q is the original graph G with nodes 1 and n removed, as well as all edges incident at these nodes. In particular, G' does not contain edges b_{ni} or b_{j1} $(i, j = 1, 2, \ldots, n)$. (See Fig. 1.2.)

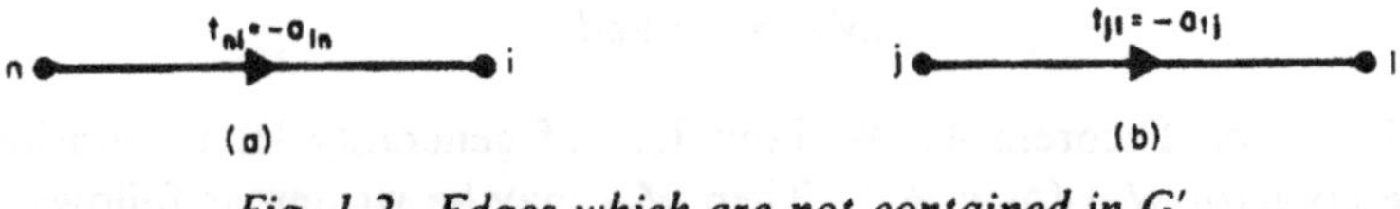

(a)   (b)

Fig. 1.2. Edges which are not contained in G'
(a) Edge b_{ni}; (b) edge b_{j1}.

The terms $t_{n1}Q$ and $t_{ni}t_{i1}Q_{i1}$, $(i = 2, \ldots, n-1)$, are of the proper form as stated by the formula, Eq. (1.19). Using the same reasoning as above, we expand the cofactor Q_{ij} $(i \neq j)$ as follows:

$$
Q_{ij} = t_{ij}V + \sum_{\substack{k,m=2 \\ k,m \neq i \text{ or } j}}^{n-1} t_{ik}t_{mj}P_{km}, \quad (1.30)
$$

where the subgraph G'' corresponding to V is the graph G' with nodes i and j removed, as well as all edges incident at these nodes. (See Fig. 1.3.)

Fig. 1.3. *Edges which are not contained in* G''
(a) Edges b_{ni} and b_{ik}; (b) edges b_{mj} and b_{jl}.

Continuing the above process, we find that a typical term in $\mathscr{A}_{n1}$ is $t_{mi}t_{ik}t_{kp} \ldots t_{qm}t_{mj}t_{j1}|\mathbf{R}|$, where $t_{ni} \ldots t_{j1}$ is the path product of path P_k from node n to node 1 and $|\mathbf{R}|$ is the determinant of the system of equations described by the graph obtained from G by removing all the nodes of P_k, and all edges incident at these nodes. This establishes the formula, Eq. (1.24). The expression for $\mathscr{A}_{ii}$ follows immediately from the fact that the submatrix obtained from $|\mathscr{A}|$ by deleting row i and column i represents the subgraph obtained from G by removing node i and all edges incident at node i. This completes the proof.

Theorems 3 and 4 yield "Mason's general gain formula," which is usually stated as follows. Let a signal-flow graph G characterize a set of equations $\mathscr{A}\mathbf{X} = \mathbf{Y}$, where $\mathscr{A}$ is the coefficient matrix. Then the response x_j due to the forcing function y_i is given by

$$\left.\frac{x_j}{y_i}\right]_{\text{all other } y\text{'s } = 0} = \frac{\mathscr{A}_{ij}}{|\mathscr{A}|}, \tag{1.31}$$

where $|\mathscr{A}|$ and $\mathscr{A}_{ij}$ are computed from the flow graph G by application of Eqs. (1.24) and (1.29).

ALTERNATIVE DEFINITION OF THE SIGNAL-FLOW GRAPH

A signal-flow graph characterizing a set of linear equations may be constructed even when the equations are not of the form of Eq. (1.1). It is often more convenient to consider equations of the following form:

$$
\begin{aligned}
(1+a_{11})x_1 + a_{12}x_2 + \ldots + a_{1n}x_n &= y_1 \\
a_{21}x_1 + (1+a_{22})x_2 + \ldots + a_{2n}x_n &= y_2 \\
&\cdots \\
a_{n1}x_1 + a_{n2}x_2 + \ldots + (1+a_{nn})x_n &= y_n.
\end{aligned}
\tag{1.32}
$$

Definition 9. The signal-flow graph associated with the set of equations of Eq. (1.32) is an oriented graph with a node corresponding to each unknown x_i ($i = 1, \ldots, n$) and a directed edge with weight $-a_{ij}$ leaving node j and entering node i. If $y_j \neq 0$, there is a directed edge with weight 1 leaving node y_j and entering node j. The graph may contain "self-loops," i.e., feedback loops formed by a single edge. This is not possible in the graph associated with Eq. (1.1).

As an example, consider the following set of equations:

$$(1-d)x_1 - fx_2 - cx_3 = y_1$$
$$-ax_1 + x_2 = 0 \qquad\qquad (1.33)$$
$$bx_2 + (1-e)x_3 = 0.$$

The signal-flow graph corresponding to the set of equations of Eq. (1.33) is shown in Fig. 1.4.

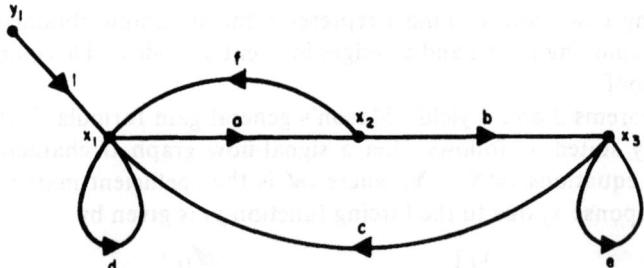

Fig. 1.4. Signal-flow graph of Eq. (1.33)

The usefulness of the alternative construction of the signal-flow graph is based on the following result.

Lemma 4. Mason's general gain formula holds without modification for the graph associated with the set of equations of Eq. (1.32).

Proof: Eq. (1.32) may be written in a simpler form as follows:

$$a_{i1}x_1 + \ldots + (1 + a_{ii})x_i + \ldots + a_{in}x_n = y_i \qquad (1.34)$$

and $i = 1, 2, \ldots, n$. If $a_{ii} \neq 0$, we transform Eq. (1.34) into the equivalent form

$$a_{i1}x_1 + \ldots + x_i + a_{ii}x'_i + \ldots + a_{in}x_n = y_i, \qquad (1.35)$$

where $x'_i = x_i$.

In matrix form, Eq. (1.35) becomes

$$
\begin{bmatrix}
1 & a_{12} \ldots & a_{1n} & a_{11} & 0 & \ldots 0 \\
a_{21} & 1 & \ldots & a_{2n} & 0 & a_{22} \ldots 0 \\
\cdot & \cdot & \cdot & \cdot & \cdot & \cdot \\
a_{n1} & a_{n2} \ldots & 1 & 0 & 0 & \ldots a_{nn} \\
-1 & 0 & \ldots 0 & 1 & 0 & \ldots 0 \\
0 & -1 & \ldots 0 & 0 & 1 & \ldots 0 \\
\cdot & \cdot & \cdot & \cdot & \cdot & \cdot \\
0 & 0 & \ldots -1 & 0 & 0 & \ldots 1
\end{bmatrix}
\begin{bmatrix}
x_1 \\ x_2 \\ \cdot \\ x_n \\ x'_1 \\ x'_2 \\ \cdot \\ x'_n
\end{bmatrix}
\begin{bmatrix}
y_1 \\ y_2 \\ \cdot \\ y_n \\ 0 \\ 0 \\ \cdot \\ 0
\end{bmatrix} . \quad (1.36)
$$

The signal-flow graph corresponding to Eq. (1.36) is related to the graph corresponding to Eq. (1.32) by the trivial modification shown in Fig. 1.5.

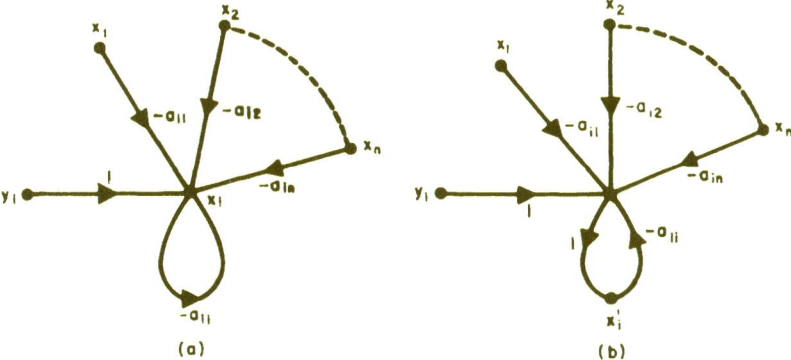

Fig. 1.5. *Modification of signal-flow graph of (a) Eq. (1.32) to fit with (b) Eq. (1.36)*

The quantities x_i/y_i may be computed from the graph of Fig. 1.5b by using Mason's formulas, Eqs. (1.19) and (1.24). It is clear that the application of the formulas to the graph of Fig. 1.5a will yield the same result. Hence, the lemma.

A similar flow-graph representation characterizing a set of linear equations has been proposed by Coates,[7] and the proof of Coates' formula has been simplified by Desoer.[8] Coates' flow graph and Mason's signal-flow graph are equivalent in the sense that one graph can be easily obtained from the other.

Coates' formula has the interesting property that no cancellation of terms can take place in the evaluation of the formula. However, for a given set of equations, Coates' graph may be more complex than Mason's since it is possible that the former contains self-loops while the latter does not. It thus appears that no definite statement which would be valid in general can be made as to the superiority of one over the other.

CHAPTER 2

Flow-Graph Analysis
of Sampled-Data Systems

A LINEAR-FEEDBACK control system may be characterized by a set of
linear algebraic equations whose unknowns are the Laplace transforms of the various system variables. Thus, Mason's "general gain
formula" applies directly to such systems. Sampled-data control
systems, on the other hand, are not amenable to a direct attack by
means of the signal-flow graph. The equations describing a sampled-
data system involve both continuous and discrete variables; the
equations must be appropriately modified before a solution can be
obtained. In this chapter a general flow-graph technique is presented
for the solution of sampled-data systems. A so-called "sampled"
signal-flow graph is constructed directly from the original system.
The sampled form of any of the desired variables may be obtained
from the sampled flow graph by means of Mason's formula.

DEFINITION OF SAMPLED-FLOW GRAPHS

It is convenient to introduce some standard nomenclature which
will be used in this and the following chapters. A "sampler" (see
Fig. 2.1.) is a switch which operates at a fixed frequency. The letter

Fig. 2.1. A sampler

T denotes the period of the sampler; the switch closes periodically
every T seconds. In most physical systems containing a sampler the
switch remains closed for an interval of time so short that the input
to the switch cannot vary appreciably over the period of closure.
For such systems the impulse modulation approximation is valid.

If $f(t)$ is the input to a sampler, then the output, denoted by $f^*(t)$, is assumed to be given by

$$f^*(t) = \sum_{n=-\infty}^{\infty} f(nT)\delta(t-nT), \tag{2.1}$$

where $\delta(t)$ is the "unit impulse or delta function."

The Laplace transform of $f(t)$ is denoted by $F(s)$, and the Laplace transform of $f^*(t)$ is denoted by $F^*(s)$, or simply by F^*. The $f^*(t)$ is called the "sampled form" of $f(t)$, and $F^*(s)$ the sampled form of $F(s)$. Starred variables are referred to as "sampled variables." The Z-transform of $f(t)$, denoted by $F(z)$, is $F^*(s)$ with the substitution $z = e^{sT}$. From Eq. (2.1)

$$F^*(s) = \sum_{n=-\infty}^{\infty} f(nT)e^{-snT} \tag{2.2a}$$

$$F(z) = \sum_{n=-\infty}^{\infty} f(nT)z^{-n}. \tag{2.2b}$$

It is clear that the sampling operation is linear, i.e.,

$$(F_1 + F_2)^* = F_1^* + F_2^*. \tag{2.3}$$

For more discussions on sampled-data control systems, see Reference 9.

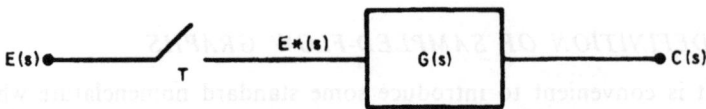

Fig. 2.2. A sampler followed by a linear system

Now consider a sampler followed by a linear system with transfer function $G(s)$, as shown in Fig. 2.2. Then,

$$C^*(s) = [E^*(s)G(s)]^* = E^*(s)G^*(s). \tag{2.4}$$

In particular, if $G(s) = 1$, then Eq. (2.4) yields

$$C^*(s) = [E^*(s)]^* = E^*(s). \tag{2.5}$$

A sampled-data control system is a control system containing linear time-invariant components and samplers. In this chapter, all samplers will be assumed to operate in synchronism; in other words, all samplers have the same frequency and close at exactly the same time.

We shall henceforth restrict ourselves to the frequency domain; all equations describing sampled-data systems will be written in terms of Laplace or Z-transforms of system variables.

Example 1. As an example of the analysis of a sampled-data system, consider the system of Fig. 2.3. In the above system a plant

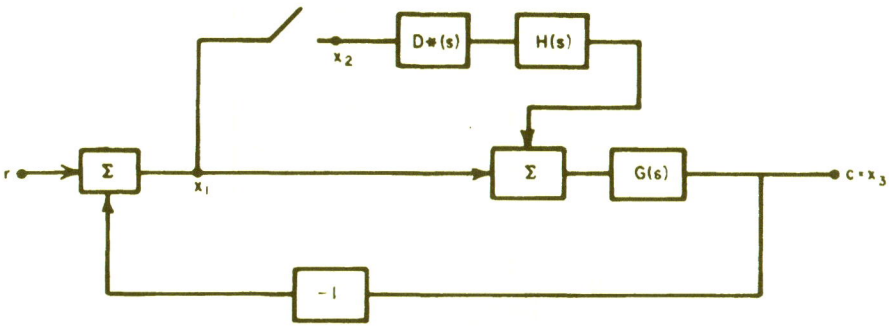

Fig. 2.3. A sampled-data system

is controlled by means of a bypass digital controller,[10] the whole structure forming a single sampled-data system. The variables x_1, x_2, x_3, and r are the Laplace transforms of system variables. It is desired to find the output x_3 in response to the input r. The system of Fig. 2.3 is characterized by the equations

$$x_1 = r - x_3$$
$$x_2 = x_1^* \qquad (2.6)$$
$$x_3 = D^*GHx_2 + Gx_1.$$

Equation (2.6) may be written in matrix form as

$$\mathscr{A}X = \mathscr{B}X^* + R, \qquad (2.7)$$

where

$$\mathscr{A} = \begin{bmatrix} 1 & 0 & 1 \\ 0 & 1 & 0 \\ -G & -D^*GH & 1 \end{bmatrix} \quad \mathscr{B} = \begin{bmatrix} 0 & 0 & 0 \\ 1 & 0 & 0 \\ 0 & 0 & 0 \end{bmatrix}$$

$$R = \begin{bmatrix} r \\ 0 \\ 0 \end{bmatrix} \quad X = \begin{bmatrix} x_1 \\ x_2 \\ x_3 \end{bmatrix} \quad X^* = \begin{bmatrix} x_1^* \\ x_2^* \\ x_3^* \end{bmatrix}. \qquad (2.8)$$

Solving Eq. (2.7) for X we obtain

$$X = \mathcal{A}^{-1}\mathcal{B}X^* + \mathcal{A}^{-1}R, \qquad (2.9)$$

where

$$\mathcal{A}^{-1} = \frac{1}{1+G}\begin{bmatrix} 1 & -D^*GH & -1 \\ 0 & 1+G & 0 \\ G & D^*GH & 1 \end{bmatrix}$$

$$\mathcal{A}^{-1}\mathcal{B} = \frac{1}{1+G}\begin{bmatrix} -D^*GH & 0 & 0 \\ 1+G & 0 & 0 \\ D^*GH & 0 & 0 \end{bmatrix} \qquad (2.10)$$

$$\mathcal{A}^{-1}R = \frac{1}{1+G}\begin{bmatrix} r \\ 0 \\ Gr \end{bmatrix}.$$

Writing out each equation of Eq. (2.9) we have

$$x_1 = -D^*\left(\frac{GH}{1+G}\right)x_1^* + \frac{r}{1+G}$$

$$x_2 = x_1^* \qquad (2.11)$$

$$x_3 = D^*\left(\frac{GH}{1+G}\right)x_1^* + \frac{Gr}{1+G}.$$

Taking the sampled form of Eq. (2.11), using Eqs. (2.3) through (2.5), we obtain

$$x_1^* = D^*\left(\frac{GH}{1+G}\right)x_1\cdot^* + \left(\frac{r}{1+G}\right)^*$$

$$x_2^* = x_1^*$$

$$x_3^* = D^*\left(\frac{GH}{1+G}\right)x_1\cdot^* + \left(\frac{Gr}{1+G}\right)^*. \qquad (2.12)$$

Equation (2.12) may be solved algebraically or by use of the signal-flow graph. We obtain

$$c^* = x_3^* = \left(\frac{r}{1+G}\right)^* \frac{D^*(GH/1+G)^*}{1+D^*(GH/1+G)^*} + \left(\frac{Gr}{1+G}\right)^*. \qquad (2.13)$$

We shall now show how the above analysis may be performed without explicitly computing the matrix $\mathcal{A}^{-1}$.

Since a sampled-data system is assumed to be linear, all system variables, whether sampled or unsampled, appear linearly in any equation describing the system. As a consequence, the equations describing a sampled-data system may be written in matrix form as:

$$\mathscr{A}\mathbf{X} = \mathscr{B}\mathbf{X}^* + \mathbf{R}, \tag{2.14}$$

where $\mathscr{A}$ and $\mathscr{B}$ are square matrices whose order is equal to the number of variables, $\mathbf{X}$ is the column vector of variables, and $\mathbf{X}^*$ is obtained from $\mathbf{X}$ by replacing each variable by its sampled form. If $\mathbf{Q}$ is any matrix, the symbol $\mathbf{Q}^*$ will be used to denote the matrix obtained from $\mathbf{Q}$ by replacing each element q_{ij} of $\mathbf{Q}$ by its sampled form. If we restrict ourselves to single-input systems (multiple-input systems are easily handled by superposition) and adopt the convention that the input variable r appears in the first equation of Eq. (2.14), then

$$\mathbf{R} = \begin{bmatrix} r \\ 0 \\ \cdot \\ \cdot \\ \cdot \\ 0 \end{bmatrix}. \tag{2.15}$$

Solving Eq. (2.14) for $\mathbf{X}$, we obtain

$$\mathbf{X} = \mathscr{A}^{-1}\mathscr{B}X^* + \mathscr{A}^{-1}\mathbf{R} \tag{2.16}$$

and the existence of $\mathscr{A}^{-1}$ will be justified in the next section.

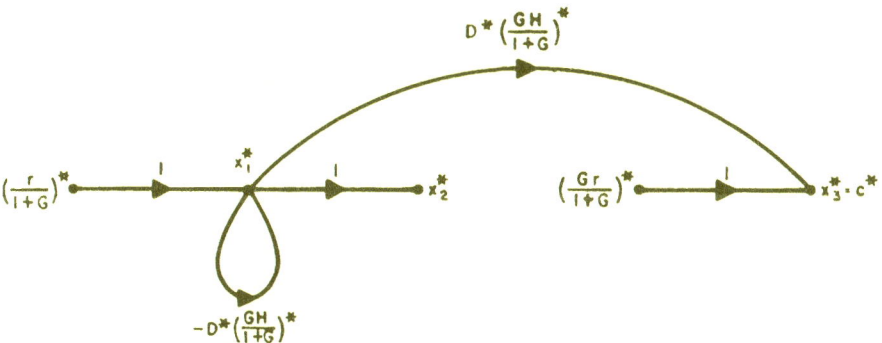

Fig. 2.4. Sampled graph for the system of Fig. 2.3

Now taking the sampled form of Eq. (2.15), using Eqs. (2.3) and (2.5), we have

$$\mathbf{X}^* = (\mathscr{A}^{-1}\mathscr{B})^*\mathbf{X}^* + (\mathscr{A}^{-1}\mathbf{R})^*. \tag{2.17}$$

Equation (2.17) represents a set of simultaneous linear equations involving only the sampled variables as unknowns. The signal-flow graph corresponding to Eq. (2.17) will be called the "sampled flow graph." For the system of Fig. 2.3, the sampled flow graph characterizes Eq. (2.12). The sampled graph is shown in Fig. 2.4.

The system output may be obtained directly from the sampled graph by using Mason's formula.

EFFICIENT CONSTRUCTION AND SIMPLIFICATION OF THE SAMPLED GRAPH

In order to construct the sampled graph of a system we need to know the elements of the matrices $\mathscr{A}^{-1}\mathscr{B}$ and $\mathscr{A}^{-1}\mathbf{R}$. In this section we shall demonstrate that it is always possible to obtain the sampled graph without the necessity of matrix inversion; the required elements of $\mathscr{A}^{-1}\mathscr{B}$ and $\mathscr{A}^{-1}\mathbf{R}$ may be found by conventional signal-flow graph techniques.

First of all, we adopt the convention of using distinct symbols for the input and output variables of a sampler. If x_i is the input and x_j the output, then $x_j = x_i^*$. This procedure simplifies the presentation to follow. We may then draw a flow-graph representation of the system in the usual way, with one modification; if x_i is the input to

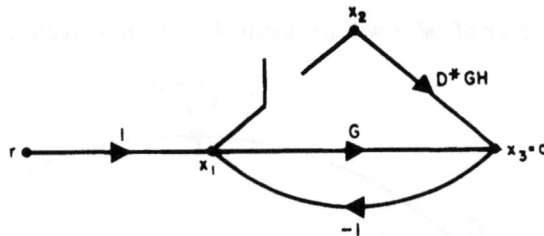

Fig. 2.5. System flow graph for the system of Fig. 2.3

a sampler and x_j the output, the sampler is shown on the diagram. The appearance of the sampler indicates the relation $x_j = x_i^*$. The flow graph just described will be called the "system flow graph" for the sampled-data system. The system flow graph for the system of Fig. 2.3 is shown in Fig. 2.5.

As a preliminary reduction of the system flow graph, we replace any configuration of two or more samplers having the same input with a single sampler. (See Fig. 2.6.) In the reduced flow graph a variable x_j cannot be the input to more than one sampler.

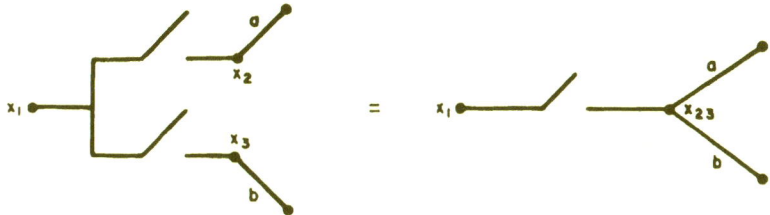

Fig. 2.6. Preliminary reduction of the system flow graph

We now describe the physical nature of $\mathscr{A}^{-1}$. Matrix $\mathscr{A}$ characterizes the set of equations

$$a_{11}x_1 + a_{12}x_2 + \ldots + a_{1n}x_n = y_1$$

$$a_{21}x_1 + a_{22}x_2 + \ldots + a_{2n}x_n = y_2$$

$$\cdot \quad \cdot \quad \cdot \quad \cdot \quad \cdot \quad \cdot \quad \cdot \quad \cdot \quad \cdot$$

$$a_{n1}x_1 + a_{n2}x_2 + \ldots + a_{nn}x_n = y_n. \qquad (2.18)$$

We may interpret these equations as describing the original control system with *all samplers open-circuited.* The forcing term y_k may be interpreted as an input connected to node k of the system flow graph. If a'_{ij} is the element in row i and column j of the matrix $\mathscr{A}^{-1}$, then

$$a'_{ij} = \frac{\mathscr{A}_{ji}}{|\mathscr{A}|} = \left. \frac{x_i}{y_j} \right]_{\text{all other } y's \, = \, 0}, \qquad (2.19)$$

where $\mathscr{A}_{ji}$ is the cofactor of a_{ji} and $|\mathscr{A}|$ the determinant of $\mathscr{A}$. Thus a'_{ij} is the response at node i of the system flow graph due to a unit input applied at node j, with all samplers open-circuited. The elements of $\mathscr{A}^{-1}$ are, therefore, physically measurable quantities; hence, the existence of $\mathscr{A}^{-1}$ for all systems of practical interest. If we denote by G_{ij} the transfer function from node i to node j of the system flow graph, with all samplers open-circuited, then

$$a'_{ij} = G_{ji}. \qquad (2.20)$$

Next we assert the following properties of matrix $\mathscr{B}$:

1. The elements of $\mathscr{B}$ are either 1 or 0;

2. The elements of b_{ij} of $\mathscr{B}$ will be 1 if and only if the variable x_j is the input to a sampler and x_i is the output of the sampler;

3. If x_j is not the input to a sampler, column j of $(\mathscr{A}^{-1}\mathscr{B})$ consists entirely of zeros;

4. $\mathscr{B}$ contains at most one nonzero element per column;

5. If $b_{ij} = 1$, then column i of $\mathscr{A}^{-1}$ appears as column j of $(\mathscr{A}^{-1}\mathscr{B})$.

Properties 1 and 2 follow from the convention of using distinct symbols for the input and output variables of a sampler. Property 3 is a direct consequence of Property 2. In the reduced system flow graph, a variable x_j is the input to not more than one sampler; this proves Property 4. Property 5 follows from Property 4 and the definition of matrix multiplication.

From the above properties and from Eq. (2.20), we may make the following statement.

If the variable x_j is the input to a sampler and the variable x_i is the output of the sampler, then column j of $(\mathscr{A}^{-1}\mathscr{B})$ appears as follows:

$$\mathscr{A}^{-1}\mathscr{B} = \begin{bmatrix} \dots G_{i1} \dots \\ \dots G_{i2} \dots \\ \cdot \quad \cdot \quad \cdot \quad \cdot \\ \dots G_{in} \dots \end{bmatrix}. \qquad (2.21)$$

If x_j is not the input to a sampler, then column j of $\mathscr{A}^{-1}\mathscr{B}$ consists entirely of zeros.

From Eq. (2.20) the matrix $(\mathscr{A}^{-1}R)$ has the form shown below:

$$\mathscr{A}^{-1}R = \begin{bmatrix} G_{11} \\ G_{12} \\ \cdot \\ \cdot \\ \cdot \\ G_{1n}{}^r \end{bmatrix}. \qquad (2.22)$$

Equations (2.21) and (2.22) allow efficient construction of the sampled graph from the original system flow graph.

A key property of the sampled graph

The flow-graph analysis of sampled-data systems is made more efficient by a special property of the sampled flow graph, namely, that for computing the sampled output we may disregard those variables which are not inputs to samplers. To establish this we first recall that the equations of a sampled-data system may be written in matrix form as

$$\mathscr{A}X = \mathscr{B}X^* + R. \tag{2.23}$$

Solving for X we obtain

$$X = (\mathscr{A}^{-1}\mathscr{B})^*X^* + (\mathscr{A}^{-1}R). \tag{2.24}$$

Taking the sampled form of Eq. (2.24) we obtain

$$X^* = (\mathscr{A}^{-1}\mathscr{B})^*X^* + (\mathscr{A}^{-1}R)^*. \tag{2.25}$$

Equation (2.25) may be rearranged as follows:

$$(U - \mathscr{A}^{-1}\mathscr{B})^*X^* = (\mathscr{A}^{-1}R)^*, \tag{2.26}$$

where U is a unit matrix whose order is equal to that of $\mathscr{A}$ and $\mathscr{B}$. The matrix $\mathscr{A}^{-1}R$ is given by Eq. (2.22).

We now make the following assertion: No edge can leave a node corresponding to a variable x_j^* of the sampled graph of a sampled-data system unless x_j is the input to a sampler of the original system.

To establish the above statement we note that the equation corresponding to node x_k^* of the sampled graph is the kth equation of Eq. (2.25), namely,

$$x_k^* = t_{1k}^*x_1^* + \ldots + t_{jk}^*x_k^* + \ldots + t_{nk}^*x_n^* + (G_{1k}r)^*, \tag{2.27}$$

where t_{jk}^* is the element in row k and column j of $(\mathscr{A}^{-1}\mathscr{B})$.

Now suppose that an edge b_{jk} of the sampled graph leaves node x_j^* and enters node x_k^*. By Eq. (2.27) the weight of b_{jk} is t_{jk}^*. However, we recall from the previous section that if x_j is not the input to a sampler, then column j of $\mathscr{A}^{-1}\mathscr{B}$ consists entirely of zeros. This implies $t_{jk}^* = 0$, contradicting the existence of b_{jk}. Thus x_j is the input to a sampler of the original system.

Note that if x_j is the input to a sampler S, then by Eq. (2.21) t_{jk} is the transfer function from the output of S to node k, with all samplers open-circuited.

Simplification of the sampled graph

It is convenient to simplify the sampled flow graph, making use of the property listed previously. Consider a sampled-data system containing p samplers $S_1, S_2, \ldots, S_p$. Suppose that the output c is the variable x_n, and the node of the system flow graph corresponding to the input r is connected to the node corresponding to x_1. Let us introduce the following notation:

> $G_{S_i S_j} \triangleq$ the transfer function from the *output* of sampler S_i to the *input* S_j, with all samplers open-circuited;
>
> $G_{1 S_i} \triangleq$ the transfer function from the *system input* to the input of sampler S_i, with all samplers open-circuited, where the input node of the system is labeled as node 1 for convenience.

Let us relabel the variables so that the input to sampler S_k is called s_k. Then, in view of the fact that $t_{jk} = 0$ if x_j is not the input to a sampler, Eq. (2.27) becomes

$$s_k* = G_{S_1 S_k}*s_1* + G_{S_2 S_k}*s_2* + \ldots + G_{S_p S_k}*s_p* + (G_{1 S_k} r)*. \qquad (2.28)$$

Using Eq. (2.28) we may draw a cross section of the sampled graph as shown in Fig. 2.7.

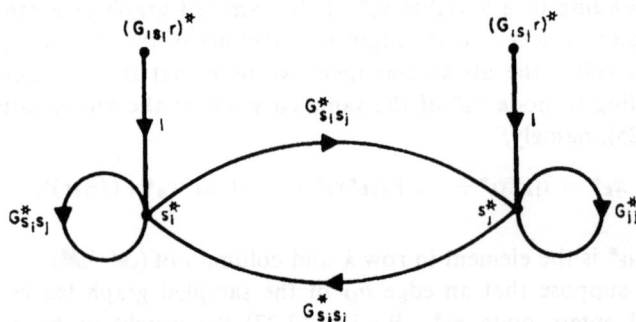

Fig. 2.7. Cross section of the sampled graph

We may assume without loss of generality that the system output c is the input to a sampler. Thus if c is the input to sampler S_n, then $c*$ is given by Eq. (2.28) with $k = n$. A similar result was obtained by Lendaris and Jury[5] using a different approach. The important point to note is that for the purpose of computing the sampled output $c*$ in response to the input r, all variables which are not input to samplers may be ignored when constructing the sampled graph, that is, it allows sampled graphs to be constructed directly without finding $\mathscr{A}^{-1}\mathscr{B}$.

We now summarize the procedure of the construction of the sampled graph and its use for the solution of a sampled-data system.

Summary of procedure

Assign a node for each sampler;

Compute $G_{S_i S_j}{}^* =$ gain from output of a sampler S_i to input of a sampler S_j with all samplers open;

Compute $(G_{1S_i} r)^*$, where $r =$ system input, and $G_{1S_i} =$ gain from system input to input of a sampler S_i provided all samplers open;

Draw sampled graph as shown in Fig. 2.7;

Apply Mason's formula.

Example 2. Let us consider a sampled multiloop system as shown in Fig. 2.8.

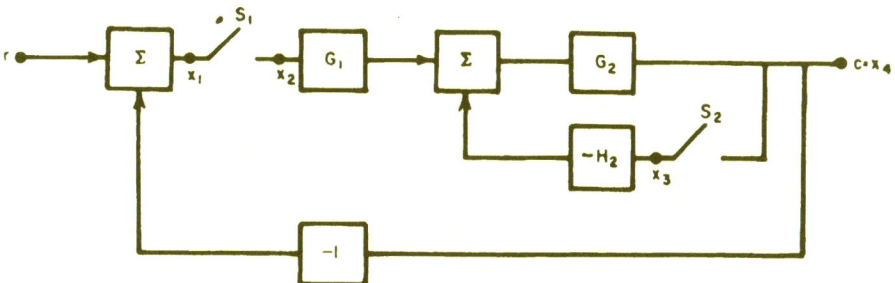

Fig. 2.8. *A multiloop sampled-data system*

First we compute

$$G_{S_1 S_1}{}^* = -(G_1 G_2)^* \qquad G_{S_1 S_2}{}^* = (G_1 G_2)$$
$$G_{S_2 S_1}{}^* = (G_2 H_2)^* \qquad G_{S_2 S_2}{}^* = -(G_2 H_2)^* \qquad (2.29)$$
$$(G_{1S_1} r)^* = r^* \qquad (G_{1S_2} r)^* = 0.$$

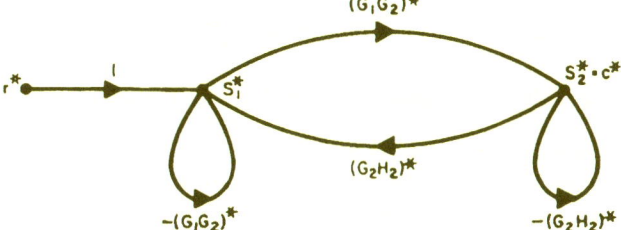

Fig. 2.9. *Sampled graph of the system of Fig. 2.8*

We therefore construct the sampled graph of the system as shown in Fig. 2.9. The output c^*, computed by Mason's formula, is

$$c^* = \frac{r^*(G_1 G_2)^*}{1 + (G_1 G_2)^* + (G_2 H_2)^*}. \qquad (2.30)$$

COMPUTATION OF THE UNSAMPLED OUTPUT

It should be clear that by using Eq. (2.16) one can find the un-sampled output c. However, the equation also suggests that a composite flow graph may be constructed which is composed of the original flow graph and the sampled graph. Thus it would *not* be necessary to compute all the sampled node variables. The idea of using the composite flow graph was suggested by Kuo.[11] For instance, the composite flow graph of the system of Fig. 2.3 is shown in Fig. 2.10. Note in Fig. 2.10 that the composite graph consists of two

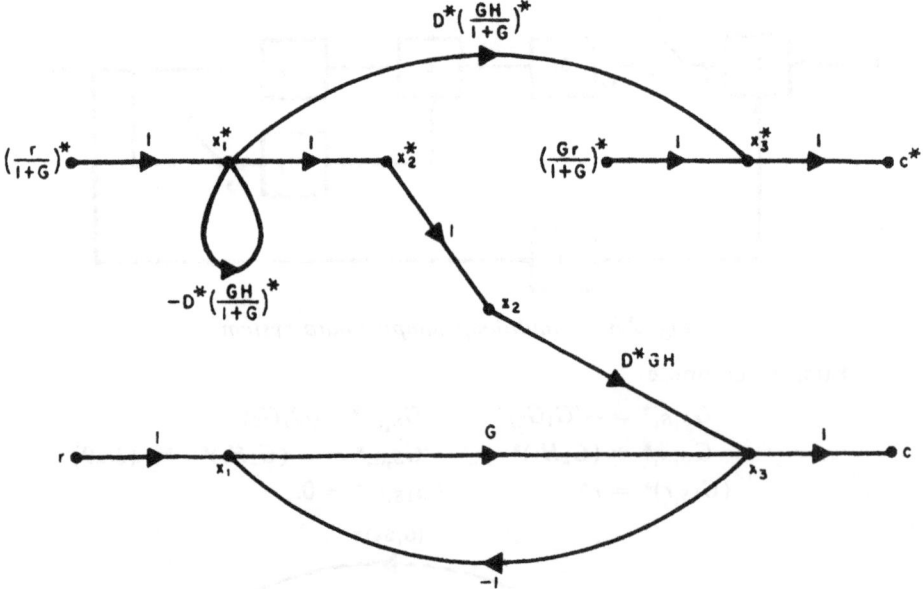

Fig. 2.10. *Composite flow graph of the system of Fig. 2.3*

graphs; one is the sampled graph of the system (Fig. 2.4) and the other the system flow graph of the system of Fig. 2.5. The two graphs are combined by an edge of unity gain from node x_2^* to node x_2 to describe the sampling operation. From the composite graph, one can compute both sampled and unsampled variables by means of Mason's formula. From Fig. 2.10, we have

$$c = x_3 = \frac{D^*GH[r/(1+G)]^*}{1+G)\{1+D^*[GH/(1+G)]^*\}} + \frac{Gr}{1+G}. \qquad (2.31)$$

A general technique for the analysis of sampled-data systems by means of the signal-flow graph has been presented. The technique is appreciably more efficient than the standard algebraic approach. The method is essentially a double application of flow-graph technique; the sampled flow graph is constructed by computing various transfer functions of the system flow graph. The sampled form of the output is found from the sampled graph. Thus the analysis of a system containing samplers is considerably more involved than the analysis of a continuous system of comparable size. In the following chapter we shall show how to adapt the flow-graph technique to analyze "multirate" systems, i.e., systems containing samplers which do not operate at the same sampling rate.

CHAPTER 3

Flow-Graph Analysis of Multirate

Sampled-Data Systems

THE TECHNIQUES DEVELOPED in the previous chapters are applicable to single-rate sampled-data systems, i.e., systems in which all switches operate in synchronism with the same sampling rate. However, many systems of practical importance are *multirate systems*, i.e., sampled-data systems containing switches which operate in synchronism but at different frequencies. Multirate systems sometimes arise because of mathematical convenience. For example, to obtain the output between sampling instants, a fictitious switch may be introduced which operates at a faster rate than the basic sampling rate of the system.[12,13] In addition, there are physical situations where multirate sampling is used to improve system performance.[14,15] A survey of available mathematical techniques for analyzing multirate systems is given by Kranc.[16] In this chapter we shall describe the analysis of such systems by means of the signal-flow graph. For this purpose, the most convenient mathematical tool is the "switch decomposition technique," due to Kranc.[16,17] By means of this technique a multirate sampled-data system is decomposed into an equivalent single-rate system. The single-rate system may then be analyzed by the flow-graph approach developed in Chapter 2.

THE SWITCH DECOMPOSITION TECHNIQUE

The method of switch decomposition is applicable to sampled-data systems containing switches whose rates are "rationally related." By this we mean that if the system to be analyzed contains switches operating with periods $T_1, T_2, \ldots, T_n$, the ratios of T_1 to T_2 to $\ldots T_n$ are expressed by rational numbers. Equivalently, there exists a number T and a set of integers $p_1, p_2, \ldots, p_n$ such that $T_1 = T/p_1$, $T_2 = T/p_2, \ldots, T_n = T/p_n$. Since an irrational number may be approximated to an arbitrary degree by a rational number, the limitation of rationally related switch rates is not serious.

The switch decomposition method involves the replacement of a switch operating with period T/p by an equivalent system containing p switches each operating with period T along with various advance and delay elements. The equivalence is illustrated in Fig. 3.1.

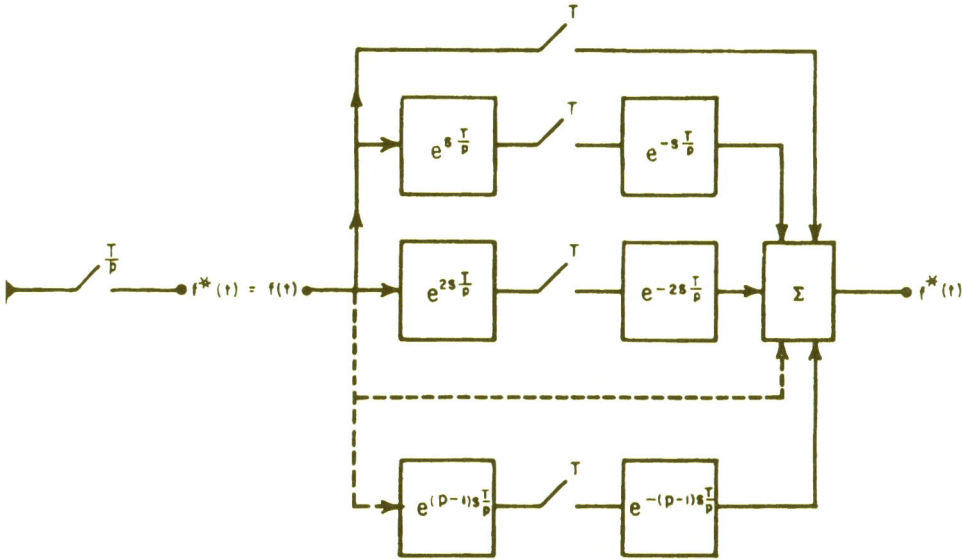

Fig. 3.1. Switch decomposition technique

In Fig. 3.1b, the system whose transfer function is $e^{-ksT/p}$ represents a "pure delay" or "right shift" of kT/p; the system whose transfer function is $e^{ksT/p}$ represents an "advance operator" or "left shift" of kT/p. An advance operator is not physically realizable since such a system is "anticipative," i.e., produces an output before the input is received. Thus the switch decomposition is only a mathematical device and has no physical counterpart.

To establish the equivalence of the two systems of Fig. 3.1, consider the system of Fig. 3.2. If $f(t)$ is the input to the system of Fig. 3.2,

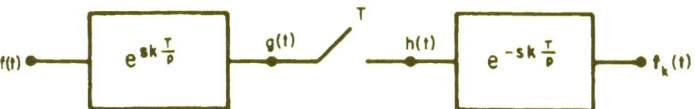

Fig. 3.2. Subsystem of the system of Fig. 3.1

then $$g(t) = f[t+(kT/p)].\tag{3.1}$$

By Eq. (2.1) in the previous chapter

$$h(t) = \sum_{n=-\infty}^{\infty} g(nT)\delta(t-nT) = \sum_{n=-\infty}^{\infty} f\left(nT+\frac{kT}{p}\right)\delta(t-nT).\tag{3.2}$$

Thus

$$f_k(t) = h\left(t-\frac{kT}{p}\right) = \sum_{n=-\infty}^{\infty} f\left(nT+\frac{kT}{p}\right)\delta\left(t-nT-\frac{kT}{p}\right).\tag{3.3}$$

The output of the system of Fig. 3.1b is then given by

$$f^*(t) = \sum_{k=0}^{p-1} f_k(t).\tag{3.4}$$

Using Eq. (3.3) we obtain

$$f^*(t) = \sum_{k=0}^{p-1} \sum_{n=-\infty}^{\infty} f\left(nT+\frac{kT}{p}\right)\delta\left(t-nT-\frac{kT}{p}\right).\tag{3.5}$$

Making the substitution $m = np+k$, Eq. (3.5) becomes

$$f^*(t) = \sum_{m=-\infty}^{\infty} f\left(\frac{mT}{p}\right)\delta\left(t-\frac{mT}{p}\right).\tag{3.6}$$

It is now clear that the expression (3.6) is the output of the system of Fig. 3.1a with the same input $f(t)$. This establishes the desired equivalence.

FLOW-GRAPH ANALYSIS OF A MULTIRATE SYSTEM

The switch decomposition technique allows the reduction of a multirate sampled-data system into an equivalent single-rate system having a more complicated structure. The flow-graph techniques previously developed can then be used without modification. As an example, consider the multirate system of Fig. 3.3. The equivalent single-rate system is shown in Fig. 3.4. Fig. 3.5 shows the system flow graph for the system of Fig. 3.4.

In order to construct the sampled graph of the system one first evaluates

$$G_{S_1S_1}{}^* = G^* \qquad G_{S_2S_1}{}^* = -H^*$$

$$G_{S_2S_1} = (Ge^{-sT/2})^* \quad G_{S_2S_2}{}^* = (-He^{sT/2})^* \tag{3.7}$$

$$(G_1 S_1 r)^* = r^* \qquad (G_{1S_2}r)^* = 0 \qquad (G_{1S_3}r)^* = (re^{sT/2})^*$$

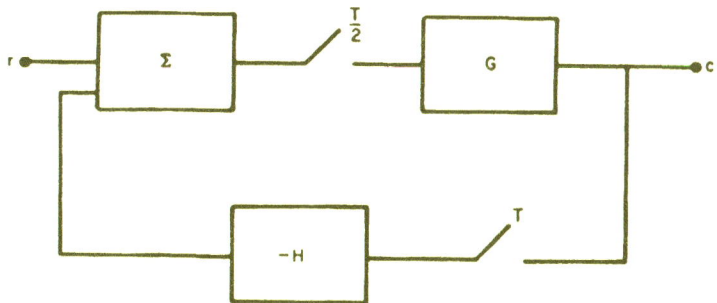

Fig. 3.3. A multirate sampled-data system

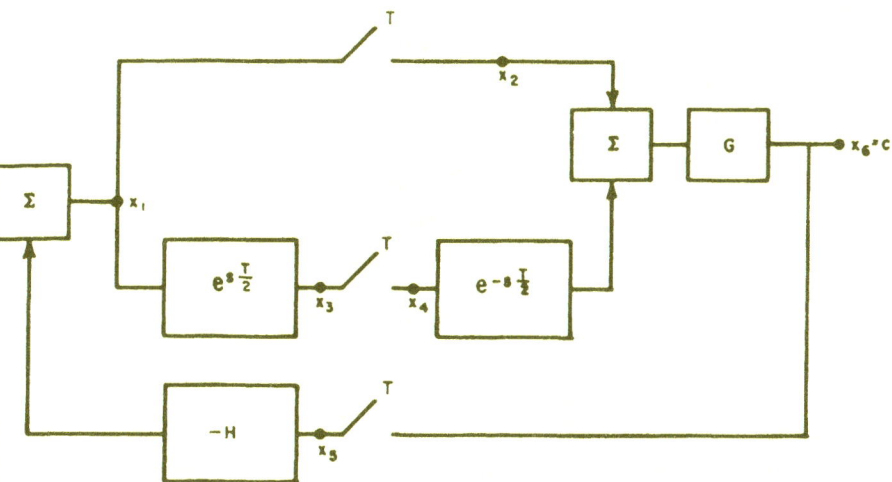

Fig. 3.4. Single-rate equivalent of the system of Fig. 3.3

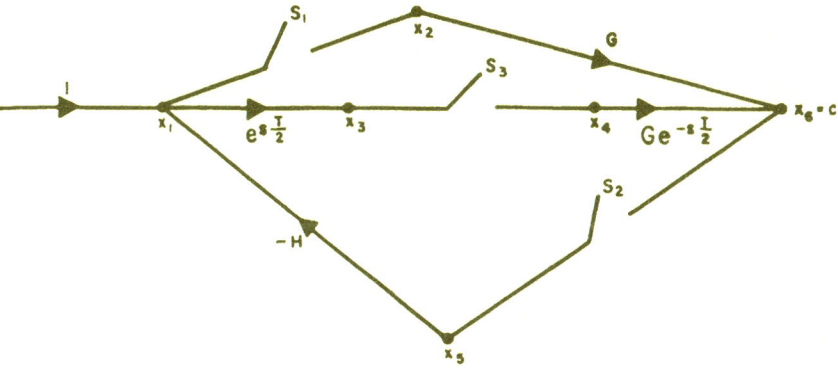

Fig. 3.5. System flow graph of the system of Fig. 3.4

and all other parameters are zero. Thus, we have the sampled graph of the system as shown in Fig. 3.6. The sampled output c^*, computed

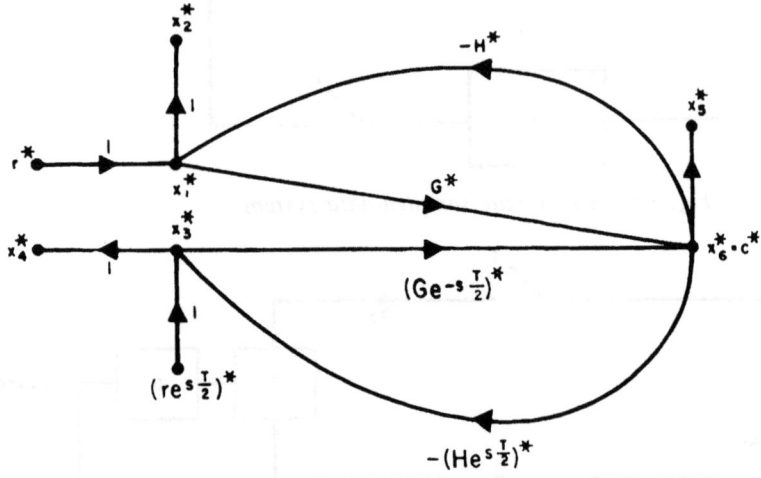

Fig. 3.6. Sampled graph for the system of Fig. 3.4

from the graph of Fig. 3.6 using Mason's formulas, is

$$c^* = \frac{r^*G^* + (re^{sT/2})^*(Ge^{-sT/2})^*}{1 + G^*H^* + (Ge^{-sT/2})^*(He^{sT/2})^*}. \tag{3.8}$$

The flow-graph technique may be applied to any multirate sampled-data system with switches having rationally related rates. This represents a decided improvement over previously used techniques which involved the solution of a very unpleasant system of simultaneous equations. However, just as the sampled flow graph of a single-rate sampled-data system is considerably more complex than the flow graph of a continuous system of comparable size, the sampled flow graph of a multirate system is quite a bit more complicated than the graph of a comparable single-rate system. A switch operating with period T/p results in p switches in the equivalent single-rate system. Each such switch adds two nodes to the sampled flow graph. Thus if a system contains two switches, one operating with period T/p and the other with period T/q, the sampled graph will contain at least $4pq$ nodes. Despite the large number of nodes, the sampled graph may frequently be constructed very easily, since in many cases a sampled-data system with all samplers open-circuited has few, if any, feedback loops.

In order to avoid the construction of the sampled graph altogether, Ash[4] proposed a general gain formula for analyzing any sampled-data system. The use of the formula is analogous to a "blindfolded computation" of the system, i.e., a computation which we would be forced to carry out if we were not allowed to see the sampled graph but were given only the various transfer functions $G_{S_i S_j}$. It is easier in general to construct the sampled graph and apply Mason's formula than to apply Ash's formula directly to a sampled-data system.

REFERENCES

1. S. J. Mason, "Feedback theory—some properties of signal flow graphs," *Proc. IRE*, 41 (1953), 1144–56.

2. S. J. Mason, "Feedback theory—further properties of signal flow graphs," *Proc. IRE*, 44 (1956), 920–26.

3. H. W. Bode, *Network Analysis and Feedback Amplifier Design*. Princeton, New Jersey, Van Nostrand, 1945.

4. R. Ash, "Topology and the solution of linear systems," *J. Franklin Inst.*, 268 (1959), 453–63. Also doctoral thesis, Department of Electrical Engineering, Columbia University, New York, 1960.

5. G. G. Lendaris and E. I. Jury, "Input–output relationships for multi-sampled loop systems," *Trans. AIEE, Part II. Application and Industry*, 46 (1960), 375–85.

6. R. B. Ash, W. H. Kim, and G. M. Kranc, "A general flow graph technique for the solution of multiloop sampled systems," *Trans. ASME, Ser. D, J. Basic Eng.*, 82 (1960), 360–66.

7. C. L. Coates, "Flow graph solutions of linear algebraic equations," *IRE Trans. PGCT-6* (1959), 170–87.

8. C. A. Desoer, "Optimum formula for the gain of a flow graph, or a simple derivation of Coates' formula," *Proc. IRE*, 48 (1960), 883–89.

9. R. R. Ragazzini and G. Franklin, *Sampled Data Control Systems*. New York, McGraw-Hill Book Co., 1958.

10. K. K. Maitra and P. E. Sarachik, "Digital compensation of continuous data feedback control systems," *Trans. AIEE, Part II. Application and Industry*, 75 (1956), 107–116.

11. B. C. Kuo, "Discussion of 'Input–output relationships for multi-sampled loop systems'," *Trans. ASME, Ser. D, J. Basic Eng.*, 82 (1960), 366–67.

12. W. K. Linvill, "Sampled-data control systems studied through comparison of sampling with amplitude modulation," *Trans. AIEE*, 70 (1951), 1779–88.

13. G. V. Lego and J. G. Truxal, "The design of sampled-data feedback systems," *Trans. AIEE*, 73 (1954), 247–53.

14. G. M. Kranc, "Compensation of an error-sampled system by a multirate controller," *Trans. AIEE*, 77 (1957), 149–59.

15. J. Sklansky, "Network compensation of error-sampled feedback systems," doctoral thesis, Department of Electrical Engineering, Columbia University, New York, 1955.

16. G. M. Kranc, "Input–output analysis of multirate feedback systems," *IRE Trans. on Automatic Control*, AC-3 (1957), 21–28.

17. G. M. Kranc, "Additional techniques for sampled-data systems problems," *IRE Wescon Convention Record*, Part 4, pp. 157-65, 1957.

PART III

Analysis and Synthesis of

Linear N-Port Networks

Introduction

NETWORKS ARE COMMONLY classified according to the number of terminals accessible for external connection, If the number of accessible terminal-pairs of a network is n, then we refer to the network as an "n-port network" or simply "n-port." Studies of the analysis and synthesis aspects of linear n-ports have frequently appeared in the literature and the many results are scattered through various publications. We shall, however, concentrate our discussion on networks consisting entirely of the ordinary two-terminal elements—resistors, inductors, and capacitors. The material which will be presented here is based on contributions made by various authors, including some pertinent up-to-date results.

Once we allow ourselves ideal transformers in addition to the ordinary elements for the realization of n-ports, Oono[1] and MacMillan[2] have shown that any symmetric positive-real matrix is realizable as the open-circuit impedance matrix of an n-port. If in our design we take negative resistors into consideration, then Sanberg[3] has shown that any arbitrary symmetric matrix of real rational functions in a complex frequency variable can be realized as the immittance matrix (open-circuit impedance and short-circuit admittance matrices) of an n-port with n negative resistors but without mutual coupling. Recently, Carlin and Youla[4] showed that any linear relation between port-voltages and port-currents, in terms of real rational functions of a complex frequency variable, can be realized with a linear n-port consisting of R, L, C elements, ideal transformers, gyrators, and negative resistors.

The realization problem however becomes extremely difficult if we restrict ourselves only to the ordinary n-ports, that is, obtaining RLC equivalent networks. This problem is significant currently for both engineering and theoretical aspects of network theory. Moreover, the problem of realizing an ordinary n-port has many applications in other areas, in addition to its importance for circuit theory. For instance, a direct extension of the realization techniques of an n-port is applied to synthesis problems of single-contact switching networks (see Part IV) and also to the realization of an optimum communication net (see Part V).

We shall open our discussion with the methods of derivation of an n-port based on port-currents and port-voltages, i.e., nodal and loop bases. Then we will investigate the realization methods of n-ports in terms of the short-circuit admittance matrices, the so-called "realizability of a Y-matrix." Finally, we shall extend the synthesis techniques of Y-matrices to realize an n-port based on the open-circuit impedance matrix, i.e., "realization of a Z-matrix." In the course of the presentation of the known results, a number of significant problems, which are still unsolved, will be pointed out as a guide to further research.

CHAPTER 1

Formulation of
N-Port Networks

FORMULATION OF AN N-PORT ON LOOP BASIS AND TRANSFORMATION OF PORT-CURRENTS

LET US CONSIDER an ordinary network N of b elements and n nodes. We shall insert an independent voltage source in series and an independent current source in parallel with each element of the network. We call such a composite element a "compositive active element" (see Fig. 1.1). If we convert every element of network N into a composite active element as defined in Fig. 1.1,† then for the

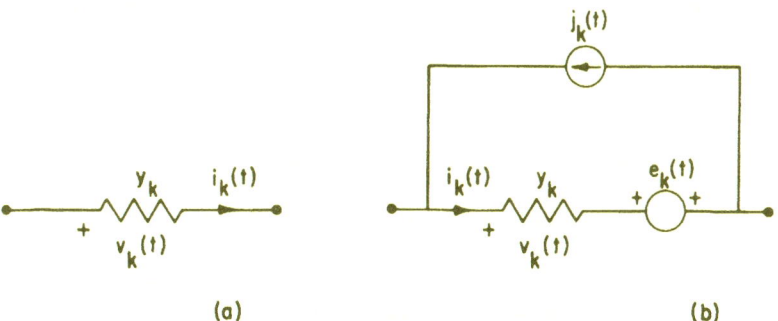

Fig. 1.1. *An ordinary element (a) is converted into a composite active element (b)*

resultant network in the steady-state condition we have Kirchhoff's equations:

$$\mathbf{BV}_e = \mathbf{BE}_e \qquad (1.1a)$$

$$\mathbf{AI}_e = \mathbf{AJ}_e, \qquad (1.1b)$$

† The stipulation for the insertion of ideal voltage and current sources is justified on the ground that the network configuration must not be altered by these sources. See Reference 5, Chapter 2, for detailed discussion.

where **B** is a loop matrix of $(b - n + 1)$ independent loops and **A** is the incidence matrix formed for the original network N; V_e and I_e are column matrices of edge-voltages and edge-currents, respectively, and E_e and J_e are also column matrices of voltage and current sources introduced into each edge, respectively. They are illustrated in the following example.

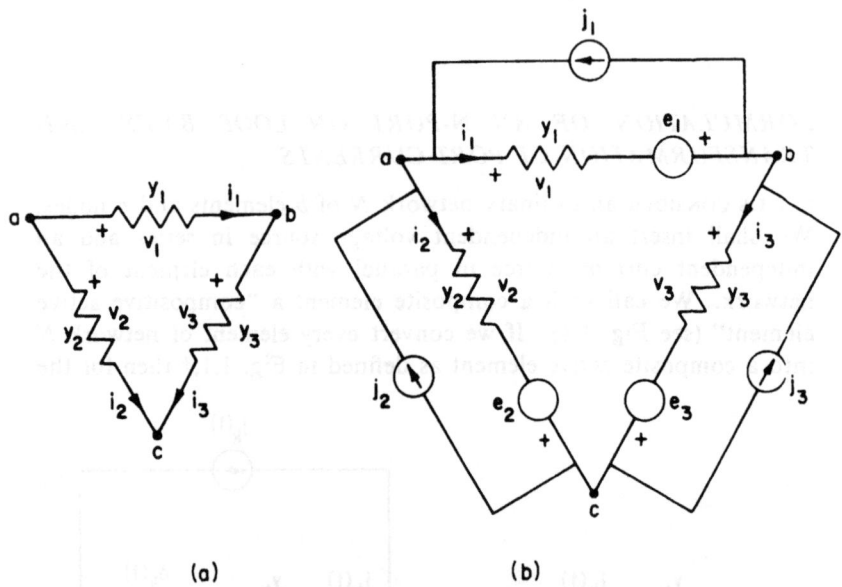

(a)   (b)

Fig. 1.2. Each ordinary element of a network (a) is replaced by a composite element (b)

Example 1. Consider an ordinary network of Fig. 1.2a. The incidence matrix and the loop matrix of the network are found to be

$$
\begin{array}{cccc}
 & i_1 & i_2 & i_3 \\
\end{array}
$$

$$
A = \begin{bmatrix} 1 & 1 & 0 \\ -1 & 0 & 1 \\ 0 & -1 & -1 \end{bmatrix} \qquad B = \begin{bmatrix} \overset{v_1}{1} & \overset{v_2}{-1} & \overset{v_3}{1} \end{bmatrix}, \qquad (1.2a)
$$

where the orientation of the loop is taken clockwise.

Now we shall replace each ordinary element of the network by a composite active element as shown in Fig. 1.2b. Then, for the resultant network we have the KCL equations:

$$i_1 + i_2 = j_1 + j_2 \qquad \text{at node } a$$
$$-i_1 + i_3 = -j_1 + j_3 \qquad \text{at node } b \tag{1.2b}$$
$$-i_1 - i_3 = -j_2 - j_3 \qquad \text{at node } c,$$

or in matrix form

$$\mathbf{AI}_e = \mathbf{AJ}_e, \tag{1.2c}$$

where

$$\mathbf{I}_e = \begin{bmatrix} i_1 \\ i_2 \\ i_3 \end{bmatrix} \qquad \mathbf{J}_e = \begin{bmatrix} j_1 \\ j_2 \\ j_3 \end{bmatrix}. \tag{1.2d}$$

And the KVL equation is

$$v_1 - v_2 + v_3 = e_1 - e_2 + e_3 \tag{1.3a}$$

or in matrix form

$$\mathbf{BV}_e = \mathbf{BE}_e, \tag{1.3b}$$

where

$$\mathbf{V}_e = \begin{bmatrix} v_1 \\ v_2 \\ v_3 \end{bmatrix} \qquad \mathbf{E}_e = \begin{bmatrix} e_1 \\ e_2 \\ e_3 \end{bmatrix}. \tag{1.3c}$$

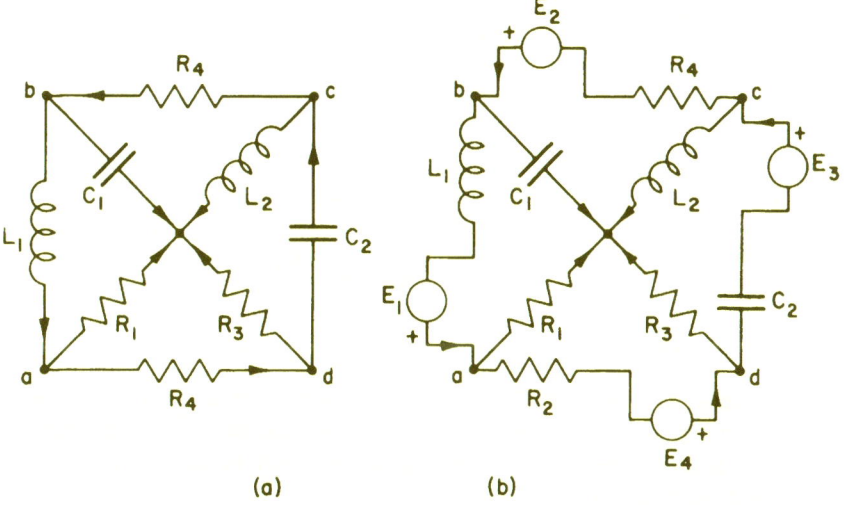

(a) (b)

Fig. 1.3. Formulation of a four-port (a) on an ordinary network (b)

In order to derive an n-port from a network in terms of the open-circuit impedance matrix, we shall insert a voltage generator in series with some of the elements of the network. However, we should assure the independence of the ports thus formed. To do this, we first choose a tree of a given network and then insert a voltage source in series with each chord of the tree. This process is called the "pliers-type connection" by Guillemin[5a] or "inscribing an n-port into a network" by Cederbaum.[6c] For instance, we are going to inscribe a four-port into the network given in Fig. 1.3a. There will be a number of ways of choosing a tree in the network. We shall choose a tree consisting of edges C_1, R_1, R_3, and L_2, and insert a voltage source in each chord of the tree as shown in Fig. 1.3b.

To evaluate the open-circuit impedances of the resultant four-port network, we first find the basic loop matrix of the original network, $\mathbf{B}_f$, with respect to the chosen tree as:

$$\mathbf{B}_f = \begin{array}{cccccccc} L_1 & R_4 & C_2 & R_2 & C_1 & R_1 & R_3 & L_2 \end{array}$$

$$\mathbf{B}_f = \begin{bmatrix} 1 & 0 & 0 & 0 & -1 & 1 & 0 & 0 \\ 0 & 1 & 0 & 0 & 1 & 0 & 0 & -1 \\ 0 & 0 & 1 & 0 & 0 & 0 & -1 & 1 \\ 0 & 0 & 0 & 1 & 0 & -1 & 1 & 0 \end{bmatrix}. \tag{1.4}$$

Then, using the topological formula derived in Part I, the open-circuit impedance matrix of the four-port, $\mathbf{Z}$, is found to be

$$\mathbf{Z} = \mathbf{B}_f \mathbf{Z}_e \mathbf{B}_f{}^t$$

$$= \begin{bmatrix} R_1+sL_1+\dfrac{C_1}{s} & \dfrac{-C_1}{s} & 0 & -R_1 \\[2ex] \dfrac{-C_1}{s} & R_4+sL_2+\dfrac{C_1}{s} & -sL_2 & 0 \\[2ex] 0 & -sL_2 & R_3+sL_2+\dfrac{C_2}{s} & -R_3 \\[2ex] -R_1 & 0 & -R_3 & R_1+R_2+R_3 \end{bmatrix}. \tag{1.5}$$

In general, the formulation of an n-port from a network of m independent loops, where $m \geqslant n$, and the evaluation of the open-circuit impedances of the resultant n-port, will be achieved by the following steps.

Choose a tree of a given network and insert a voltage source in each chord in series for any desired set of *n* chords, assuming that *n* ≤ the number of chords of the network.

Form the KVL equation for each basic loop and for all the loops of the resultant network with respect to the chosen tree. Thus, obtain the basic loop impedance matrix of the resultant network.

Apply the pivotal condensation on the determinant of the loop impedance matrix about the diagonal elements corresponding to the chords which are not included in forming the *n*-port.

We must note here that it is possible to derive an *n*-port from a network in which the number of independent loops is less than *n*, even though it was assumed that the number of ports to be formulated should be greater than the number of chords (or independent basic loops). As an illustration, let us assume that in the network of Fig. 1.3a $R_2 = \infty$ such that the network is reduced to the one shown in Fig. 1.4a which has three independent loops. However, a voltage generator E_4 with zero internal impedance, i.e., an ideal voltage source, can be introduced between nodes *a* and *b* as shown in Fig. 1.4b to form an extra port. It should, however, be clear that the topological formula for the open-circuit impedance matrix cannot be directly used because the edge corresponding to E_4 introduces an extra basic loop which was not included in the original network of Fig. 1.4a. That is, the basic loop matrix of the network of Fig. 1.4a

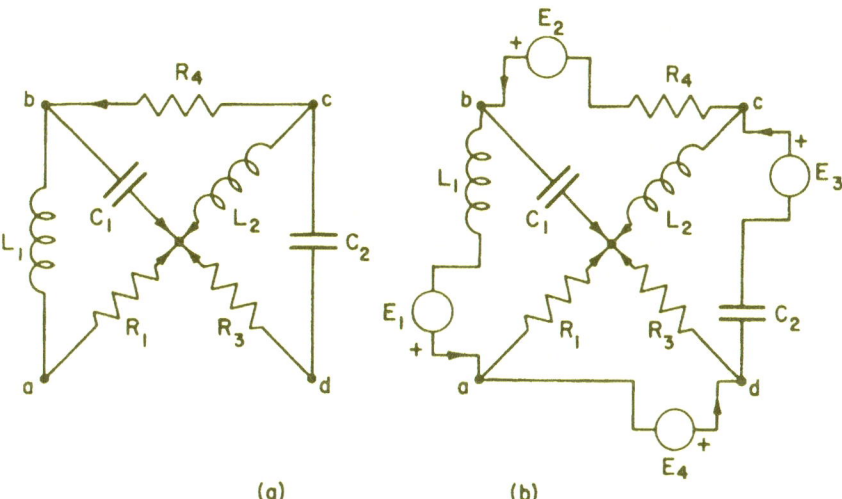

(a)　　　　(b)

Fig. 1.4. A four-port is inscribed into a network of three independent loops

will contain only three loops and will yield three KVL equations. We should therefore modify the basic loop matrix of the network of Fig. 1.4a such that the new loop is taken into consideration. The modified basic loop matrix, $\mathbf{B}_{fm}$, should be as:

$$\mathbf{B}_{fm} = \begin{array}{c} \begin{array}{cccccccc} L_1 & R_4 & C_2 & E_4 & C_1 & R_1 & R_3 & L_2 \end{array} \\ \begin{bmatrix} 1 & 0 & 0 & 0 & 1 & -1 & 0 & 0 \\ 0 & 1 & 0 & 0 & -1 & 0 & 0 & 1 \\ 0 & 0 & 1 & 0 & 0 & 0 & 1 & 1 \\ 0 & 0 & 0 & 1 & 0 & 1 & -1 & 0 \end{bmatrix} \end{array}. \qquad (1.6)$$

The basic loop matrix $\mathbf{B}_f$ of the original network is given by

$$\mathbf{B}_f = \begin{array}{c} \begin{array}{ccccccc} L_1 & R_4 & C_2 & C_1 & R_1 & R_3 & L_2 \end{array} \\ \begin{bmatrix} 1 & 0 & 0 & 1 & -1 & 0 & 0 \\ 0 & 1 & 0 & -1 & 0 & 0 & 0 \\ 0 & 0 & 1 & 0 & 0 & 1 & 1 \end{bmatrix} \end{array}. \qquad (1.7)$$

Note that in the modified loop matrix of Eq. (1.6) the extra column and row corresponding to edge E_4 and the fourth loop completed by E_4, respectively, are introduced.

In forming the edge-impedance matrix of all the edges which are involved in the modified loop matrix, the edge corresponding to E_4 is assumed to be associated with zero impedance. Thus we have the modified edge-impedance matrix, $\mathbf{Z}_{em}$,

$$\mathbf{Z}_{em} = \begin{bmatrix} L_1 s & 0 & 0 & 0 & 0 & 0 & 0 & 0 \\ 0 & R_4 & 0 & 0 & 0 & 0 & 0 & 0 \\ 0 & 0 & \dfrac{1}{C_2 s} & 0 & 0 & 0 & 0 & 0 \\ 0 & 0 & 0 & 0 & 0 & 0 & 0 & 0 \\ 0 & 0 & 0 & 0 & \dfrac{1}{C_1 s} & 0 & 0 & 0 \\ 0 & 0 & 0 & 0 & 0 & R_1 & 0 & 0 \\ 0 & 0 & 0 & 0 & 0 & 0 & R_3 & 0 \\ 0 & 0 & 0 & 0 & 0 & 0 & 0 & L_2 s \end{bmatrix}, \qquad (1.8)$$

where the fourth column of $\mathbf{Z}_{em}$ consisting of all zeros corresponds to edge E_4.

Finally, the open-circuit impedance matrix **Z** of the resultant four-port network of Fig. 1.4b is found to be

$$\mathbf{Z} = \mathbf{B}_{fm}\mathbf{Z}_{em}\mathbf{B}_{fm}{}^t$$

$$= \begin{bmatrix} L_1s + \dfrac{1}{C_1s} + R_1 & -\dfrac{1}{C_1s} & 0 & -R_1 \\[2mm] \dfrac{-1}{C_1s} & L_2s + \dfrac{1}{C_1s} + R_4 & -L_2s & 0 \\[2mm] 0 & -L_2s & L_2s + \dfrac{1}{C_2s} + R_3 & -R_3 \\[2mm] -R_1 & 0 & -R_3 & R_1 + R_3 \end{bmatrix}. \quad (1.9)$$

There is an alternate way of finding the open-circuit impedance matrix of the inscribed four-port network using the edge impedance matrix of the original network, $\mathbf{Z}_e$ which is given by

$$\mathbf{Z}_e = \begin{bmatrix} L_1s & 0 & 0 & 0 & 0 & 0 & 0 \\[1mm] 0 & R_4 & 0 & 0 & 0 & 0 & 0 \\[1mm] 0 & 0 & \dfrac{1}{C_2s} & 0 & 0 & 0 & 0 \\[1mm] 0 & 0 & 0 & \dfrac{1}{C_1s} & 0 & 0 & 0 \\[1mm] 0 & 0 & 0 & 0 & R_1 & 0 & 0 \\[1mm] 0 & 0 & 0 & 0 & 0 & R_3 & 0 \\[1mm] 0 & 0 & 0 & 0 & 0 & 0 & L_2s \end{bmatrix}. \quad (1.10)$$

In order to have the product of the modified basic loop matrix and the edge-impedance matrix of the original network exist, and at the same time to include the loop completed by edge E_4, we will delete the column from the modified basic loop matrix corresponding to edge E_4, which we shall denote by $\mathbf{B}_{fm_{-4}}$ since the fourth column of $\mathbf{B}_{fm}$ corresponds to edge E_4. Thus

$$\mathbf{B}_{fm_{-4}} = \begin{array}{c} \begin{array}{ccccccc} L_1 & R_4 & C_2 & C_1 & R_1 & R_3 & L_2 \end{array} \\ \begin{bmatrix} 1 & 0 & 0 & 1 & -1 & 0 & 0 \\ 0 & 1 & 0 & -1 & 0 & 0 & 1 \\ 0 & 0 & 1 & 0 & 0 & 1 & 1 \\ 0 & 0 & 0 & 0 & 1 & -1 & 0 \end{bmatrix} \end{array} \quad (1.11)$$

and the open-circuit impedance matrix of the four-port is then

$$Z = B_{fm} {}_{,}Z_e B_{fm-{}_{,}}{}^t.$$ (1.12)

In the case of derivation of a multiport, care must be exercised so that no loop is formed with only voltage sources.

In general, we shall call a modified basic loop matrix with the columns corresponding to the fictitious chords deleted a "special basic loop matrix" and denoted by B_s. The fictitious chords represent the impedanceless voltage generators. We therefore have

$$Z = B_s Z_e B_s{}^t,$$ (1.13)

where in general more than one column will be deleted from B_{fm} resulting in B_s.

Thus far we have derived a multiport from a network using the port-currents as independent variables which we again identified with the currents through the elements corresponding to a set of chords of a tree of the network. We shall examine the properties of transformation relating two different systems of multiports based on the same parent network. Moreover, we will develop the relationship of the open-circuit impedance matrix of one system to that of the other.

On an ordinary network of b elements and $n+1$ nodes let us choose two different trees, t_k and t_r, and let the basic loop matrices of the network with respect to the trees be denoted by B_{fk} and B_{fr}, respectively. Then, from the relationship derived in Part I, we have

$$I_e = B_{fk}{}^t I_{kc} = B_{fr}{}^t I_{rc},$$ (1.14)

where I_e is the column matrix of all the edge-currents of the original network, and I_{kc} and I_{rc} are the column matrices for each set of the chord-currents with respect to trees t_k and t_r respectively. It is of course required to arrange the columns of both loop matrices in the same order. If we denote the submatrices of $B_{fk}{}^t$ and $B_{fr}{}^t$ by M and N, respectively, such that the rows of the submatrices correspond to each set of chords of trees t_r and t_k, respectively, then from Eq. (1.14) we deduce

$$I_{rc} = MI_{kc}$$ (1.15a)

$$I_{kc} = NI_{rc},$$ (1.15b)

where M and N are nonsingular submatrices as proved in Chapter 1 of Part I. We therefore have

$$M = N^{-1}.$$ (1.16)

Here again we should note that $\mathbf{M}$ and $\mathbf{N}$ are nonsingular unimodular matrices, i.e., the determinant of each matrix takes the value of 1 or -1, since the basic loop matrix of a connected graph is a unimodular matrix (see Chapter 2, Part I).

The substitution of Eq. (1.15) into Eq. (1.14) yields

$$\mathbf{I}_e = \mathbf{B}_{fk}{}^t \mathbf{N} \mathbf{I}_{rc} = \mathbf{B}_{fr}{}^t \mathbf{I}_{rc} \tag{1.17a}$$

$$\mathbf{I}_e = \mathbf{B}_{fk}{}^t \mathbf{I}_{kc} = \mathbf{B}_{fr}{}^t \mathbf{M} \mathbf{I}_{kc}. \tag{1.17b}$$

From Eqs. (1.16) and (1.17)

$$\mathbf{B}_{fr} = \mathbf{N}^t \mathbf{B}_{fk} \tag{1.18a}$$

$$\mathbf{B}_{fk} = \mathbf{M}^t \mathbf{B}_{fr}. \tag{1.18b}$$

Thus, we found that two systems of chord-currents are related by a nonsingular unimodular matrix. The relationship was derived based on the assumption that the state of the network under consideration is not altered in derivation of multiports, i.e., keeping the same edge-currents and edge-voltages. Corresponding to each set of the chord-currents, $\mathbf{I}_{rc}$ and $\mathbf{I}_{kc}$, we have the open-circuit impedance matrices $\mathbf{Z}_k$ and $\mathbf{Z}_r$ as

$$\mathbf{Z}_k = \mathbf{B}_{fk} \mathbf{Z}_e \mathbf{B}_{fk}{}^t \tag{1.19a}$$

$$\mathbf{Z}_r = \mathbf{B}_{fr} \mathbf{Z}_e \mathbf{B}_{fr}{}^t. \tag{1.19b}$$

If we substitute Eq. (1.18) into Eq. (1.19)

$$\mathbf{Z}_k = \mathbf{M}^t (\mathbf{B}_{fr} \mathbf{Z}_e \mathbf{B}_{fr}{}^t) \mathbf{M} = \mathbf{M}^t \mathbf{Z}_r \mathbf{M} \tag{1.20a}$$

$$\mathbf{Z}_r = \mathbf{N}^t (\mathbf{B}_{fk} \mathbf{Z}_b \mathbf{B}_{fk}{}^t) \mathbf{N} = \mathbf{N}^t \mathbf{Z}_k \mathbf{N}. \tag{1.20b}$$

Equation (1.20) gives the relation between the two impedance matrices. It is of interest to note that the determinants of the two impedance matrices are equal. This is so, because

$$\det \mathbf{Z}_k = (\det \mathbf{M}^t)(\det \mathbf{Z}_r)(\det \mathbf{M}) = (\det \mathbf{Z}_r)(\det \mathbf{M})^2$$

$$= \det \mathbf{Z}_r, \tag{1.21}$$

since $\mathbf{M}$ is a nonsingular unimodular matrix, i.e., $\det \mathbf{M} = \pm 1$.

FORMULATION OF AN N-PORT ON NODAL BASIS AND TRANSFORMATION OF PORT-VOLTAGES

On an ordinary network with b elements and $(n+1)$ nodes, it was found that

$$\mathbf{V}_e = \mathbf{A}_1{}^t \mathbf{V}_n \tag{1.22a}$$

$$\mathbf{Y}_n = \mathbf{A}_1 \mathbf{Y}_e \mathbf{A}_1{}^t, \tag{1.22b}$$

where V_n is a column matrix with n rows representing the node-datum voltages, i.e., node-pair voltages, with respect to the reference node. Y_n is the node admittance matrix and Y_e the diagonal matrix of edge-admittances.

Here, however, we are looking at the network as an n-port, thus it is assumed that expression (1.22b) represents the short-circuit admittance matrix of an n-port with exactly $(n+1)$ nodes and all ports having a common node. That is, the structure of the ports forms a "star-tree." The voltages across the ports with respect to the reference node (the common node) are chosen as independent variables. It is therefore clear that an n-port can be derived from a network of $(n+1)$ nodes and b elements by selecting any star-tree with n edges and by connecting current generators between the various nodes and the common node. The connection of current generators between node-pairs of a network is called a "soldering-iron type of connection" by Guillemin and "describing an n-port into a network" by Cederbaum.

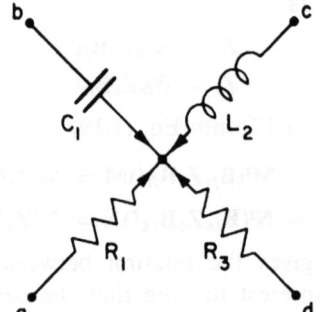

Fig. 1.5. A star-tree of the network of Fig. 1.3a

Let us consider the network given in Fig. 1.5a, where the orientation of each element indicates the direction of the polarity of the voltage across the element. We choose a tree of the network consisting of R_1, C_1, L_2, and R_3 as shown in Fig. 1.5b. Then, the incidence matrix of the network A_1 is found to be

$$A_1 = \begin{array}{c} \\ a \\ b \\ c \\ d \end{array} \begin{array}{c} L_1 \quad R_4 \quad C_2 \quad R_2 \quad R_1 \quad C_1 \quad L_2 \quad R_3 \\ \begin{bmatrix} -1 & 0 & 0 & 1 & 1 & 0 & 0 & 0 \\ 1 & -1 & 0 & 0 & 0 & 1 & 0 & 0 \\ 0 & 1 & -1 & 0 & 0 & 0 & 1 & 0 \\ 0 & 0 & 1 & 1 & 0 & 0 & 0 & 1 \end{bmatrix} \end{array}, \quad (1.23)$$

where node r is the reference node and the corresponding row was deleted in forming the incident matrix.

Now we introduce a current generator with zero internal admittance across each branch of the chosen tree as shown in Fig. 1.6. Note

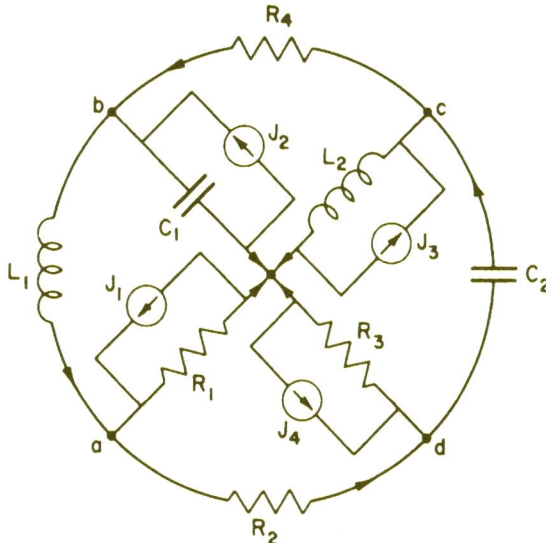

Fig. 1.6. A four-port derived from the network of Fig. 1.3a

here that each current source was connected so that the edge-voltages of the network are not affected. The short-circuit admittance matrix of the resultant four-port network, $\mathbf{Y}$, is then given by

$$\mathbf{Y} = \mathbf{A}_1 \mathbf{Y}_e \mathbf{A}_1^t$$

$$= \begin{bmatrix} \dfrac{1}{R_1}+\dfrac{1}{R_2}+\dfrac{1}{L_1 s} & -\dfrac{1}{L_1 s} & 0 & -\dfrac{1}{R_2} \\[2ex] -\dfrac{1}{L_1 s} & C_1 s+\dfrac{1}{L_1 s}+\dfrac{1}{R_4} & -\dfrac{1}{R_4} & 0 \\[2ex] 0 & -\dfrac{1}{R_4} & C_2 s+\dfrac{1}{L_2 s}+\dfrac{1}{R_4} & -C_2 s \\[2ex] -\dfrac{1}{R_2} & 0 & -C_2 s & C_2 s+\dfrac{1}{R_2}+\dfrac{1}{R_3} \end{bmatrix}$$

$$(1.24)$$

and

$$\mathbf{Y}_e = \begin{bmatrix} \dfrac{1}{L_1 s} & & & & & & & 0 \\ & \dfrac{1}{R_4} & & & & & & \\ & & C_2 s & & & & & \\ & & & \dfrac{1}{R_2} & & & & \\ & & & & \dfrac{1}{R_1} & & & \\ & & & & & C_1 s & & \cdot \\ & & & & & & \dfrac{1}{L_2 s} & \\ & & \cdot & & & & & \dfrac{1}{R_3} \end{bmatrix} . \quad (1.25)$$

In general we may derive an n-port from a network of m nodes, where $m > n+1$. In this case we first choose a node as the reference node, but all the remaining $(m-1)$ nodes are not to be used to connect current generators, instead we utilize only n of the remaining nodes.

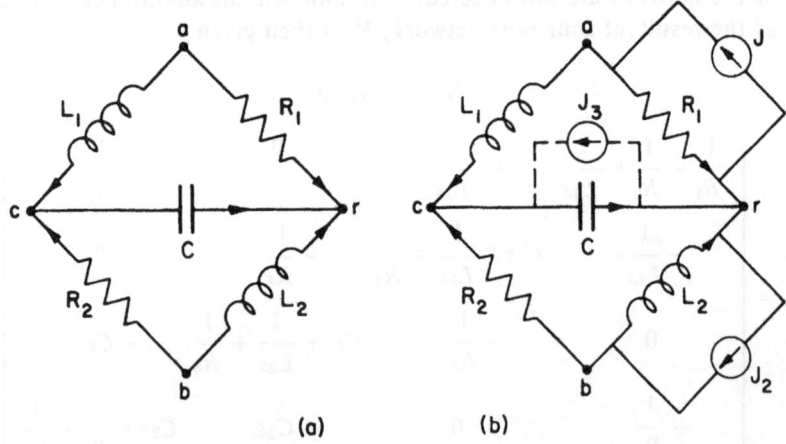

Fig. 1.7. Derivation of a two-port

However, in order to find the short-circuit admittance coefficients of the resultant n-port, the short-circuit admittance matrix of the $(m-1)$-port is first evaluated. The pivotal condensation is then applied to the determinant of the admittance matrix of the $(m-1)$-port with respect to the diagonal elements corresponding to the ports which are left opened. This is illustrated in the following example.

Example 2. Given a network in Fig. 1.7a, we are going to derive a two-port with respect to terminal-pairs (a, r) and (c, r) where node r is taken as the reference node. First, the short-circuit admittance matrix of the three-port with respect to nodes a, b, and c, i.e., the node admittance matrix is found as:

$$\mathbf{Y} = \begin{array}{c} a \\ b \\ c \end{array} \begin{bmatrix} \dfrac{1}{L_1s} + \dfrac{1}{R_1} & 0 & -\dfrac{1}{L_1s} \\[2ex] 0 & \dfrac{1}{L_2s} + \dfrac{1}{R_2} & -\dfrac{1}{R_2} \\[2ex] -\dfrac{1}{L_1s} & -\dfrac{1}{R_2} & Cs + \dfrac{1}{L_1s} + \dfrac{1}{R_2} \end{bmatrix}. \quad (1.26)$$

In order to find the admittance coefficients of the two-port, the matrix of Eq. (1.26) will be condensed about the elements in (3, 3)-position. If we denote the condensed matrix by $\mathbf{Y}^*$, then

$$\mathbf{Y}^* = \frac{1}{Cs + (1/L_1s) + (1/R_2)}$$

$$\times \begin{bmatrix} \dfrac{Cs}{R_1} + \dfrac{1}{L_1R_1s} + \dfrac{C}{L_1} + \dfrac{1}{R_1R_2} + \dfrac{1}{R_2L_1s} & \dfrac{-1}{L_1R_2s} \\[2ex] \dfrac{-1}{L_1R_2s} & \dfrac{Cs}{R_2} + \dfrac{1}{L_1R_2s} + \dfrac{1}{L_1L_2s^2} + \dfrac{C}{L_2} + \dfrac{1}{R_2L_2s} \end{bmatrix}.$$

$$(1.27)$$

Thus, the short-circuit admittance matrix of the derived two-port is found.

Once the short-circuit admittance coefficients of an n-port are found, each port-current which would be a chord-current can be obtained by solving the equation

$$\mathbf{V}_n = \mathbf{Y}^{-1}\mathbf{I}_c. \quad (1.28)$$

In order to assure that there always exists a set of the nontrivial solutions for Eq. (1.27), **Y** must be nonsingular.†

We have thus far discussed the formulation of an n-port based on the star-tree port-structure, i.e., every port has the common grounded node. However, one may select a port-structure which is not a star-tree. We can actually chose a tree of any structure and its branch-voltages are used as the port-voltages of an n-port. The short-circuit admittance coefficients of the resultant n-port are then no longer evaluated by the topological formula which was used for an n-port of the star-tree port-structure. We therefore introduce the generalized approach to the evaluation of the admittance coefficients of an n-port.

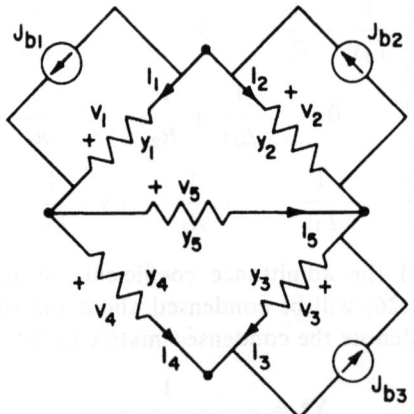

Fig. 1.8. A described four-port network

Consider a described three-port network as shown in Fig. 1.8. Then, with respect to the set of edges y_1, y_2, and y_3 which form a tree of the parent network, we have KVL equations,

$$\mathbf{B}_f\mathbf{V}_e = \begin{array}{c} \\ \end{array} \overset{\displaystyle y_4 \quad y_5 \quad y_1 \quad y_2 \quad y_3}{\begin{bmatrix} 1 & 0 & -1 & -1 & -1 \\ 0 & 1 & -1 & -1 & 0 \end{bmatrix}} \begin{bmatrix} V_4 \\ V_5 \\ V_1 \\ V_2 \\ V_3 \end{bmatrix} = 0. \qquad (1.29)$$

---

† In other words, there should be no node in a formulated n-port with only current sourses connected to it.

If we form the basic cut-sets with respect to the same tree, then they are given by the basic cut-set matrix C_f

$$
C_f = \begin{array}{c} \begin{array}{ccccc} y_4 & y_5 & y_1 & y_2 & y_3 \end{array} \\ \begin{bmatrix} 1 & 1 & 1 & 0 & 0 \\ 1 & 1 & 0 & 1 & 0 \\ 1 & 0 & 0 & 0 & 1 \end{bmatrix} \end{array}. \tag{1.30}
$$

Thus, KCL equations for the basic cut-sets are found to be

$$
C_f I_e = C_f J_e, \tag{1.31a}
$$

where

$$
I_e = \begin{bmatrix} i_4 \\ i_5 \\ i_1 \\ i_2 \\ i_3 \end{bmatrix} \quad \text{and} \quad J_e = \begin{bmatrix} 0 \\ 0 \\ j_{b1} \\ j_{b2} \\ j_{b3} \end{bmatrix}. \tag{1.31b}
$$

The right-hand expression of Eq. (1.31a) is a column matrix in which each row represents the algebraic sum of the current sources connected across the branches of the tree. We therefore replace the expression by a column matrix such that

$$
J = C_f J_e, \tag{1.32}
$$

where an element of J, j_k, is the current source connected across the kth node-pair.

Referring to the same network, one also can readily verify the relationship between the edge-voltages and branch-voltages, as shown previously that

$$
V_e = C_f{}^t V_b \tag{1.33a}
$$

and

$$
V_e = \begin{bmatrix} V_4 \\ V_5 \\ V_1 \\ V_2 \\ V_3 \end{bmatrix} \quad \text{and} \quad V_b = \begin{bmatrix} V_1 \\ V_2 \\ V_3 \end{bmatrix}. \tag{1.33b}
$$

Using Eqs. (1.31), (1.32), and (1.33) with Ohm's law, we obtain

$$
J = C_f I_e = C_f Y_e V_e = C_f Y_e C_f{}^t V_b \tag{1.34a}
$$

and

$$\mathbf{Y}_e = \begin{bmatrix} y_4 & 0 & 0 & 0 & 0 \\ 0 & y_5 & 0 & 0 & 0 \\ 0 & 0 & y_1 & 0 & 0 \\ 0 & 0 & 0 & y_2 & 0 \\ 0 & 0 & 0 & 0 & y_3 \end{bmatrix}. \tag{1.34b}$$

It is thus clear that the short-circuit admittance matrix of the network, **Y**, is given by:

$$\mathbf{Y} = \mathbf{C}_f \mathbf{Y}_e \mathbf{C}_f{}^t$$

$$= \begin{bmatrix} 1 & 1 & 1 & 0 & 0 \\ 1 & 1 & 0 & 1 & 0 \\ 1 & 0 & 0 & 0 & 1 \end{bmatrix} \begin{bmatrix} y_4 & 0 & 0 & 0 & 0 \\ 0 & y_5 & 0 & 0 & 0 \\ 0 & 0 & y_1 & 0 & 0 \\ 0 & 0 & 0 & y_2 & 0 \\ 0 & 0 & 0 & 0 & y_3 \end{bmatrix} \begin{bmatrix} 1 & 1 & 1 \\ 1 & 1 & 0 \\ 1 & 0 & 0 \\ 0 & 1 & 0 \\ 0 & 0 & 1 \end{bmatrix}$$

$$= \begin{bmatrix} y_1 + y_4 + y_5 & y_4 + y_5 & y_4 \\ y_4 + y_5 & y_2 + y_4 + y_5 & y_4 \\ y_4 & y_4 & y_3 + y_4 \end{bmatrix}. \tag{1.35}$$

Thus far, we have considered the derivation of multiports based on the short-circuit admittances. Let us consider an ordinary network with b elements and $(n+1)$ nodes. Then, with respect to trees t_k and t_r of the network, let $\mathbf{C}_{fk}$ and $\mathbf{C}_{fr}$ be the basic cut-set matrices, respectively. We then have, using Eq. (2.17c) of Part I,

$$\mathbf{V}_e = \mathbf{C}_{fk}{}^t \mathbf{V}_{bk} \tag{1.36a}$$

$$\mathbf{V}_e = \mathbf{C}_{fr}{}^t \mathbf{V}_{br}, \tag{1.36b}$$

where the elements of the network are arranged in the same order for both systems. $\mathbf{V}_{bk}$ and $\mathbf{V}_{br}$ are column matrices of the branch-voltages, i.e., port-voltages of the n-ports formed based on trees t_k and t_r, respectively.

If we again denote the submatrices of $\mathbf{C}_{fk}{}^t$ and $\mathbf{C}_{fr}{}^t$ corresponding to each set of the branches of the chosen trees t_r and t_k, respectively, by α and β, then the following relationships, which are similar to those derived for the transformation of port-currents, can be directly obtained

$$\mathbf{V}_{bk} = \alpha\mathbf{V}_{br} \tag{1.37a}$$

$$\mathbf{V}_{br} = \beta\mathbf{V}_{bk} \tag{1.37b}$$

$$\alpha = \beta^{-1} \tag{1.37c}$$

$$\mathbf{C}_{fk} = \alpha^t\mathbf{C}_{fr} \tag{1.37d}$$

$$\mathbf{C}_{fr} = \beta^t\mathbf{C}_{fk} \tag{1.37e}$$

and α and β are nonsingular unimodular matrices, therefore,

$$\det\alpha = \det\beta = \pm1. \tag{1.38}$$

Thus we have shown that the transformation matrix between two complete sets of independent port-voltages is a nonsingular unimodular matrix with its determinant equal to ±1.

For the short-circuit admittance matrices $\mathbf{Y}_k$ and $\mathbf{Y}_r$ of both n-ports we have

$$\mathbf{Y}_k = \mathbf{C}_{fk}\mathbf{Y}_e\mathbf{C}_{fk}{}^t \tag{1.39a}$$

$$\mathbf{Y}_r = \mathbf{C}_{fr}\mathbf{Y}_e\mathbf{C}_{fr}{}^t. \tag{1.39b}$$

Substituting Eqs. (1.37d) and (1.37e) into Eq. (1.39) we obtain

$$\mathbf{Y}_k = \alpha^t\mathbf{Y}_r\alpha \tag{1.40a}$$

$$\mathbf{Y}_r = \beta^t\mathbf{Y}_k\beta. \tag{1.40b}$$

This set of equations is dual to Eq. (1.20) which relates two sets of independent chord-currents, i.e., port-currents, and their corresponding open-circuit impedance matrices.

Thus far we have discussed the derivation of an n-port from a network of $(n+1)$ nodes on the basis of basic cut-sets with respect to a tree of a network. The short-circuit admittance matrix $\mathbf{Y}$ of the n-port is given by

$$\mathbf{Y} = \mathbf{C}_f\mathbf{Y}_e\mathbf{C}_f{}^t, \tag{1.41}$$

where $\mathbf{Y}_e$ is the edge admittance matrix.

It may not always be necessary, however, to identify each port-voltage with a branch-voltage of a tree of the parent network in order to form an n-port. In other words, as long as the port-structure is chosen so that the port-voltages are independent of each other, then each port-voltage is not necessarily identical with a branch-voltage of a tree of a network into which the n-port is described. For instance, let us consider the network given in Fig. 1.9a, where the orientation of each edge indicates the direction of an edge-current, and let us describe a four-port into the network as indicated in

Fig. 1.9b. Then, since there exists no edge connected between nodes 1 and 2 on the parent network, the port-structure of the four-port is *not* based on an actual tree of the original network, even though

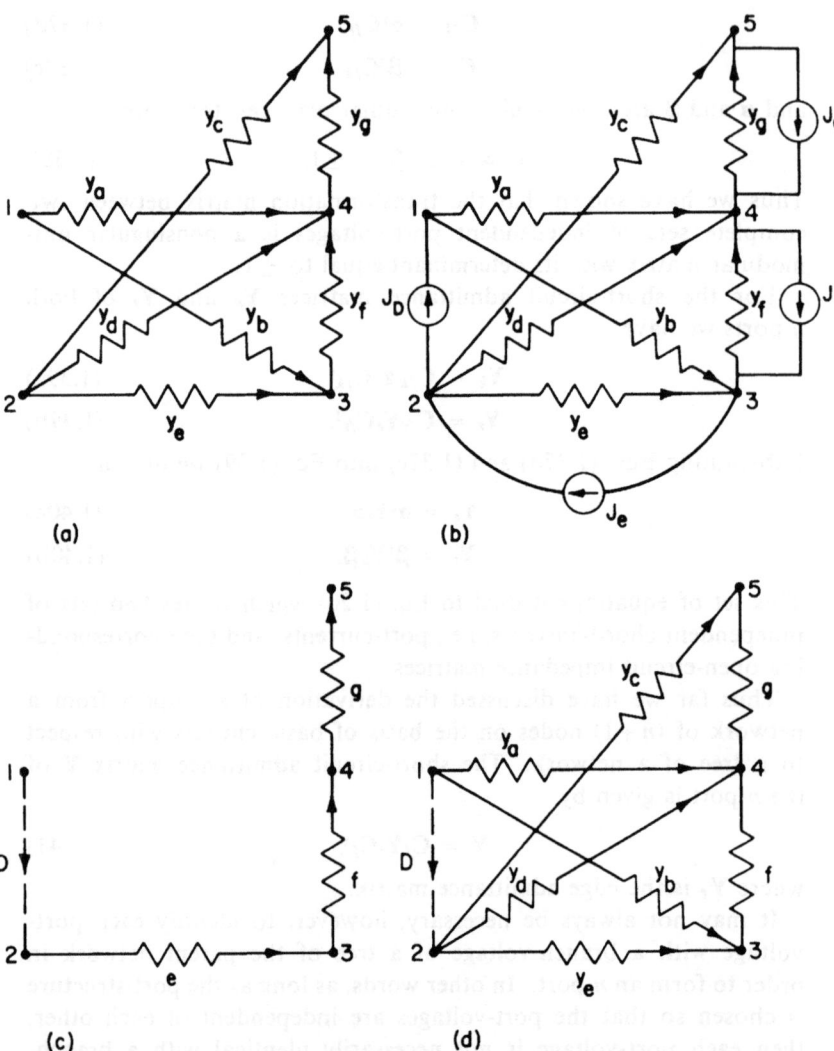

(a)

(b)

(c)

(d)

Fig. 1.9. Illustration of the derivation of a multiport based on a fictitious tree of a network

(a) An ordinary network of five nodes; (b) a four-port is described into the network of (a); (c) the fictitious tree on which the four-port of (b) is derived; (d) the network of (a) is modified with a fictitious edge.

the ports form a "tree-like structure." This port-structure is, however, based on the tree-like structure as shown in Fig. 1.9c, which contains a fictitious edge (indicated by a dotted line segment) corresponding to a current generator. This structure will be referred to as a "modified" or "fictitious tree" of the parent network. Then, the basic cut-set matrix based on the modified tree, the "modified basic cut-set matrix," denoted by C_{fm}, of the network of Fig. 1.9a, or the basic cut-set matrix of the modified network of Fig. 1.9d, is found to be

$$
C_{fm} = \begin{array}{c}
\begin{array}{cccccccc} a & b & c & d & D & e & f & g \end{array} \\
\left[\begin{array}{cccccccc}
1 & 1 & 0 & 0 & 1 & 0 & 0 & 0 \\
1 & 1 & 1 & 1 & 0 & 1 & 0 & 0 \\
1 & 0 & 1 & 1 & 0 & 0 & 1 & 0 \\
0 & 0 & 1 & 0 & 0 & 0 & 0 & 1
\end{array}\right]
\end{array},
\tag{1.42}
$$

and the short-circuit admittance matrix of the four-port, Y, is then given by

$$
Y = C_{fm}Y_{em}C_{fm}{}^t,
\tag{1.43}
$$

where Y_{em} is the edge-admittance of the modified network. That is, Y_{em} includes a zero diagonal element corresponding to the fictitious edge which is a current generator with zero admittance.

If the zero diagonal element in the modified edge-admittance matrix Y_{em} should be eliminated in Eq. (1.43), the column corresponding to edge D must also be removed from C_{fm}. We therefore have C_{fm} with the fifth column deleted,

$$
C_{fm_{-3}} = \begin{array}{c}
\begin{array}{ccccccc} a & b & c & d & e & f & g \end{array} \\
\left[\begin{array}{ccccccc}
1 & 1 & 0 & 0 & 0 & 0 & 0 \\
1 & 1 & 1 & 1 & 1 & 0 & 0 \\
1 & 0 & 1 & 1 & 0 & 1 & 0 \\
0 & 0 & 1 & 0 & 0 & 0 & 1
\end{array}\right]
\end{array}
\tag{1.44a}
$$

and Eq. (1.43) is reduced to

$$
Y = C_{fm_{-3}}Y_eC_{fm_{-3}}{}^t,
\tag{1.44b}
$$

where Y_e is the edge-admittance of the network of Fig. 1.9a.

We now investigate the matrix $C_{fm_{-3}}$ more carefully. The first, third, and fourth rows of the matrix correspond to cut-sets of the original network. The second row, however, corresponds to a union

156 *Analysis and Synthesis of Linear N-Port Networks*

of two cut-sets; edges *a*, *b*, and edges *c*, *d*, *e*. It is thus clear that the formula stated in Eq. (1.41) is not always applicable to an *n*-port described into a network of $(n+1)$ nodes, unless the port-structure of the *n*-port is based on a tree of the parent network. We therefore need to introduce an extended concept of a cut-set and cut-set matrix which includes the union of cut-sets. The following definitions and concepts were due to Reed.[7]

Definition 1. A "seg" is a set of edges in a connected graph *G*, such that:

The removal of the edges in the set separates *G* into two disjoint groups of nodes, N_1 and N_2;

No edge in the set has both of its nodes in either N_1 or N_2;

N_1 or N_2 is not necessarily a connected subgraph of *G*.

Definition 2. A "seg matrix", $C_s = [c_{s_{ij}}]$ is defined by a set of edges and segs of a graph such that:

$c_{s_{ij}} = 1$, if edge *j* is included in seg *i* and the orientations of both edge *j* and seg *i* coincide;

$c_{s_{ij}} = -1$, if edge *j* is included in seg *i* but the orientation of edge *j* and seg *i* are opposite;

$c_{s_{ij}} = 0$, if edge *j* is *not* included in seg *i*, where the orientation of each seg is arbitrarily assigned.

From Definitions 1 and 2, it should be clear that a cut-set and cut-set matrix are the special cases of a seg and a seg matrix. With these new concepts, we can write an expression for the short-circuit admittance matrix, **Y**, of an *n*-port described into a network of $(n+1)$ nodes and *b* edges as:

$$\mathbf{Y} = \mathbf{C}_s \mathbf{Y}_e \mathbf{C}_s{}^t, \qquad (1.45)$$

where the order of $\mathbf{C}_s$ is *n* by *b*. And, in general, the seg matrix of Eq. (1.45) may be obtained from a modified basic cut-set matrix with more than one column deleted, and with each row of the seg matrix corresponding to a simple cut-set or an edge-disjoint union of simple cut-sets,† although a seg matrix may be interpreted to include an edge-joint union of cut-sets.

† The matrix $\mathbf{C}_s$ in the formulation of Eq. (1.45) is called a "node-pair-to-edge matrix" by Cederbaum.[6c]

CHAPTER 2

Analytic Properties of
Y- and Z-Matrices

WE SHALL INVESTIGATE the analytic properties of the short-circuit admittance and the open-circuit impedance matrices of an ordinary n-port network. In particular, the properties of a "tapered matrix," "dominant matrix," and "paramount matrix" are investigated at length because they play an important role in the realization of multiport networks. We shall study also the "linear congruent transformation" of a matrix, since it is actually a synthesis technique which we will discuss in subsequent chapters.

STAR-TREE PORT-STRUCTURE

Let us now consider a very common port-structure which gives rise to the nodal admittance matrix. Figure 2.1 shows a completely

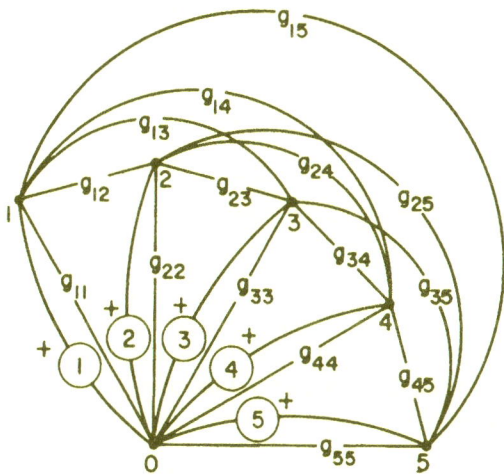

Fig. 2.1. A star-tree port-structure of a completely connected network

connected five-port network with a star-tree port-structure and six nodes. For simplicity, let us assume that all edges represent resistive elements; conductance g_{ij} is connected between nodes i and j and conductance g_{ii} is connected between the reference node O and node i. The orientation of a port is the same as the orientation of each branch with which the n-port is associated. In general, the graph may be oriented arbitrarily.

In general, the basic cut-set matrix C_f with respect to a star-tree which forms the port-structure of an n-port is of the form

$$C_f = [U_n \quad C_{f_{12}}], \qquad (2.1)$$

where the columns of a unit matrix U_n correspond to the branches and the columns of $C_{f_{12}}$ to a set of the chords of the tree. Since the tree under consideration is of a star-like structure, i.e., all ports have a common reference node, and the orientation of each port is defined by that of a branch of the tree, each column of $C_{f_{12}}$ will contain at most one 1 and one -1. That is, the basic cut-set matrix of a star-like tree of a graph is actually the incidence matrix of the graph. It is of course assumed that the graph is nonseparable.

For the n-port formed based on a star-tree, the short-circuit admittance matrix Y of order n is then found by

$$Y = [y_{ij}] = C_f G_e C_f^t \qquad (2.2)$$

and G_e is a diagonal matrix of edge-conductances.

It is now obvious that in the congruence of Eq. (2.2) we have:

$$y_{ii} = \sum_{j=1}^{n} g_{ij}, \qquad \text{for } i = 1, 2 \ldots, n \qquad (2.3a)$$

$$y_{ij} = -g_{ij}, \qquad \text{for } i \neq j \qquad (2.3b)$$

and

$$g_{ii} = \sum_{j=1}^{n} y_{ij}, \qquad \text{for } i = 1, 2 \ldots, n \qquad (2.4a)$$

$$g_{ij} = -y_{ij}, \qquad \text{for } i \neq j. \qquad (2.4b)$$

Equations (2.3) and (2.4) show that matrix Y corresponding to a star-tree has the property that the main diagonal elements are not less than the sum of the absolute values of all other elements on the same row; that is,

$$y_{ii} \geq \sum_{\substack{j=1 \\ i \neq j}}^{n} |y_{ij}|. \qquad (2.5)$$

A Y-matrix satisfying condition (2.5) is called a "dominant matrix," and we have the following theorem.

Theorem 1. The short-circuit admittance matrix of a resistive n-port, every port having a common node as the reference node, is a dominant matrix with nonpositive off-diagonal elements.

Proof: If the network contains only $(n+1)$ nodes, then the short-circuit admittance matrix of the network is actually the node admittance matrix and the theorem is therefore true. If the network contains more than $(n+1)$ nodes, then star-mesh transformations in the unused nodes will transform the network into an equivalent network with only $(n+1)$ nodes. Thus, the theorem is proved.

The short-circuit admittance matrix of an *RLC* network with $(n+1)$ nodes and with a star-tree port-structure also gives rise to a dominant matrix when s takes a positive, real value. That is, in general, we have every element y_{ij} of the short-circuit admittance matrix Y of the form

$$y_{ij} = C_{ij}s + g_{ij} + (1/L_{ij}s), \tag{2.6a}$$

and Y itself is given by

$$Y = sC + G + (1/s)\Gamma; \tag{2.6b}$$

C is the capacitance matrix, G the conductance matrix, and $\Gamma = [1/L]$. All three matrices are dominant. If we let s take a positive real value, say k, then an *RLC* network is equivalent to a resistive network, and $Y(k)$ is therefore dominant because of Theorem 1.

Dominant matrices play an important role in the theory of n-port networks because the dominant property is a sufficient condition for the realizability of a Y-matrix. The study of a dominant matrix and its application to the synthesis of multiport networks are presented in detail in later chapters.

LINEAR-TREE PORT-STRUCTURE

In the previous section we examined the star-tree port-structure and the dominant short-circuit admittance matrix associated with it. In this section we will deal with another important port-structure, a so-called "linear-tree" port-structure. Most of the material in this section is based on the work of Guillemin,[5b] and also of Biorci and Civalleri.[8a]

Let us again consider a completely connected resistive network with six nodes as shown in Fig. 2.2. The network shown in Fig. 2.2

is identical to the one of Fig. 2.1, except that the port-structure, that is, the port-structure presently under consideration, is a "linear-tree."

First, we note that if the ports are numbered and oriented as shown in Fig. 2.2, all elements of the basic cut-set matrix correspond-

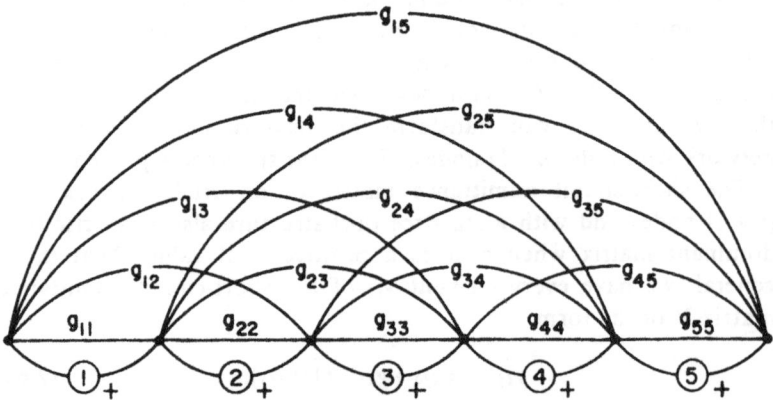

Fig. 2.2. A linear-tree port-structure of a completely connected network

ing to the linear-tree of the port-structure are positive. The basic cut-set matrix C_f with respect to the tree is then

$$
C_f = 3 \quad
\begin{array}{c}
1 \\ 2 \\ 3 \\ 4 \\ 5
\end{array}
\begin{array}{c}
 g_{11}\; g_{12}\; g_{13}\; g_{14}\; g_{15}\; g_{22}\; g_{23}\; g_{24}\; g_{25}\; g_{33}\; g_{34}\; g_{35}\; g_{44}\; g_{45}\; g_{55} \\
\left[
\begin{array}{ccccccccccccccc}
1 & 1 & 1 & 1 & 1 & 0 & 0 & 0 & 0 & 0 & 0 & 0 & 0 & 0 & 0 \\
0 & 1 & 1 & 1 & 1 & 1 & 1 & 1 & 1 & 0 & 0 & 0 & 0 & 0 & 0 \\
0 & 0 & 1 & 1 & 1 & 0 & 1 & 1 & 1 & 1 & 1 & 1 & 0 & 0 & 0 \\
0 & 0 & 0 & 1 & 1 & 0 & 0 & 1 & 1 & 0 & 1 & 1 & 1 & 1 & 0 \\
0 & 0 & 0 & 0 & 1 & 0 & 0 & 0 & 1 & 0 & 0 & 1 & 0 & 1 & 1
\end{array}
\right]
\end{array}. \quad (2.7)
$$

If we form the congruence $Y = C_f G_e C_f{}^t$ where G_e is a diagonal matrix of edge-conductances, then for the elements of Y we have

$$y_{11} = g_{15} + g_{14} + g_{13} + g_{12} + g_{11}$$

$$y_{12} = g_{15} + g_{14} + g_{13} + g_{12}$$

$$y_{13} = g_{15} + g_{14} + g_{13}$$

$$y_{14} = g_{15} + g_{14}$$

$$y_{15} = g_{15} \tag{2.8}$$

$$y_{22} = g_{15} + g_{14} + g_{25} + g_{24} + g_{23} + g_{13} + g_{22} + g_{12}$$

$$y_{23} = g_{15} + g_{14} + g_{25} + g_{24} + g_{23} + g_{13}$$

$$y_{24} = g_{15} + g_{14} + g_{25} + g_{24}$$

$$y_{25} = g_{15} + g_{25},$$

and similarly for other rows.

Here we observe the following properties of Y:
The elements of Y are non-negative.
The following relationships are valid:

$$y_{11} \geqslant y_{12} \geqslant \ldots\ldots\ldots \geqslant y_{1n}$$

$$y_{22} \geqslant y_{23} \geqslant \ldots \geqslant y_{2n}$$

$$\cdot \quad \cdot \quad \cdot \quad \cdot \quad \cdot \quad \cdot \quad \cdot \quad \cdot \quad \cdot \quad \cdot$$

$$y_{n-1,n-1} \geqslant y_{n-1,n} \tag{2.9a}$$

and

$$y_{nn} \geqslant y_{n-1,n} \geqslant \ldots\ldots\ldots \geqslant y_{1n}$$

$$y_{n-1,n-1} \geqslant y_{n-2,n-1} \geqslant \ldots \geqslant y_{1,n-1}$$

$$\cdot \quad \cdot \quad \cdot \quad \cdot \quad \cdot \quad \cdot \quad \cdot \quad \cdot \quad \cdot \quad \cdot$$

$$y_{22} \geqslant y_{12}. \tag{2.9b}$$

A matrix with properties of Eq. (2.9) is called a "tapered matrix," and we may now state the results in the form of a theorem.

Theorem 2. The short-circuit admittance matrix Y of a resistive n-port network with a linear-tree port-structure numbered and oriented in one direction (as shown in Fig. 2.2) is a tapered matrix.[5b,8a,b.] That is,

$$y_{ij} \geqslant 0, \quad \text{for } all \text{ } i \tag{2.10a}$$

$$y_{ij} \geqslant y_{i,j+1}, \quad \text{for } j \geqslant i \tag{2.10b}$$

$$y_{ij} \geqslant y_{i-1,j}, \quad \text{for } 1 < i \leqslant j. \tag{2.10c}$$

Proof: If an n-port network under consideration contains $n+1$ nodes, then the theorem follows directly from the previous discussion. If the network contains more than $n+1$ nodes, then star-mesh transformation applied to the unused nodes will result in an equivalent network with $n+1$ nodes. This completes the proof.

If any of the linear-tree ports have an opposite orientation than that shown in Fig. 2.2, then the corresponding row and column of $\mathbf{Y}$ is multiplied by -1. Also, if the sequence of the ports is different from the sequence shown in Fig. 2.2, then the corresponding rows and columns of $\mathbf{Y}$ are transposed. For example, if we interchange the positions of ports i and j, then the ith row and ith column are transposed with the jth row and column.

In order to investigate further the properties of a tapered matrix appropriate to a linear-tree, let us transform matrix $\mathbf{Y}$ corresponding to a linear-tree to another matrix corresponding to a star-tree. In Fig. 2.3, we show the new port-structure as well as the old one. The

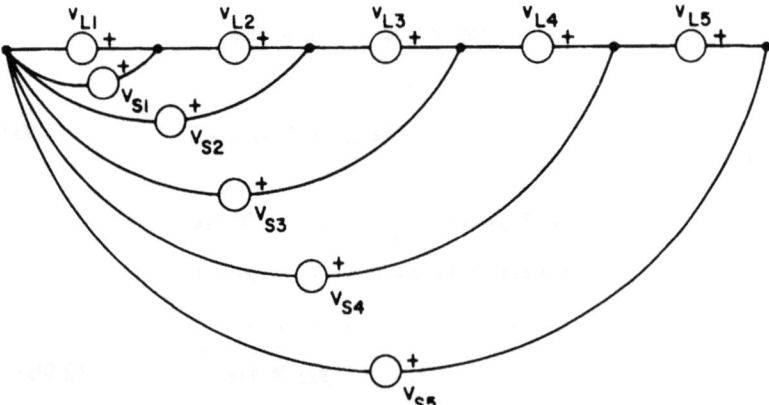

Fig. 2.3. Transformation of a star-tree port-structure into a linear-tree port-structure

subscripts "L" and "S" stand for "linear-tree-port" and "star-tree-port," respectively. Then, we have

$$\mathbf{Y}\mathbf{V}_L = \mathbf{J}_L, \qquad (2.11)$$

where $\mathbf{Y}$ is the tapered matrix, and $\mathbf{V}_L$ and $\mathbf{J}_L$ are column matrices of the linear-tree port-voltages and port-currents, respectively.

For the same parent network, we may choose a star-tree upon which to base a five-port as shown in Fig. 2.3. Let us denote the short-circuit admittance matrix of the resultant five-port by $\overline{\mathbf{Y}}$. Then, as shown in the previous section, $\overline{\mathbf{Y}}$ is a dominant matrix and

$$\overline{\mathbf{Y}}\mathbf{V}_S = \mathbf{J}_S, \qquad (2.12)$$

where $\mathbf{V}_S$ and $\mathbf{J}_S$ are column matrices of the star-tree port-voltages and port-currents, respectively.

Since the two systems of the five-ports were derived from the same network, the port-voltages and port-currents of the two systems should be related by

$$\mathbf{V}_L = \mathbf{T}\mathbf{V}_S \qquad (2.13)$$

and **T** is a nonsingular transformation matrix. Then, as was shown in the linear transformation of port-voltages of the previous chapter, the short-circuit admittance matrix of the five-port of the star-tree structure is found to be

$$\overline{\mathbf{Y}} = \mathbf{T}^t\mathbf{Y}\mathbf{T}, \qquad (2.14)$$

where **T** was found in the previous chapter to be a nonsingular unimodular matrix.

As an illustration, let us consider two systems of five-ports which are related by

$$\begin{bmatrix} V_{L1} \\ V_{L2} \\ V_{L3} \\ V_{L4} \\ V_{L5} \end{bmatrix} = \begin{bmatrix} 1 & 0 & 0 & 0 & 0 \\ -1 & 1 & 0 & 0 & 0 \\ 0 & -1 & 1 & 0 & 0 \\ 0 & 0 & -1 & 1 & 0 \\ 0 & 0 & 0 & -1 & 1 \end{bmatrix} \begin{bmatrix} V_{S1} \\ V_{S2} \\ V_{S3} \\ V_{S4} \\ V_{S5} \end{bmatrix}. \qquad (2.15)$$

Using Eq. (2.14) one obtains the short-circuit admittance coefficients of the five-port with star-tree port-structure in terms of those of the network with linear-tree port-structure. They are shown in Eq. (2.16):

$$\underline{Y} = \begin{bmatrix} 1 & -1 & 0 & 0 & 0 \\ 0 & 1 & -1 & 0 & 0 \\ 0 & 0 & 1 & -1 & 0 \\ 0 & 0 & 0 & 1 & -1 \\ 0 & 0 & 0 & 0 & 1 \end{bmatrix} \begin{bmatrix} y_{11} & y_{12} & y_{13} & y_{14} & y_{15} \\ y_{21} & y_{22} & y_{23} & y_{24} & y_{25} \\ y_{31} & y_{32} & y_{33} & y_{34} & y_{35} \\ y_{41} & y_{42} & y_{43} & y_{44} & y_{45} \\ y_{51} & y_{52} & y_{53} & y_{54} & y_{55} \end{bmatrix} \begin{bmatrix} 1 & 0 & 0 & 0 & 0 \\ -1 & 1 & 0 & 0 & 0 \\ 0 & -1 & 1 & 0 & 0 \\ 0 & 0 & -1 & 1 & 0 \\ 0 & 0 & 0 & -1 & 1 \end{bmatrix}$$

$$= \begin{bmatrix}
(y_{11}-y_{21})-(y_{12}-y_{22}) & (y_{12}-y_{22})-(y_{13}-y_{23}) & (y_{13}-y_{23})-(y_{14}-y_{24}) & (y_{14}-y_{24})-(y_{15}-y_{25}) & y_{15}-y_{25} \\
(y_{21}-y_{31})-(y_{22}-y_{32}) & (y_{22}-y_{32})-(y_{23}-y_{33}) & (y_{23}-y_{33})-(y_{24}-y_{34}) & (y_{24}-y_{34})-(y_{25}-y_{35}) & y_{25}-y_{35} \\
(y_{31}-y_{41})-(y_{32}-y_{42}) & (y_{32}-y_{42})-(y_{33}-y_{43}) & (y_{33}-y_{43})-(y_{34}-y_{44}) & (y_{34}-y_{44})-(y_{35}-y_{45}) & y_{35}-y_{45} \\
(y_{41}-y_{51})-(y_{42}-y_{52}) & (y_{42}-y_{52})-(y_{43}-y_{53}) & (y_{43}-y_{53})-(y_{44}-y_{54}) & (y_{44}-y_{54})-(y_{45}-y_{55}) & y_{45}-y_{55} \\
(y_{51}-y_{52}) & (y_{52}-y_{53}) & y_{53}-y_{54} & y_{54}-y_{55} & y_{55}
\end{bmatrix}. \quad (2.16)$$

Since the matrix Y is based on a star-tree its elements are very simply related to the edge-conductances. That is,

$$g_{11} = y_{11} - y_{12}$$
$$g_{12} = y_{12} - y_{13}$$
$$g_{13} = y_{13} - y_{14}$$
$$g_{14} = y_{14} - y_{15} \qquad (2.17)$$
$$g_{15} = y_{15}$$
$$g_{22} = y_{22} + y_{13} - y_{12} - y_{23}$$
$$g_{23} = y_{23} + y_{14} - y_{13} - y_{24},$$

and so on for the remaining conductances. In general, we can write with stipulation that for $(i-1) = 0$, $y_{i-1,j} = y_{i-1,j+1} = 0$ and for $(j+1)$ greater than the order of the Y-matrix then $y_{i-1,j+1} = y_{i,j+1} = 0$,

$$g_{ij} = y_{ij} + y_{i-1,j+1} - y_{i-1,j} - y_{i,j+1}, \qquad i \geqslant j. \qquad (2.18)$$

Conductance g_{ij} is connected between the node of positive polarity of port i and the node of negative polarity of port j, and is therefore a non-negative number. We have now shown that a Y-matrix based on a linear-tree satisfies a condition stronger than the tapered condition. We state the above result as a theorem since it is very important for the realization of multiport networks.

Theorem 3. The short-circuit admittance matrix Y of a resistive n-port, with a linear-tree port-structure in which the polarity of the port-voltage and ordering of each port is aligned in one direction, is a "uniformly tapered matrix."[5b,8a,b] That is, in addition to the conditions required for Theorem 2; Eqs. (2.10a), (2.10b), and (2,10c),

$$y_{ij} + y_{i-1,j+1} \geqslant y_{i-1,j} + y_{i,j+1}, \qquad \text{for all } i \geqslant j. \qquad (2.19)$$

Proof: If the network under consideration contains $n+1$ nodes, then the theorem follows from Eq. (2.18), because all conductances, i.e., g_{ij} for all i and j, are non-negative. If the network contains more than $n+1$ nodes, then star-mesh transformations applied to the unused nodes will result in an equivalent network with $n+1$ nodes. Thus, the theorem is proved.

It is obvious that Theorems 2 and 3 are valid for *RLC* n-ports for positive real values of s.

Example 3. Let us assume that the network of Fig. 2.2 is formed by unit conductances. Then, the short-circuit admittance matrix of the network, **Y**, is reduced to

$$\mathbf{Y} = \begin{bmatrix} 5 & 4 & 3 & 2 & 1 \\ 4 & 8 & 6 & 4 & 2 \\ 3 & 6 & 9 & 6 & 3 \\ 2 & 4 & 6 & 8 & 4 \\ 1 & 2 & 3 & 4 & 5 \end{bmatrix}. \tag{2.20}$$

This matrix is uniformly tapered. However, if the sequence of the ports is changed to (2, 1, 3, 4, 5) from (1, 2, 3, 4, 5), then, by interchanging the first and second rows and columns in **Y**, we obtain

$$\mathbf{Y}_{21345} = \begin{bmatrix} 8 & 4 & 6 & 4 & 2 \\ 4 & 5 & 3 & 2 & 1 \\ 6 & 3 & 9 & 6 & 3 \\ 4 & 2 & 6 & 8 & 4 \\ 2 & 1 & 3 & 4 & 5 \end{bmatrix}. \tag{2.21}$$

This new matrix is not tapered because the branches of the tree are not arranged in an ordered sequence.

PROPERTIES OF UNIMODULAR CONGRUENT TRANSFORMATIONS

In the previous sections we derived some of the properties of the short-circuit admittance and open-circuit impedance matrices of *RLC* networks without ideal transformers. For our investigation of *Z*- and *Y*-matrices it is essential at this point to consider some of the properties of unimodular congruent transformations and also the properties of paramount matrices which result from these transformations. The material presented in this section is based on the work of Cederbaum.[6]

We have shown that the **Y**-matrix of an *RLC* n-port with $(n+1)$ nodes is of the form

$$\mathbf{Y} = \mathbf{C}_f \mathbf{Y}_e \mathbf{C}_f^t, \tag{2.22}$$

where $\mathbf{C}_f$ is the basic cut-set matrix corresponding to a tree port-structure, and $\mathbf{Y}_e$ is a diagonal matrix of edge-admittances. Similarly, for the Z-matrix of an *RLC* network with *n* independent loops we have

$$\mathbf{Z} = \mathbf{B}_f \mathbf{Z}_e \mathbf{B}_f^t, \tag{2.23}$$

where $\mathbf{B}_f$ is the basic loop matrix with respect to the tree port-structure, and $\mathbf{Z}_e$ is a diagonal matrix of edge-impedances.

Equations (2.22) and (2.23) are called "unimodular congruences" because the matrices $\mathbf{C}_f$ and $\mathbf{B}_f$ are unimodular or *E*-matrices. If we consider only resistive networks, then the diagonal elements of $\mathbf{Y}_e$ and $\mathbf{Z}_e$ are always positive and real numbers. In order to extend our discussions to *RLC* networks in general, we shall assume that the complex variable *s* takes only positive real values.

We can now examine properties of the congruence $\mathbf{K} = \mathbf{PDP}^t$, where $\mathbf{P}$ is a unimodular matrix and $\mathbf{D}$ is a diagonal matrix with positive real elements.

Theorem 4. Each principal minor of the matrix $\mathbf{K} = \mathbf{PDP}^t$, where $\mathbf{D}$ is a diagonal matrix with positive elements and $\mathbf{P}$ is a unimodular matrix, is not less than the absolute value of any other minor built from the same rows (or columns).[6b]

Proof: Let us denote the elements of a unimodular matrix $\mathbf{P}$, of order *n* by *b*, by p_{ij} such that $p_{ij} = \pm 1$ or 0 for $i = 1, 2, \ldots, n$ and $j = 1, 2, \ldots, b$. Also denote the diagonal elements of $\mathbf{D}$ by $d_1, d_2, \ldots, d_b$. Then elements of $\mathbf{K}$ are found to be

$$k_{ii} = p_{i1}^2 d_1 + p_{i2}^2 d_2 + \ldots + p_{ib}^2 d_b \tag{2.24a}$$

$$k_{ij} = p_{i1}p_{j1}d_1 + p_{i2}p_{j2}d_2 + \ldots + p_{ib}p_{jb}d_b. \tag{2.24b}$$

From Eqs. (2.24) we see that

$$|k_{ij}| = |p_{i1}p_{j1}d_1 + \ldots + p_{ib}p_{jb}d_b| \leqslant |p_{i1}p_{j1}d_1| + \ldots + |p_{ib}p_{jb}d_b|$$

$$\leqslant |p_{i1}^2 d_1 + p_{i2}^2 d_2 + \ldots + p_{ib}^2 d_b| = k_{ii}. \tag{2.25}$$

Hence we obtain

$$k_{ii} \geqslant |k_{ij}|, \qquad \text{for } i, j = 1, 2 \ldots, n. \tag{2.26}$$

Now we use the Binet–Cauchy theorem which states that the compound of a product of matrices is equal to the product of the compounds. We therefore have for the *h*th compound

$$\mathbf{K}^{(h)} = \mathbf{P}^{(h)} \mathbf{D}^{(h)} \mathbf{P}^{t(h)}. \tag{2.27}$$

The elements of $\mathbf{D}^{(h)}$ are obviously positive. The elements of $\mathbf{P}^{(h)}$ are submatrices of a unimodular matrix, hence they are equal to ± 1 or 0. Using these facts, we can deduce that the main diagonal elements of $K^{(h)}$ are not smaller than the absolute value of any other element on the same row and column. An element on the main diagonal of $K^{(h)}$ is a principal minor, of order h, of K, and an element on the same row (column) of $K^{(h)}$ is a minor of order h built from the same rows (columns) of K which were included in $K^{(h)}$. This completes the proof.

Definition 3. A symmetric matrix with constant real elements satisfying the property that any principal minor is not smaller than the modulus of any other minor built from the same rows (or columns) is called a "paramount matrix" (or M-matrix).

We have just shown that paramountcy is a necessary condition for a symmetric matrix K to be expressible in the form $K = PDP^t$. This condition is not sufficient because there exist paramount matrices which are not presentable in the same form. Another property which we must now consider is the essential uniqueness of congruent transformations.

Theorem 5. Two decompositions $K = P_1D_1P_1{}^t$ and $K = P_2D_2P_2{}^t$, where D_1 and D_2 are diagonal matrices with positive elements, and P_1 and P_2 are square nonsingular unimodular matrices, may differ from each other only in the order and the sign of the columns of P and the order of elements of D.

The proof of this theorem is rather lengthy and may be found in Reference 6b.

If P is a rectangular unimodular matrix in general, then Theorem 5 is valid but the following operations may be performed on the transformation:

Columns with only one nonzero element may be added to matrix P and zero elements may be placed in the corresponding place on the diagonal of D without upsetting the matrix K;

Any column of P may be repeated and the corresponding element of D decomposed into two elements whose sum remains constant;

Columns of zeros may be added to P and arbitrary positive elements may be placed in the corresponding place of D. These steps are illustrated in the following example.

Example 4. Given a decomposition of K,

$$K \underset{=}{\Delta} \begin{bmatrix} 1 & 1 & 1 \\ 1 & 3 & 2 \\ 1 & 2 & 3 \end{bmatrix} = \begin{bmatrix} 0 & 0 & 0 & 1 \\ 1 & 0 & 1 & 1 \\ 0 & 1 & 1 & 1 \end{bmatrix} \begin{bmatrix} 1 & 0 & 0 & 0 \\ 0 & 1 & 0 & 0 \\ 0 & 0 & 1 & 0 \\ 0 & 0 & 0 & 1 \end{bmatrix} \begin{bmatrix} 0 & 1 & 0 \\ 0 & 0 & 1 \\ 0 & 1 & 1 \\ 1 & 1 & 1 \end{bmatrix}. \tag{2.28}$$

Since the decomposition matrix of Eq. (2.28) is a rectangular matrix, we shall introduce one more column to the matrix with only one nonzero element in order to obtain another decomposition. Let us insert a column with only one 1 as the first column of the resultant matrix. And we should augment the diagonal matrix with a row of all zeros as the first row. Thus

$$\mathbf{K} \triangleq \begin{bmatrix} 1 & 1 & 1 \\ 1 & 3 & 2 \\ 1 & 2 & 3 \end{bmatrix} = \begin{bmatrix} 1 & 0 & 0 & 0 & 1 \\ 0 & 1 & 0 & 1 & 1 \\ 0 & 0 & 1 & 1 & 1 \end{bmatrix} \begin{bmatrix} 0 & 0 & 0 & 0 & 0 \\ 0 & 1 & 0 & 0 & 0 \\ 0 & 0 & 1 & 0 & 0 \\ 0 & 0 & 0 & 1 & 0 \\ 0 & 0 & 0 & 0 & 1 \end{bmatrix} \begin{bmatrix} 1 & 0 & 0 \\ 0 & 1 & 0 \\ 0 & 0 & 1 \\ 0 & 1 & 1 \\ 1 & 1 & 1 \end{bmatrix}. \quad (2.29)$$

One may also decompose a diagonal element of the diagonal matrix of Eq. (2.28) so that the first diagonal element shall be decomposed into two elements. Then the first column of the decomposition matrix should be repeated. We therefore have

$$\mathbf{K} \triangleq \begin{bmatrix} 1 & 1 & 1 \\ 1 & 3 & 2 \\ 1 & 2 & 3 \end{bmatrix} = \begin{bmatrix} 0 & 0 & 0 & 0 & 1 \\ 1 & 1 & 0 & 1 & 1 \\ 0 & 0 & 1 & 1 & 1 \end{bmatrix} \begin{bmatrix} \frac{1}{2} & 0 & 0 & 0 & 0 \\ 0 & \frac{1}{2} & 0 & 0 & 0 \\ 0 & 0 & 1 & 0 & 0 \\ 0 & 0 & 0 & 1 & 0 \\ 0 & 0 & 0 & 0 & 1 \end{bmatrix} \begin{bmatrix} 0 & 1 & 0 \\ 0 & 1 & 0 \\ 0 & 0 & 1 \\ 0 & 1 & 1 \\ 1 & 1 & 1 \end{bmatrix}. \quad (2.30)$$

We may add a column of all zeros to the decomposition matrix and a positive element d_1 to the diagonal matrix in the corresponding position as

$$\mathbf{K} \triangleq \begin{bmatrix} 1 & 1 & 1 \\ 1 & 3 & 2 \\ 1 & 2 & 3 \end{bmatrix} = \begin{bmatrix} 0 & 0 & 0 & 0 & 1 \\ 0 & 1 & 0 & 1 & 1 \\ 0 & 0 & 1 & 1 & 1 \end{bmatrix} \begin{bmatrix} d_1 & 0 & 0 & 0 & 0 \\ 0 & 1 & 0 & 0 & 0 \\ 0 & 0 & 1 & 0 & 0 \\ 0 & 0 & 0 & 1 & 0 \\ 0 & 0 & 0 & 0 & 1 \end{bmatrix} \begin{bmatrix} 0 & 0 & 0 \\ 0 & 1 & 0 \\ 0 & 0 & 1 \\ 0 & 1 & 1 \\ 1 & 1 & 1 \end{bmatrix}. \quad (2.31)$$

Definition 4. A unimodular matrix is called "nonredundant" if it has no columns with all zero elements and no two columns in which the patterns of zero and nonzero elements are identical.

Thus far, we have shown that the unimodular congruent transformation results in a paramount matrix. Tellegen[9] first considered paramount matrices of order three; however, Cederbaum[6a,e] obtained the following properties for a general form of a paramount matrix.

Properties of paramount matrices

Each principal submatrix of a paramount matrix is paramount; this follows from the definition of the paramount matrix.

The inverse of a nonsingular paramount matrix is paramount; this can be shown using Jacobi's theorem and is left as an exercise.

If $\mathbf{K}$ is paramount and if $\mathbf{K}^*$ results by applying pivotal condensation with respect to any one of the main diagonal elements of $\mathbf{K}$, then $\mathbf{K}^*$ is paramount; proof of this property can be found in Reference 6e.

Two more properties follow from the definition:

A paramount matrix is positive semidefinite;

A nonsingular paramount matrix is positive definite.

In many instances we will be concerned with the problem of whether a given matrix of real elements is paramount. The test of the paramountcy, of course, follows from the definition. That is, first we have to form all principal minors and all nonprincipal minors corresponding to the same rows. The evaluation of all these minors is very laborious and it is, however, not yet known if a simpler test can be formulated. Further work must be done in this direction. In the case that a matrix to be tested is a symmetric matrix of rational functions of s, then the paramount character of the matrix must be tested for all positive real values of s. Slepian and Weinberg have proposed a test which involves testing a rational function of s for non-negative character. Reference 10 may be consulted for this test.

We have so far found that a necessary condition for a symmetric real matrix $\mathbf{K}$ to be of the form $\mathbf{PDP}^t$, where $\mathbf{P}$ is a unimodular matrix and $\mathbf{D}$ a diagonal matrix with positive elements, is that $\mathbf{K}$ be paramount. For synthesis applications it is necessary for us to know a systematic procedure for decomposition of a paramount matrix into the unimodular congruent form, so that the unimodular matrix $\mathbf{P}$ will be realized as the basic cut-set or loop matrix of a connected graph. Cederbaum[6d] proposed a simple straightforward algorithm which results in finding the pair of matrices $\mathbf{P}$ and $\mathbf{D}$ for a given paramount matrix $\mathbf{K}$ if the decomposition is possible. Before investigating the decomposition algorithm the following theorem is necessary.

Theorem 6 (sign rule theorem). All elements of the matrix $(\mathbf{PDP}^t)$. where $\mathbf{P}$ is a unimodular matrix and $\mathbf{D}$ is a diagonal matrix with positive real diagonal elements d_i for $i = 1, 2, \ldots$, are of the form $\pm(d_r + d_s + \ldots)$ where r, s are some of the indices $1, 2, \ldots, m$.[6d]

Proof: Let

$$\mathbf{K} = \mathbf{PDP}^t.$$

Then we have

$$k_{ij} = p_{i1}p_{j1}d_1 + p_{i2}p_{j2}d_2 + \ldots + p_{im}p_{jm}d_m. \tag{2.32}$$

All products $p_{ig}p_{jg}(g = 1, 2, \ldots, m)$ are equal to ± 1 or 0 because P is a unimodular matrix.

Let $p_{iq}p_{jq}$ and $p_{ik}p_{jk}$ be different from zero and let it be assumed that contrary to the theorem they are of different sign. Thus

$$p_{iq}p_{jq} = -p_{ik}p_{jk}. \tag{2.33}$$

Let us now consider the following subdeterminant of $\mathbf{P}$:

$$\det \mathbf{P}_{ij} = \begin{vmatrix} p_{iq} & p_{ik} \\ p_{jq} & p_{jk} \end{vmatrix} = p_{iq}p_{jk} - p_{jq}p_{ik}. \tag{2.34}$$

Now, multiply Eq. (2.34) by $p_{jq}p_{jk}$ which is equal to ± 1, because $p_{iq}p_{jq}$ and $p_{ik}p_{jk}$ are nonvanishing. We have, using Eq. (2.33),

$$p_{jq}p_{jk}(\det \mathbf{P}_{ij}) = p_{jk}{}^2p_{iq}p_{jq} - p_{jq}{}^2p_{jk}p_{ik}$$
$$= 2p_{iq}p_{jq} = \pm 2. \tag{2.35a}$$

Therefore

$$\det \mathbf{P}_{ij} = \pm 2. \tag{2.35b}$$

Equation (2.35b) contradicts the assumption that the matrix $\mathbf{P}$ is unimodular. Hence, all coefficients of d_i of Eq. (2.32) for $i = 1, 2, \ldots, m$ have the same sign.

Theorem 7.[†] Let $\mathbf{K} = \mathbf{PDP}^t$ be a decomposition of a matrix $\mathbf{K}$ of an arbitrary order, where $\mathbf{P}$ is a nonredundant unimodular matrix and $\mathbf{D}$ is a diagonal matrix with real positive elements. If $k_{ij}, j \neq i$, is the off-diagonal nonzero element with the minimum absolute value (or one of them, or the only one), then in matrix $\mathbf{P}$ the rows i and j have nonzero elements simultaneously exactly in one column.

Corollary 1. Let $\mathbf{K}$ be decomposed into $\mathbf{K} = \mathbf{PDP}^t$. If k_{ij} for $i \neq j$ is the nonzero off-diagonal element of $\mathbf{K}$ with the minimum absolute value, then k_{ij} is equal to one of the diagonal elements of $\mathbf{D}$. Based on Theorem 7 and Corollary 1, let us proceed with the algorithm.

† The proofs of Theorem 7 and Corollary 1 can be found in Reference 6d.

Decomposition of a paramount matrix **K** *of order n* (*Cederbaum's algorithm*)

Let us assume that the decomposition into the form of **PDP**^{t} is possible, and let us proceed to find, step by step, all the columns of **P** and the diagonal elements of **D**.

Case 1: Matrix **K** is a diagonal matrix: Then, the decomposition of **K** is simply **K** = **UKU** where **U** is the unit matrix of order n.

Case 2: Matrix **K** is diagonal with some zero diagonal elements: Since in the standard form of the decomposition we do not allow zero main diagonal elements in matrix **D**, we therefore cross out all rows $q, g, \ldots$, and columns $q, g, \ldots$ of **K** corresponding to the zero elements $k_{qq}, k_{gg}, \ldots$, and then cross out the columns $q, g, \ldots$, of the left unit matrix and the rows $q, g, \ldots$, of the right unit matrix of the decomposition. This is illustrated in the following example.

Example 5.

$$\mathbf{K} \underset{=}{\Delta} \begin{bmatrix} 2 & 0 & 0 \\ 0 & 1 & 0 \\ 0 & 0 & 0 \end{bmatrix} = \begin{bmatrix} 1 & 0 \\ 0 & 1 \\ 0 & 0 \end{bmatrix} \begin{bmatrix} 2 & 0 \\ 0 & 1 \end{bmatrix} \begin{bmatrix} 1 & 0 & 0 \\ 0 & 1 & 0 \end{bmatrix}. \qquad (2.36)$$

Case 3: Matrix **K** has some nonzero off-diagonal elements: Let us select a nonzero off-diagonal element of **K** with the smallest absolute value, say k_{12}. Then, by Corollary 1, there must be a diagonal element of **D** which is equal to k_{12}. For convenience, let us assume that

$$d_1 = |k_{12}|. \qquad (2.37)$$

If we denote the first column of **P** by p_1, then the first two elements of p_1 should have nonzero elements since k_{12} was assigned to d_1 which is in the first column of a diagonal matrix **D** due to Theorem 6. The sign-rule theorem also indicates that the second element of p_1 has the same sign as the sign of k_{12}, provided we make the first element $+1$. It is of course required that according to Theorem 6 the first two rows of matrix **P** *cannot* have nonzero elements simultaneously in any but the first column. We therefore arbitrarily assign $+1$ to the first element of p_1 such that

$$\mathbf{p}_1 = \begin{bmatrix} 1 \\ \pm 1 \\ \cdot \\ \cdot \\ \cdot \end{bmatrix}. \tag{2.38}$$

It should be clear here that, in general, if element $k_{\alpha\beta}$ was chosen as an element of the smallest value and taken as d_r, then the αth and βth rows of $\mathbf{P}$ will have nonzero elements simultaneously in the rth column, but not in any other columns.

In order to obtain the other elements of $\mathbf{p}_1$ we use the following rules. If

$$k_{12}k_{1j}k_{2j} > 0, \qquad \text{for } j \neq 1, 2, \tag{2.39}$$

then the element of the jth row of the column is nonzero and the sign of this element is the same as the sign of k_{1j}, and if

$$k_{12}k_{1j}k_{2j} \leqslant 0, \qquad \text{for } j \neq 1, 2, \tag{2.40}$$

then the element on the jth row of the first column is zero. Thus, we can find all elements of the first column.

In the next step we must find the second column $\mathbf{p}_2$ of $\mathbf{P}$. Let us delete the first column from $\mathbf{P}$ and also the first row and column of $\mathbf{D}$.† This will leave us with the matrices $\mathbf{P}_{-1}$ and $\mathbf{D}_{-1}$. We have

$$\mathbf{K}_{-1} = \mathbf{P}_{-1}\mathbf{D}_{-1}\mathbf{P}_{-1}{}^t = \mathbf{K} - \mathbf{p}_1[d_1]\mathbf{p}_1{}^t. \tag{2.41}$$

If the elements of $\mathbf{p}_1$ are $p_{11}, p_{21}, \ldots, p_{n1}$ and if $k_{ij}{}^1$ is an element of $\mathbf{K}_{-1}$, then

$$k_{ij}{}^1 = k_{ij} - p_{i1}p_{j1}d_1. \tag{2.42}$$

From Eq. (2.42) we see that all zero elements of $\mathbf{K}$ remain unchanged in $\mathbf{K}_{-1}$ (by virtue of the sign rule theorem) and no nonzero element k_{ij} can change its sign or increase in absolute value.

The resulting matrix $\mathbf{K}_{-1}$ has more zero elements than matrix $\mathbf{K}$. In some cases $\mathbf{K}_{-1}$ may be a diagonal matrix. If $\mathbf{K}_{-1}$ has some non-zero off-diagonal elements, then the procedure discussed above is repeated by selecting the element (or one of them) of $\mathbf{K}_{-1}$ with the smallest absolute value and then proceeding to find the second

† It is assumed that matrix $\mathbf{P}$ exists in order to derive a relationship necessary for the development of an algorithm for finding $\mathbf{P}$.

column p_2 and a new diagonal element d_2. This leads to a new matrix K_{-2}, and

$$K_{-2} = P_{-2}D_{-2}P_{-2}{}^t = K_{-1} - p_2[d_2]p_2{}^t. \qquad (2.43)$$

In every step of the algorithm of finding K_{-1}, K_{-2}, ..., K_{-r}, the number of zero off-diagonal elements increases by at least one. The maximum number of steps for which K_{-t} has all zero off-diagonal elements is equal to $[n(n-1)]/2$. Matrix K_{-t} will fall into either of Cases 1 and 2 discussed previously, and, hence, may be decomposed. If the algorithm diverges, that is, if some of the off-diagonal elements of K_{-t} start increasing in absolute value or changing signs, then the decomposition is impossible. In some cases the decomposition may be possible but P may not be a unimodular matrix or D may have negative diagonal elements or both.

Let us now illustrate the algorithm. The following matrix has been proposed by Slepian and Weinberg[10] for realization as a short-circuit admittance matrix of a resistive network containing exactly five nodes:

$$K = \begin{bmatrix} 7 & 1 & 2 & 3 \\ 1 & 12 & 4 & 5 \\ 2 & 4 & 15 & 6 \\ 3 & 5 & 6 & 18 \end{bmatrix}. \qquad (2.44)$$

Then, we should try to interchange the rows and columns to see if the matrix will become a tapered form. Let us interchange the second row and column with the fourth row and column, respectively. Then we will have

$$K_{1432} = \begin{bmatrix} 7 & 3 & 2 & 1 \\ 3 & 18 & 6 & 5 \\ 2 & 6 & 15 & 4 \\ 1 & 5 & 4 & 12 \end{bmatrix}. \qquad (2.45)$$

It is clear then that K_{1432} is not tapered and therefore not realizable by a network of five nodes with a linear-tree port-structure.

Now, let us see if we will come to the same conclusion for the given matrix by taking another approach. That is, we are going to apply the decomposition technique just described. First we choose the smallest element of K, say the element of $(1, 2)$-position which has the value of 1. Thus, the first diagonal element d_1 is determined as $d_1 = 1$. Then, we apply criteria (2.39) and (2.40) and determine the first column, $\mathbf{p}_1$ of $\mathbf{P}$. Thus

$$\mathbf{p}_1 = \begin{bmatrix} 1 \\ 1 \\ 1 \\ 1 \end{bmatrix}. \tag{2.46}$$

We now form the product

$$\mathbf{p}_1[d_1]\mathbf{p}_1^t = \begin{bmatrix} 1 \\ 1 \\ 1 \\ 1 \end{bmatrix}[1][1 \quad 1 \quad 1 \quad 1] = \begin{bmatrix} 1 & 1 & 1 & 1 \\ 1 & 1 & 1 & 1 \\ 1 & 1 & 1 & 1 \\ 1 & 1 & 1 & 1 \end{bmatrix}. \tag{2.47}$$

Therefore,

$$\mathbf{K}_{-1} = \mathbf{K} - \mathbf{p}_1[d_1]\mathbf{p}_1 = \begin{bmatrix} 6 & 0 & 1 & 2 \\ 0 & 11 & 3 & 4 \\ 1 & 3 & 14 & 5 \\ 2 & 4 & 5 & 17 \end{bmatrix}. \tag{2.48}$$

We now choose the smallest element in $\mathbf{K}_{-1}$, say the element of (1, 3)-position, as d_2. Then, again we apply criteria (2.39) and (2.40) to $\mathbf{K}_{-1}$ to determine the second column $\mathbf{p}_2$ of $\mathbf{P}$. Once all elements of $\mathbf{p}_2$ are found, we obtain $\mathbf{K}_{-2}$. This process is repeated until we reach a diagonal matrix $\mathbf{K}_{-t}$. Thus, we have

$$\mathbf{p}_2[d_2]\mathbf{p}_2^t = \begin{bmatrix} 1 & 0 & 1 & 1 \\ 0 & 0 & 0 & 0 \\ 1 & 0 & 1 & 1 \\ 1 & 0 & 1 & 1 \end{bmatrix} \qquad \mathbf{K}_{-2} = \mathbf{K}_{-1} - \mathbf{p}_2[d_2]\mathbf{p}_2' = \begin{bmatrix} 5 & 0 & 0 & 1^* \\ 0 & 11 & 3 & 4 \\ 0 & 3 & 13 & 4 \\ 1 & 4 & 4 & 16 \end{bmatrix}$$

$$\mathbf{p}_3[d_3]\mathbf{p}_3^t = \begin{bmatrix} 1 & 0 & 0 & 1 \\ 0 & 0 & 0 & 0 \\ 0 & 0 & 0 & 0 \\ 1 & 0 & 0 & 1 \end{bmatrix} \qquad \mathbf{K}_{-3} = \mathbf{K}_{-2} - \mathbf{p}_3[d_3]\mathbf{p}_3' = \begin{bmatrix} 4 & 0 & 0 & 0 \\ 0 & 11 & 3^* & 4 \\ 0 & 3 & 13 & 4 \\ 0 & 4 & 4 & 15 \end{bmatrix}$$

$$\mathbf{p}_4[d_4]\mathbf{p}_4^t = \begin{bmatrix} 0 & 0 & 0 & 0 \\ 0 & 3 & 3 & 3 \\ 0 & 3 & 3 & 3 \\ 0 & 3 & 3 & 3 \end{bmatrix} \qquad \mathbf{K}_{-4} = \mathbf{K}_{-3} - \mathbf{p}_4[d_4]\mathbf{p}_4^t = \begin{bmatrix} 4 & 0 & 0 & 0 \\ 0 & 8 & 0 & 1^* \\ 0 & 0 & 10 & 1 \\ 0 & 1 & 1 & 12 \end{bmatrix}$$

$$\mathbf{p}_5[d_5]\mathbf{p}_5{}^t = \begin{bmatrix} 0 & 0 & 0 & 0 \\ 0 & 1 & 0 & 1 \\ 0 & 0 & 0 & 0 \\ 0 & 1 & 0 & 1 \end{bmatrix} \qquad \mathbf{K}_{-5} = \mathbf{K}_{-4} - \mathbf{p}_5[d_5]\mathbf{p}_5{}^t = \begin{bmatrix} 4 & 0 & 0 & 0 \\ 0 & 7 & 0 & 0 \\ 0 & 0 & 10 & 1^* \\ 0 & 0 & 1 & 11 \end{bmatrix}$$

$$\mathbf{p}_6[d_6]\mathbf{p}_6{}^t = \begin{bmatrix} 0 & 0 & 0 & 0 \\ 0 & 0 & 0 & 0 \\ 0 & 0 & 1 & 1 \\ 0 & 0 & 1 & 1 \end{bmatrix} \qquad \mathbf{K}_{-6} = \mathbf{K}_{-5} - \mathbf{p}_6[d_6]\mathbf{p}_6{}^t = \begin{bmatrix} 4 & 0 & 0 & 0 \\ 0 & 7 & 0 & 0 \\ 0 & 0 & 9 & 0 \\ 0 & 0 & 0 & 10 \end{bmatrix},$$

$$(2.49)$$

where the starred element in each matrix is used for the determination of each column of **P**. Then

$$\mathbf{K} = \mathbf{P}\mathbf{D}\mathbf{P}^t = \begin{bmatrix} 1 & 1 & 1 & 0 & 0 & 0 & 1 & 0 & 0 & 0 \\ 1 & 0 & 0 & 1 & 1 & 0 & 0 & 1 & 0 & 0 \\ 1 & 1 & 0 & 1 & 0 & 1 & 0 & 0 & 1 & 0 \\ 1 & 1 & 1 & 1 & 1 & 1 & 0 & 0 & 0 & 1 \end{bmatrix} \mathbf{D} \begin{bmatrix} 1 & 1 & 1 & 1 \\ 1 & 0 & 1 & 1 \\ 1 & 0 & 0 & 1 \\ 0 & 1 & 1 & 1 \\ 0 & 1 & 0 & 1 \\ 0 & 0 & 1 & 1 \\ 1 & 0 & 0 & 0 \\ 0 & 1 & 0 & 0 \\ 0 & 0 & 1 & 0 \\ 0 & 0 & 0 & 1 \end{bmatrix}, \quad (2.50\text{a})$$

where

$$\mathbf{D} = \begin{bmatrix} 1 & 0 & 0 & 0 & 0 & 0 & 0 & 0 & 0 & 0 \\ 0 & 1 & 0 & 0 & 0 & 0 & 0 & 0 & 0 & 0 \\ 0 & 0 & 1 & 0 & 0 & 0 & 0 & 0 & 0 & 0 \\ 0 & 0 & 0 & 3 & 0 & 0 & 0 & 0 & 0 & 0 \\ 0 & 0 & 0 & 0 & 1 & 0 & 0 & 0 & 0 & 0 \\ 0 & 0 & 0 & 0 & 0 & 1 & 0 & 0 & 0 & 0 \\ 0 & 0 & 0 & 0 & 0 & 0 & 4 & 0 & 0 & 0 \\ 0 & 0 & 0 & 0 & 0 & 0 & 0 & 7 & 0 & 0 \\ 0 & 0 & 0 & 0 & 0 & 0 & 0 & 0 & 9 & 0 \\ 0 & 0 & 0 & 0 & 0 & 0 & 0 & 0 & 0 & 10 \end{bmatrix}. \quad (2.50\text{b})$$

Now, it is easily observed that matrix $\mathbf{P}$ in Eq. (2.50a) is not a unimodular matrix, from the fact that it contains a submatrix whose determinant takes a value other than ± 1 or 0, such as

$$\det \begin{bmatrix} 1 & 1 & 0 & 0 \\ 1 & 0 & 1 & 0 \\ 1 & 0 & 0 & 1 \\ 1 & 1 & 1 & 1 \end{bmatrix} = 2. \tag{2.51}$$

Thus the given matrix is not realizable as the short-circuit admittance matrix of a network of five nodes. It is of importance to point out here that Cederbaum's decomposition (if possible) is unique only if $\mathbf{P}$ is unimodular and $\mathbf{D}$ has only positive real elements in the main diagonal. If these two conditions are not satisfied, there may exist more than one decomposition. As a matter of interest the matrix of Eq. (2.44) may also be decomposed in the following form:

$$\mathbf{K} = \begin{bmatrix} 1 & 0 & 0 & 0 & 1 & 1 & 1 & 0 & 0 \\ 0 & 0 & 0 & 1 & 0 & 0 & 1 & 1 & 1 \\ 0 & 0 & 1 & 0 & 0 & 1 & 1 & 1 & 1 \\ 0 & 1 & 0 & 0 & 1 & 1 & 1 & 1 & 0 \end{bmatrix} \mathbf{D}_1 \begin{bmatrix} 1 & 0 & 0 & 0 \\ 0 & 0 & 0 & 1 \\ 0 & 0 & 1 & 0 \\ 0 & 1 & 0 & 0 \\ 1 & 0 & 0 & 1 \\ 1 & 0 & 1 & 1 \\ 1 & 1 & 1 & 1 \\ 0 & 1 & 1 & 1 \\ 0 & 1 & 1 & 0 \end{bmatrix}, \tag{2.52a}$$

where

$$\mathbf{D}_1 = \begin{bmatrix} 4 & 0 & 0 & 0 & 0 & 0 & 0 & 0 & 0 \\ 0 & 11 & 0 & 0 & 0 & 0 & 0 & 0 & 0 \\ 0 & 0 & 10 & 0 & 0 & 0 & 0 & 0 & 0 \\ 0 & 0 & 0 & 8 & 0 & 0 & 0 & 0 & 0 \\ 0 & 0 & 0 & 0 & 1 & 0 & 0 & 0 & 0 \\ 0 & 0 & 0 & 0 & 0 & 1 & 0 & 0 & 0 \\ 0 & 0 & 0 & 0 & 0 & 0 & 1 & 0 & 0 \\ 0 & 0 & 0 & 0 & 0 & 0 & 0 & 4 & 0 \\ 0 & 0 & 0 & 0 & 0 & 0 & 0 & 0 & -1 \end{bmatrix}. \tag{2.52b}$$

In this decomposition, the decomposition matrix is realizable as a basic cut-set matrix; however, the diagonal matrix K_1 contains one negative element requiring one negative resistor for the realization of **K**. The above decomposition cannot be obtained using the algorithm which we have studied because it may result in negative elements only in the last step where a diagonal matrix K_{-i} is obtained. Hence a negative element d_i may correspond only to a column p_i with only one nonzero element. In the example previously considered, the negative resistance element corresponds to a column with two non-zero elements. We are going to consider some further implications of the above comments when we discuss the realization of n-ports.

At this stage, one must have a clear understanding that the decomposition algorithm may diverge. That is, if the number of zeros in the off-diagonal positions in K_{-i} is smaller than those of K_{-j} for $j < i$, then the algorithm does not work. In such a case, one concludes that the matrix **K** is not expressible in terms of unimodular congruence.

We have now established a number of necessary conditions for a Z- and Y-matrix. The necessary and sufficient condition will now be stated, which is due to Cederbaum[6e] and Bryant.[11]

Theorem 8. A necessary and sufficient condition for a given symmetric matrix of order n, **Z** (or **Y**) to be the open-circuit impedance (or the short-circuit admittance) matrix of a nonsingular *RLC* n-port is that **Z** (or **Y**) is a principal submatrix of the inverse of the matrix $C_f Y_e C_f{}^t$ (or $B_f Z_e B_f{}^t$). Where Y_e (or Z_e) is a diagonal matrix in which each diagonal element is of the form of a, bs, or $1/cs$ for $a, b, c > 0$, C_f is a realizable basic cut-set matrix and B_f is a realizable basic loop matrix.†

The proof of the theorem is not given here because it is presently of little use in synthesis application, since no systematic procedure exists for the decomposition of a Z- or Y-matrix as the principal minor of the inverse of the desired congruent transformation. This problem remains unsolved.

---

† The theorem is also valid when the seg and special loop matrices are under consideration.

CHAPTER 3

Realization of a Y-Matrix

with $(n+1)$ Nodes

WE SHALL FIRST investigate the various known methods for the realization of a Y-matrix, and we shall then extend them to synthesize a multiport network based on a given Z-matrix.

For a given Y-matrix, there may exist many possible structures of a realized network. That is, to form n independent ports, a network must have at least $(n+1)$ terminal nodes and up to $2n$ terminals, because a terminal node should be connected to one or more voltage or current generators. Thus, if a given Y-matrix is not realizable with $(n+1)$ terminal nodes, then it may be realizable with $(n+2)$ nodes, or with $(n+3)$ nodes, and so on. However, except for a few trial-and-error techniques available at present, there has been no systematic technique developed for realizing a Y-matrix with any arbitrary number of terminal nodes, say $n+k$ and $k = 1, 2, \ldots, n$. The problem of realizing a Y-matrix with an n-port of $(n+1)$ nodes has been completely solved by the contributions of various authors. We therefore open this discussion with the realization of a Y-matrix with $(n+1)$ nodes and then extend our study to a network of more than $(n+1)$ nodes in the next chapter.

REALIZATION OF A Y-MATRIX OF ORDER n WITH A RESISTIVE n-PORT OF $(n+1)$ NODES

There exist two major approaches to the synthesis of an n-port with $(n+1)$ nodes. We shall first summarize the two approaches and then investigate each of the methods in detail.

Approach 1 (method of determination of port-structure): In this approach, the first step involves the determination of the port-structure (if one exists) and the second step involves the transformation of the existing port-structure into a star-tree port-structure. If the transformation results in a dominant matrix, the realization can be accomplished by inspection. This approach is recommended in case a Y-matrix contains no zero elements. The reason for this preference is that when $\mathbf{Y}$ contains many zero elements, the derivation of the port-structure becomes exceedingly laborious. The work of Guillemin,[5b] Biorci and Civalleri,[8a,d,12] and Halkias, Cederbaum, and Kim[13a] may use this approach.

Approach 2 (decomposition): The second approach is based on the decomposition (if such decomposition is possible) of a Y-matrix into the triple product $\mathbf{Y} = \mathbf{C}_s \mathbf{Y}_e \mathbf{C}_s{}^t$ where $\mathbf{C}_s$ is a seg matrix and $\mathbf{Y}_e$ is the diagonal matrix with positive elements representing edge-conductances. The realization of $\mathbf{Y}$ is accomplished if $\mathbf{C}_s$ can be realized into a connected graph. The second approach is recommended when a Y-matrix contains some zero elements. The work by Cederbaum[6d] and Halkias[13b] may be classified under this approach.

Determination of port-structure

Suppose we are given the short-circuit admittance matrix of a resistive n-port network with $n+1$ nodes. Then the problem of determining the port-structure, that is, the sequence, numbering, and polarity of the n-ports, should be first considered.

Since the n-ports of a network with $n+1$ nodes form a tree-like structure, we shall consider a tree with n branches corresponding to ports. We shall call two branches of a tree "series-connected" if they have a common node and no other branch is incident with this common node. Two tree-branches are "star-connected" if they are end-branches and have a common node. The following topological properties of a tree, which will be useful to further discussion, were proposed by Halkias.[13b]

Lemma 1. Every tree formed by $n \geqslant 4$ branches must have at least four distinct branches, i, j, p, and q, such that the branches of each of the two pairs (i, j) and (p, q) are either series- or star-connected.

Proof: Let us consider a tree T of n branches.

Case 1: T is a star-tree. If T is a star-tree then all branches are star-connected and the lemma follows.

Case 2: T is not a star-tree. Remove all end-branches, and suppose there are $m \geqslant 1$ branches left corresponding to a subtree T'. For $m = 1$, let us label this branch b_1 and its two nodes v_1 and v_2, respectively. Node v_1 must have been connected in T to at least one end-branch of T. (Otherwise b_1 would have been an end-branch itself.)

If v_1 were connected to only one end-branch, say branch b_2, then b_1 and b_2 would form a series-connection. If v_1 were connected to more than one of the end-branches, then these end-branches would form at least one star-connection. A similar argument can be made for node v_2 and the lemma is proven for $m = 1$.

For $m > 1$, T' must have at least two end-branches and the previous situation holds for each one of these end-branches. This completes the proof of the lemma.

Theorem 9. If **Y** is a short-circuit admittance matrix of an n-port resistive network with $n + 1$ nodes and if the pair of ports i and j are series-connected, then for an arbitrary index $k = 1, 2, \ldots, n$, we have[13a]

$$y_{ij}y_{ki}y_{kj} \geqslant 0. \tag{3.1}$$

Proof: Let $\mathbf{C_s}$ be the seg matrix corresponding to the tree of the port-structure of a network under consideration. (This matrix may have rows which correspond to an edge-disjoint union of the simple cut-sets.) Then

$$\mathbf{Y} = \mathbf{C_s Y_e C_s^t}. \tag{3.2}$$

If $k = i$ or $k = j$, the expression of (3.1) is trivial for each pair of indices i and j. We denote the modified tree on which $\mathbf{C_s}$ is formed by t, and assume that in t two branches i and j, corresponding to ports i and j, are connected in series. For convenience we shall call each seg based on branches i and j "seg i" and "seg j," respectively. We now choose another seg k with respect to branch k. Then, it should be clear that there exists a path joining branches $i, j,$ and k since branches i and j are series-connected. For each branch k, $k = 1, 2, \ldots, n$, there may exist, therefore, a chord with respect to tree t of the network such that the chord is included in segs $i, j,$ and k. Consequently, there exists a column in $\mathbf{C_s}$ having nonzero elements in rows $i, j,$ and k. Thus, in product $y_{ij}y_{ki}y_{kj}$ its sign with the terms with negative sign cannot be negative. If there exists no such chord, then the product vanishes. This completes the proof of the theorem.

Theorem 10. If **Y** is an admittance matrix of an n-port resistive network with $n + 1$ nodes and if the pair of ports i and j are star-connected, then for an arbitrary number $k = 1, 2, \ldots, n, k \neq i, j,$[13a]

$$y_{ij}y_{ki}y_{kj} \leqslant 0. \tag{3.3}$$

Proof: Let us suppose that, on the contrary, for some index $k \neq i, j,$

$$y_{ij}y_{ki}y_{kj} > 0. \tag{3.4}$$

Then, referring to Eq. (3.2) and the discussions used in the proof of

Theorem 9, there should exist a chord belonging to all three segs i, j, and k which is impossible since i and j are star-connected. Thus, the theorem is proved.

Corollary 2. If $\mathbf{Y}$ is an admittance matrix of an n-port resistive network with $n + 1$ $(n \geq 4)$ nodes, then there must exist at least *two pairs* of different indices (i, j) and (k, m) such that either

$$y_{ij}y_{pi}y_{pj} \geq 0, \qquad \text{for all } p = 1, 2, \ldots, n$$
$$y_{km}y_{qk}y_{qm} \geq 0, \qquad \text{for all } q = 1, 2, \ldots, n \tag{3.5a}$$

(ports i, j series-connected; ports k, m series-connected) or

$$y_{ij}y_{pi}y_{pj} \leq 0, \qquad \text{for all } p \neq i, j,$$
$$y_{km}y_{qk}y_{qm} \leq 0, \qquad \text{for all } q \neq k, m \tag{3.5b}$$

(ports i, j star-connected; ports k, m star-connected) or

$$y_{ij}y_{pi}y'_{pj} \geq 0, \qquad \text{for all } p = 1, 2, \ldots, n,$$
$$y_{km}y_{qk}y_{qm} \leq 0, \qquad \text{for all } q \neq k, m \tag{3.5c}$$

(ports i, j series-connected; ports k, m star-connected).[13a]

Since the proof of the corollary is evident from Lemma 1 and Theorems 9 and 10, it is left to the reader.

The application of the corollary requires knowledge only of the signs of the elements of a Y-matrix. In both pairs of columns (i, j) and (k, m) in $\mathbf{Y}$ the elements in each row are required to be associated with the same or always opposite signs. The main diagonal elements and their counterparts in the other column may be exempted from this rule since $p \neq i, j$ and $q \neq k, m$.

On the basis of the previous theorems we now give a systematic procedure for obtaining a tree-port-structure (if one exists) appropriate to the short-circuit admittance matrix $\mathbf{Y}$ of a resistive n-port with $n + 1$ nodes.

Let us assume that matrix $\mathbf{Y}$ contains no zero elements.

Step 1: By applying Theorems 9 and 10 or Corollary 2, first find the pairs of ports (or branches) which are series- or star-connected. There must be at least two such pairs; if not, the procedure can be terminated here.

Step 2: Start at any one pair of series-connected or star-connected branches of the tree and short-circuit one of the branches which is equivalent to crossing out the corresponding row and column of matrix $\mathbf{Y}$. Apply Corollary 2 and determine if the remaining branch is series-connected or star-connected to any other branch. Repeat the process until arriving at another pair of series- or star-connected branches going through all branches. If the tree has more than two such pairs, then one may arrive eventually at a branch which is neither series- nor star-connected to any of the remaining branches.

Step 3: Start at another series- or star-connected pair and repeat Step 2. By repeated applications of Steps 2 and 3, one should eventually go through every tree-branch. If it is impossible to continue this process until one goes through every tree-branch, then the sign pattern is nonrealizable and the procedure may be discontinued. If the sign pattern is realizable, then proceed with Step 4.

Step 4: Arbitrarily assign the direction of one port, say port i; then the direction of any other port k is obtained from the sign of the element y_{ik}.

The following examples will illustrate the previous procedure.

Example 6. Determine the port-structure appropriate to the following sign pattern of a short-circuit admittance matrix:

$$
\text{Sign pattern of } \mathbf{Y} =
\begin{bmatrix}
+ & - & + & - & + & - & - \\
- & + & + & + & + & - & - \\
+ & + & + & - & + & - & - \\
- & + & - & + & + & - & + \\
+ & + & + & + & + & - & + \\
- & - & - & - & - & + & - \\
- & - & - & + & + & - & +
\end{bmatrix}. \quad (3.6)
$$

Then, applying Step 1, we see that for only ports 5 and 6 we have $y_{56}y_{5k}y_{6k} > 0$. Thus ports 5 and 6 are series-connected. We need not proceed any further because it is clear from Corollary 2 that no tree-port-structure exists corresponding to the sign pattern of Eq. (3.6).

Example 7. Determine the port-structure appropriate to the sign pattern of a matrix $\mathbf{Y}$ of order six, given in Eq. (3.7):

$$
\text{Sign pattern of } \mathbf{Y} =
\begin{bmatrix}
+ & + & + & + & + & + \\
+ & + & - & - & - & + \\
+ & - & + & - & - & - \\
+ & - & - & + & + & - \\
+ & - & - & + & + & - \\
+ & + & - & - & - & +
\end{bmatrix}. \quad (3.7)
$$

In order to make the procedure of finding the port-structure more systematic, it may be necessary to build a table of signs for $y_{ij}y_{ik}y_{jk}$ as shown in Table 1. In Table 1, the signs of the product $y_{ij}y_{ik}y_{jk}$ are given for $i, j, k = 1, 2, 3, 4, 5, 6$. In the positions corresponding to $k = i$ or $k = j$ an asterisk is placed because we are not interested in those signs.

By Step 1, we find that ports (2, 6) and (4, 5) are series-connected and that ports (1, 3) are star-connected. Now, short-circuit port 1. This is equivalent to deleting the first column of Table 1. Then

Table 1. Signs of the triple product $(y_{ij}y_{ik}y_{jk})$

i, j \ k	1	2	3	4	5	6	
1, 2	*	*	−	−	−	+	
1, 3	*	−	*	−	−	−	Star
1, 4	*	−	−	*	+	−	
1, 5	*	−	−	+	*	−	
1, 6	*	+	−	−	−	*	
2, 3	−	*	*	−	−	+	
2, 4	−	*	−	*	+	+	
2, 5	−	*	−	+	*	+	
2, 6	+	*	+	+	+	*	Series
3, 4	−	−	*	*	+	−	
3, 5	−	−	*	+	*	−	
3, 6	−	+	*	−	−	*	
4, 5	+	+	+	*	*	+	Series
4, 6	−	+	−	*	+	*	
5, 6	−	+	−	+	*	*	

port 3 is neither series- nor star-connected to any other port.

As the last step: (a) Short-circuit ports 1 and 6; i.e., consider Table 1 with columns 1 and 6 deleted and do not take into account

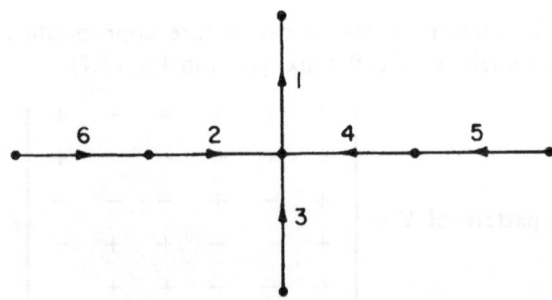

Fig. 3.1. Port-structure corresponding to the sign pattern of Eq. (3.7)

the rows which contain 1 or 6. Then ports 2 and 3 are star-connected; (b) Short-circuit port 2; then ports 3, 4, and 5 are series-connected. Hence, the tree may be constructed as shown in Fig. 3.1.

It must be noted that the sequence of ports (6, 2) and (4, 5) cannot be determined solely on the basis of the sign pattern. The submatrix corresponding to ports 6, 2, 4 and 5 must be realized with a linear-tree using known methods.

Thus far, we have developed the procedure of deriving the port-structure. Once the port-structure is known we use the results in the previous chapter to transform the tree into a star-tree. If the resulting matrix is dominant the realization can be accomplished by inspection. Let us illustrate the final realization of a given matrix.

Example 8. Realize the following matrix with a resistive network of seven nodes:

$$\mathbf{Y} = \begin{bmatrix} 12 & 5 & 1 & 5 & 2 & 2 \\ 5 & 13 & -5 & -2 & -2 & 6 \\ 1 & -5 & 12 & -5 & -2 & -2 \\ 5 & -2 & -5 & 13 & 6 & -2 \\ 2 & -2 & -2 & 6 & 7 & -2 \\ 2 & 6 & -2 & -2 & -2 & 7 \end{bmatrix}. \tag{3.8}$$

The sign pattern of the matrix is the same as the sign pattern of Example 7; thus the port-structure is also the same. The order of ports 4 and 5 can be obtained by considering the submatrix

$$\mathbf{Y}_{145} = \begin{bmatrix} 12 & 5 & 2 \\ 5 & 13 & 6 \\ 2 & 6 & 7 \end{bmatrix}. \tag{3.9}$$

This matrix is uniformly tapered and the sequence of ports 1, 4, and 5 is as shown in Fig. 3.2. The same procedure can be followed for ports 1, 2, and 6. Now we arbitrarily select a reference node and obtain the star-tree port-structure with respect to this node, which is shown by dotted lines in Fig. 3.2. We proceed to write the transformation matrix between the two sets of port-voltages:

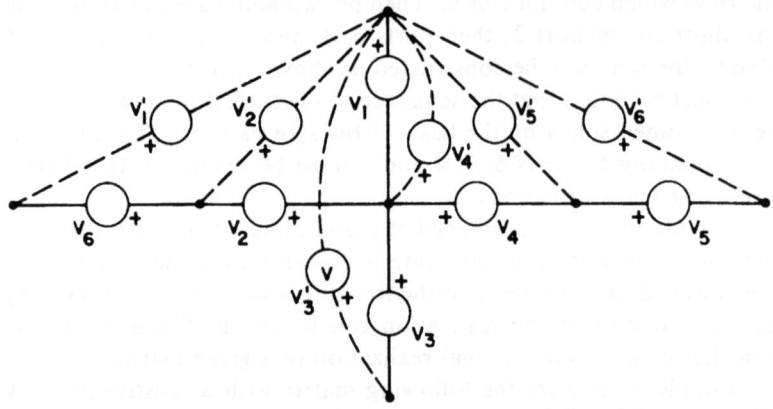

Fig. 3.2. Transformation of the port-structure of Fig. 3.1 into a star-tree port-structure

$$
\begin{bmatrix} V_1 \\ V_2 \\ V_3 \\ V_4 \\ V_5 \\ V_6 \end{bmatrix} = \begin{bmatrix} 0 & 0 & 0 & -1 & 0 & 0 \\ 0 & -1 & 0 & 1 & 0 & 0 \\ 0 & 0 & -1 & 1 & 0 & 0 \\ 0 & 0 & 0 & 1 & -1 & 0 \\ 0 & 0 & 0 & 0 & 1 & -1 \\ -1 & 1 & 0 & 0 & 0 & 0 \end{bmatrix} \begin{bmatrix} V_1' \\ V_2' \\ V_3' \\ V_4' \\ V_5' \\ V_6' \end{bmatrix}, \quad (3.10a)
$$

where V_i is the port-voltage of port i of the tree while V_k' is the port-voltage of port k of the star-tree. In matrix form we have

$$
V = TV' \tag{3.10b}
$$

and T is the coefficient matrix of Eq. (3.10 a).

We therefore transform matrix Y to $\bar{Y}$ corresponding to the star-tree by T:

$$
\bar{Y} = T'YT = \begin{bmatrix} 7 & -1 & -2 & 0 & 0 & -2 \\ -1 & 8 & -3 & -1 & 0 & 0 \\ -2 & -3 & 12 & -1 & -3 & -2 \\ 0 & -1 & -1 & 4 & -1 & 0 \\ 0 & 0 & -3 & -1 & 8 & -1 \\ -2 & 0 & -2 & 0 & -1 & 7 \end{bmatrix}. \tag{3.11}
$$

Now we can see that matrix $\bar{Y}$ is dominant with negative off-diagonal elements and can therefore be realized by inspection. (See

Fig. 3.3.) The ports were determined in Fig. 3.2. This completes the realization of the given network.

In the previous example we considered a matrix with no zero elements; from our discussion of a star-tree and a linear-tree port-structure two special classes of matrices can be realized almost by inspection. The first class is the class of dominant matrices with

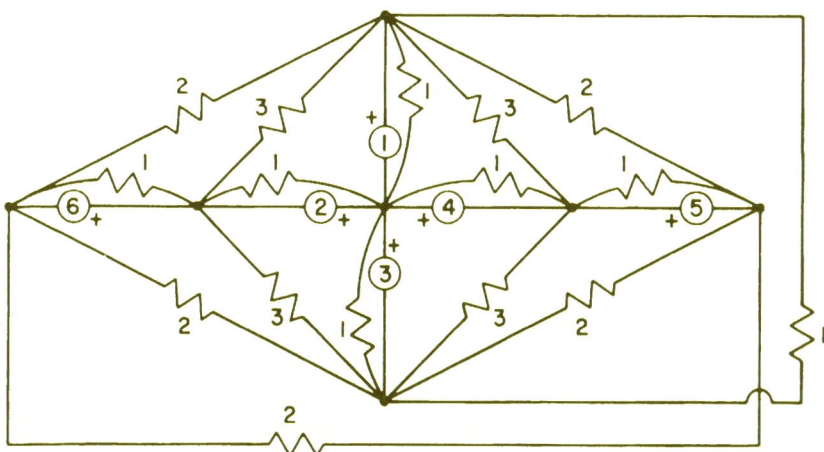

Fig. 3.3. Final realization of the Y-matrix of Eq. (3.8)

negative off-diagonal elements. This is the ordinary node admittance matrix and its realization is obvious. The second class is the class of matrices with all positive elements. If such a matrix can be made uniformly tapered then the conductance of each edge of a network to be realized can be obtained using the equation

$$g_{ij} = y_{ij} + y_{i-1,j+1} - y_{i-1,j} - y_{i,j+1}. \tag{3.12}$$

Example 9. Realize the matrix of Eq. (3.13), which is uniformly tapered, with a resistive network of six nodes:

$$\mathbf{Y} = \begin{bmatrix} 5 & 4 & 3 & 2 & 1 \\ 4 & 8 & 6 & 4 & 2 \\ 3 & 6 & 9 & 6 & 3 \\ 2 & 4 & 6 & 8 & 4 \\ 1 & 2 & 3 & 4 & 5 \end{bmatrix}. \tag{3.13}$$

Using Eq. (3.12), we obtain the matrix of edge-conductances

$$\begin{bmatrix} g_{11} & g_{12} & g_{13} & g_{14} & g_{15} \\ g_{21} & g_{22} & g_{23} & g_{24} & g_{25} \\ g_{31} & g_{32} & g_{33} & g_{34} & g_{35} \\ g_{41} & g_{42} & g_{43} & g_{44} & g_{45} \\ g_{51} & g_{52} & g_{53} & g_{54} & g_{55} \end{bmatrix} = \begin{bmatrix} 1 & 1 & 1 & 1 & 1 \\ 1 & 1 & 1 & 1 & 1 \\ 1 & 1 & 1 & 1 & 1 \\ 1 & 1 & 1 & 1 & 1 \\ 1 & 1 & 1 & 1 & 1 \end{bmatrix}. \quad (3.14)$$

The realization is shown in Fig. 3.4, where all conductances are 1 mho.

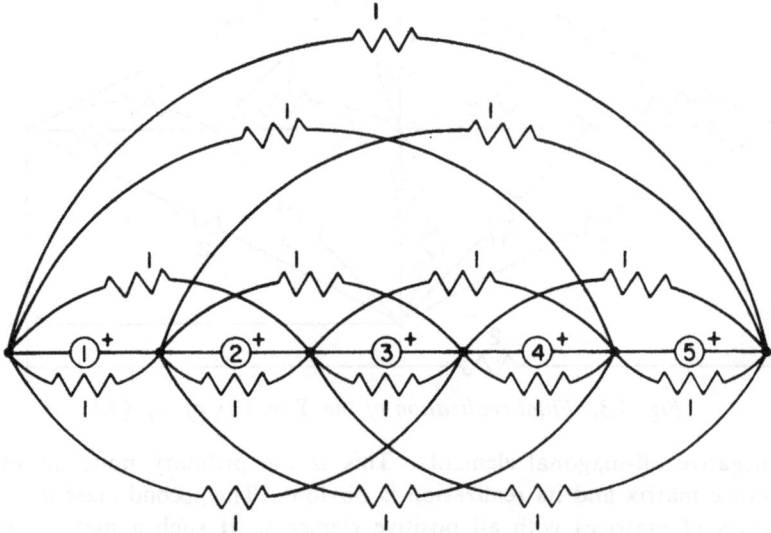

Fig. 3.4. Realization of the Y-matrix of Eq. (5.12)

Decomposition method

Let us now consider the second approach. If a matrix contains k zero elements above the main diagonal then 2^k sign patterns are possible. A number of different port-structures may be appropriate to some of the sign patterns and the n-port network, if it exists, may sometimes be realizable with more than one port-structure. The method presented here is based on the decomposition $\mathbf{Y} = \mathbf{PDP}^t$. Suppose that a column of matrix $\mathbf{P}$ has nonzero entries in the rows $i, j, \ldots, k$; this implies that ports $i, j, \ldots, k$ are linked by the branch corresponding to the column, and, hence, if all other ports are short-circuited, the submatrix corresponding to rows and columns $i, j, \ldots, k$ must be realizable with a linear-tree. The ordering of the ports is obtained by tapering this submatrix. In this way, we proceed to realize the *n*-port network using column after column of matrix **P**. The procedure is illustrated in the following examples.

Example 10. Here we show an example of a matrix which is realizable with $(n + 1)$ nodes and with two distinct port-structures. Consider

$$\mathbf{Y} = \begin{bmatrix} 4 & 2 & 0 \\ 2 & 6 & 3 \\ 0 & 3 & 5 \end{bmatrix} = \begin{bmatrix} 1 & 0 & 1 & 0 & 0 \\ 1 & 1 & 0 & 1 & 0 \\ 0 & 1 & 0 & 0 & 1 \end{bmatrix} \begin{bmatrix} 2 & 0 & 0 & 0 & 0 \\ 0 & 3 & 0 & 0 & 0 \\ 0 & 0 & 2 & 0 & 0 \\ 0 & 0 & 0 & 1 & 0 \\ 0 & 0 & 0 & 0 & 2 \end{bmatrix} \begin{bmatrix} 1 & 1 & 0 \\ 0 & 1 & 1 \\ 1 & 0 & 0 \\ 0 & 1 & 0 \\ 0 & 0 & 1 \end{bmatrix} . (3.15)$$

The first column of the cut-set matrix shows that a conductance of 2 mho links ports 1 and 2. (See Fig. 3.5a.) The second column shows that a conductance of 3 mho links ports 2 and 3. (See Fig. 3.5b.)

The connection of ports (1, 2) and (2, 3) can be made in either one of two ways, as shown in Fig. 3.6.

Two different realizations of the given *Y*-matrix with a network of

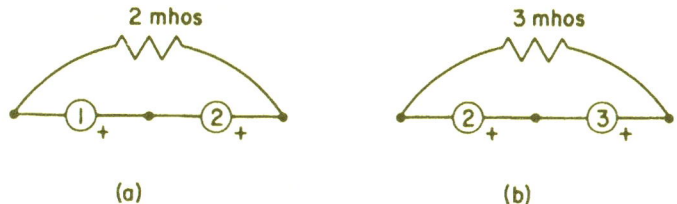

Fig. 3.5. *Realization of the first (a) and second (b) columns of the cut-set matrix of Eq. (3.15)*

Fig. 3.6. *Possible connections of ports 1, 2, and 3 for the Y-matrix of Eq. (3.15)*

(a) Series-connection; (b) star-connection.

four nodes, based on the tree-port-structures of Fig. 3.6 are shown in Fig. 3.7.

It may be noted that the realizations of Fig. 3.7 correspond to assuming the signs (+) and (−), respectively, to the zero element of matrix **Y**.

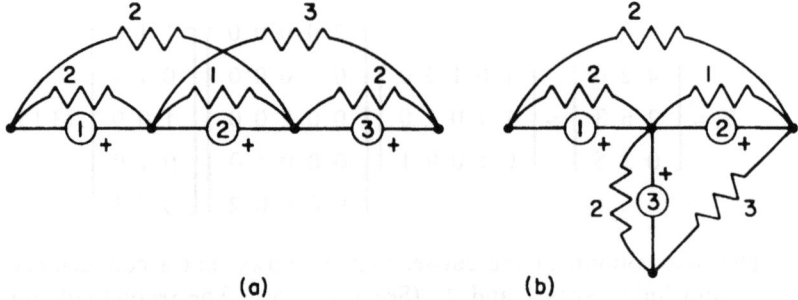

(a)   (b)

Fig. 3.7. Realizations of the Y-matrix of Eq. (3.15), where each edge-admittance is in mhos

Example 11. We shall try to realize matrix **Y** of Eq. (3.16) with a large number of zero elements:

$$\mathbf{Y} = \begin{bmatrix} 3 & 0 & 3 & 0 & 0 & 0 & -1 & 0 \\ 0 & 7 & 7 & 0 & 0 & 0 & 0 & -4 \\ 3 & 7 & 19 & 0 & 0 & 0 & -1 & -4 \\ 0 & 0 & 0 & 36 & 11 & 15 & -5 & -8 \\ 0 & 0 & 0 & 11 & 11 & 0 & -5 & 0 \\ 0 & 0 & 0 & 15 & 0 & 15 & 0 & -8 \\ -1 & 0 & -1 & -5 & -5 & 0 & 6 & 0 \\ 0 & -4 & -4 & -8 & 0 & -8 & 0 & 12 \end{bmatrix}. \qquad (3.16)$$

The given matrix contains sixteen zero elements above the main diagonal. It is obvious that great effort would be required to investigate all possible sign patterns for the appropriate tree-port-structure. The given matrix may be easily decomposed into the triple product as

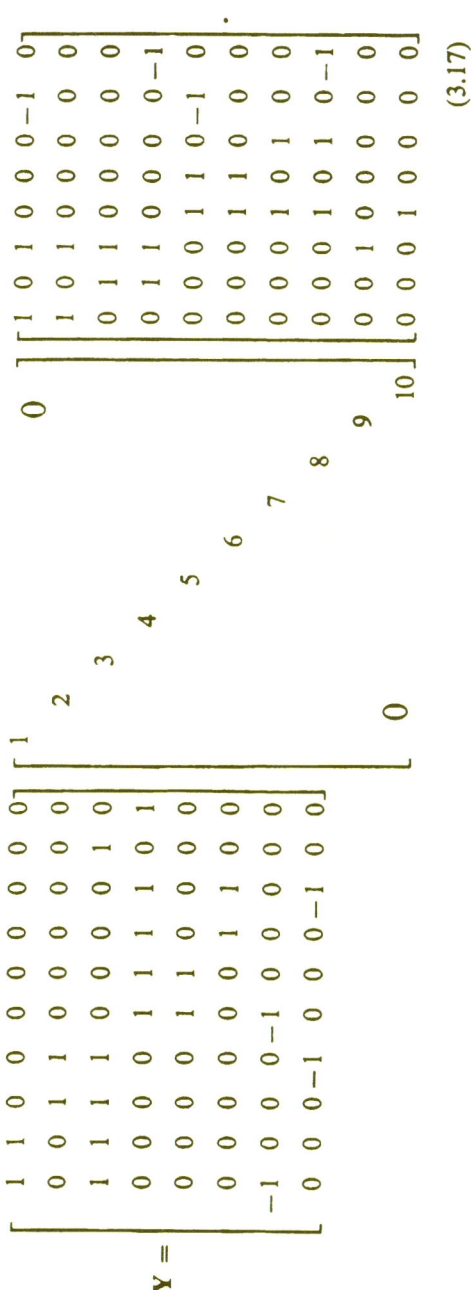

$$\mathbf{Y} = \qquad (3.17)$$

Then, first, we see that conductance 1 mho, corresponding to the first column of matrix **P**, links the ports 1, 3, and 7. They follow in sequence because the submatrix containing the rows and columns 1, 3, and 7, representing the admittance matrix of the *n*-port with all other ports short-circuited, appears in tapered form (see Fig. 3.8):

$$\mathbf{Y}_{137} = \begin{bmatrix} 3 & 3 & -1 \\ 3 & 19 & -1 \\ -1 & -1 & 6 \end{bmatrix}. \qquad (3.18)$$

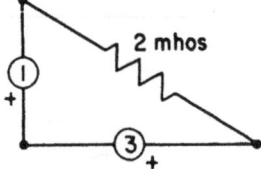

Fig. 3.8. Partial port-structure corresponding to $\mathbf{Y}_{137}$ *of Eq. (3.18)*

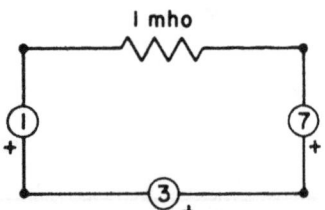

Fig. 3.9. Partial port-structure corresponding to the second column of **P** *of Eq. (3.17)*

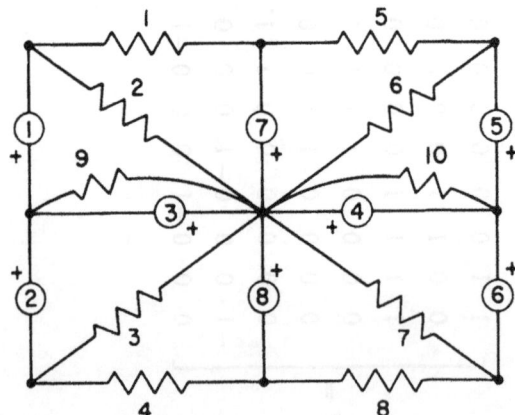

Fig. 3.10. Final realization of the Y-matrix of Eq. (3.16), in ohms

Second, we know that conductance 2 mho corresponds to the second column of **P** and links the ports 1 and 3 as shown in Fig. 3.9. Continuing in this way, we finally realize an 8-port network as shown in Fig. 3.10.

REALIZATION OF A Y-MATRIX OF ORDER n WITH AN RLC n-PORT WITH n+1 NODES

Let us consider a connected graph with $n + 1$ nodes and b edges. Let each edge represent either a resistor, an inductor, or a capacitor. In such a case the short-circuit admittance matrix **Y** of an n-port described on this network is of the form

$$\mathbf{Y} = \mathbf{C}_s \mathbf{Y}_e \mathbf{C}_s^t, \tag{3.19}$$

where $\mathbf{Y}_e$ is the diagonal matrix with main diagonal elements of the form g, sC, $1/sL$. A necessary condition for the realization of **Y** with $(n + 1)$ nodes is that each element of this matrix be of the form

$$y_{ij} = \pm [g + sC + (1/sL)], \tag{3.20}$$

where g, C, and L are non-negative real numbers. Equation (3.20) follows from Eq. (3.19) and the sign-rule theorem. The realization of **Y** can now be based on the following observation: For real and positive values of the complex variable s an *RLC* n-port behaves as a resistive network.

We arbitrarily assign, then, a real positive value to the variable s resulting in a matrix $\bar{\mathbf{Y}}$, and proceed as in the previous section. Of course, we must find the value of the various resistors, inductors, and capacitors required for the realization. This can be done in two different ways depending on the nature of the matrix $\bar{\mathbf{Y}}$.

Case 1. $\bar{\mathbf{Y}}$ contains no zero elements. If matrix $\bar{\mathbf{Y}}$ contains no zero elements, then Approach 1 illustrated in the previous section can be used to obtain the port-structure. The transformation to a star-tree port-structure can now be applied to $\bar{\mathbf{Y}}$ and the resulting matrix $\bar{\mathbf{Y}}^*$ must be realizable as the node admittance matrix. This is illustrated in the following example.

Example 12. Realize the following Y-matrix with an *RLC* network with exactly seven nodes:

$$\mathbf{Y} = \begin{bmatrix} 1+6s+\dfrac{5}{s} & 3s+\dfrac{2}{s} & s & 2s+\dfrac{3}{s} & 2s & \dfrac{2}{s} \\[2mm] 3s+\dfrac{2}{s} & 7s+\dfrac{6}{s} & -2s-\dfrac{3}{s} & -2s & -2s & 4s+\dfrac{2}{s} \\[2mm] s & -2s-\dfrac{3}{s} & 1+5s+\dfrac{6}{s} & -2s-\dfrac{3}{s} & -2s & -2s \\[2mm] 2s+\dfrac{3}{s} & -2s & -2s-\dfrac{3}{s} & 1+6s+\dfrac{6}{s} & 6s & -2s \\[2mm] 2s & -2s & -2s & 6s & 7s & -2s \\[2mm] \dfrac{2}{s} & 4s+\dfrac{2}{s} & -2s & -2s & -2s & 1+4s+\dfrac{2}{s} \end{bmatrix}.$$

$$(3.21)$$

Let $s = 1$, then matrix $\mathbf{Y}$ becomes

$$\overline{\mathbf{Y}} = \begin{bmatrix} 12 & 5 & 1 & 5 & 2 & 2 \\ 5 & 13 & -5 & -2 & -2 & 6 \\ 1 & -5 & 12 & -5 & -2 & -2 \\ 5 & -2 & -5 & 13 & 6 & -2 \\ 2 & -2 & -2 & 6 & 7 & -2 \\ 2 & 6 & -2 & -2 & -2 & 7 \end{bmatrix}.$$

$$(3.22)$$

The matrix of Eq. (3.22) is the same as the matrix which was considered in Example 8. The port-structure of the network corresponding to the matrix is therefore the same as the port-structure obtained in that example (see Fig. 3.2). The transformation of the port-structure into a star-tree port-structure applied to the matrix results in the following matrix:

$$\overline{\mathbf{Y}}^* = \begin{bmatrix} 1+4s+\dfrac{2}{s} & -1 & -2s & 0 & 0 & -2s \\[2mm] -1 & 1+3s+\dfrac{4}{s} & -\dfrac{3}{s} & -\dfrac{1}{s} & 0 & 0 \\[2mm] -2s & -\dfrac{3}{s} & 1+5s+\dfrac{6}{s} & -1 & -\dfrac{3}{s} & -2s \\[2mm] 0 & -\dfrac{1}{s} & -1 & 3+\dfrac{1}{s} & -1 & 0 \\[2mm] 0 & 0 & -\dfrac{3}{s} & -1 & 1+s+\dfrac{6}{s} & -s \\[2mm] -2s & 0 & -2s & 0 & -s & 7s \end{bmatrix} . \quad (3.23)$$

The matrix $\overline{\mathbf{Y}}^*$ is the ordinary node admittance matrix and can be realized by inspection with the port-structure shown in Fig. 3.3. The final realization is shown in Fig. 3.11.

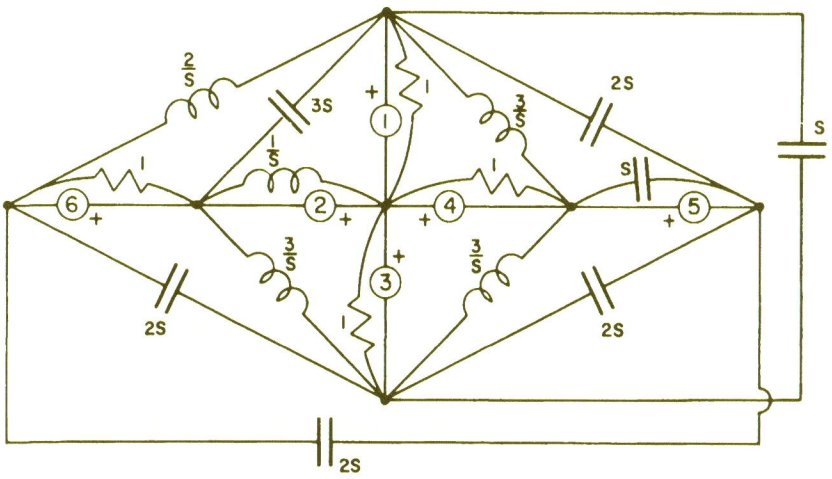

Fig. 3.11. Final realization of the Y-matrix of Eq. (3.21)

Case 2: $\overline{\mathbf{Y}}$ contains some zero elements. If $\overline{\mathbf{Y}}$ contains zero elements, then Approach 2 of the previous section should be used to obtain a seg matrix $\mathbf{C}_s$. Thus, we have

$$\mathbf{Y} = \mathbf{C}_s \mathbf{D} \mathbf{C}_s{}^t \text{ and } \overline{\mathbf{Y}} = \mathbf{C}_s \overline{\mathbf{D}} \mathbf{C}_s{}^t, \quad (3.24)$$

where $\bar{\mathbf{D}}$ is obtained from $\mathbf{D}$ by replacing the complex variable s with a positive real number. In other words, first derive $\bar{\mathbf{Y}}$ from $\mathbf{Y}$ by the substitution of a positive real number in s, and then apply Cederbaum's decomposition technique to get $\mathbf{C}_s$. This procedure is illustrated in the following example.

Example 13. Realize the following Y-matrix with an RLC network with four nodes:

$$\mathbf{Y} = \begin{bmatrix} 2+\dfrac{2}{s} & 2s & 0 \\[2mm] 2s & 2s+\dfrac{4}{s} & \dfrac{3}{s} \\[2mm] 0 & -\dfrac{3}{s} & 2s+\dfrac{3}{s} \end{bmatrix}. \tag{3.25}$$

Let $s = 1$; then $\mathbf{Y}$ becomes

$$\bar{\mathbf{Y}} = \begin{bmatrix} 4 & 2 & 0 \\ 2 & 6 & 3 \\ 0 & 3 & 5 \end{bmatrix}. \tag{3.26}$$

However, this matrix was considered in Example 10. Thus we have

$$\mathbf{Y} = \begin{bmatrix} 2+2s & 2s & 0 \\[2mm] 2s & 2s+\dfrac{4}{s} & \dfrac{3}{s} \\[2mm] 0 & \dfrac{3}{s} & 2s+\dfrac{3}{s} \end{bmatrix} = \begin{bmatrix} 1 & 0 & 1 & 0 & 0 \\ 1 & 1 & 0 & 1 & 0 \\ 0 & 1 & 0 & 0 & 1 \end{bmatrix} \begin{bmatrix} d_1 & 0 & 0 & 0 & 0 \\ 0 & d_2 & 0 & 0 & 0 \\ 0 & 0 & d_3 & 0 & 0 \\ 0 & 0 & 0 & d_4 & 0 \\ 0 & 0 & 0 & 0 & d_5 \end{bmatrix} \begin{bmatrix} 1 & 1 & 0 \\ 0 & 1 & 1 \\ 1 & 0 & 0 \\ 0 & 1 & 0 \\ 0 & 0 & 1 \end{bmatrix}.$$

$$\tag{3.27}$$

From Eq. (3.37) we solve for the unknowns d_1, d_2, d_3, d_4, and d_5. Thus

$$2+2s = d_1+d_3$$
$$2s = d_1$$
$$2s+(4/s) = d_1+d_2+d_4 \tag{3.28}$$
$$3/s = d_2$$
$$2s+(3/s) = d_2+d_5.$$

Hence

$$d_1 = 2s$$
$$d_2 = 3/s$$
$$d_3 = 2 \qquad (3.29)$$
$$d_4 = 1/s$$
$$d_5 = 2s.$$

Two distinct realizations are shown in Figs. 3.12a and b.

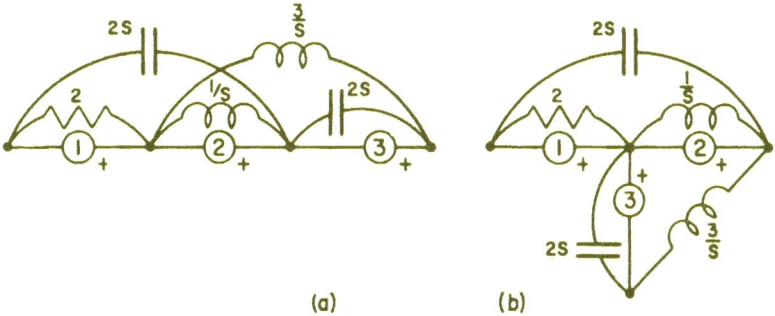

(a)   (b)

Fig. 3.12. Two distinct realizations of the Y-matrix of Eq. (3.25)
With (a) a linear-tree port-structure and (b) a star-tree port-structure.

CHAPTER 4

Generalized Realization
of a Y- and Z-Matrix

THERE EXIST a number of trial and error procedures for the realization of a Y-matrix with more than $n+1$ nodes and of a Z-matrix. The materials presented in this chapter are the contributions of Cederbaum,[6a] Slepian and Weinberg,[10] Tellegen,[9] and Halkias.[13b]

REALIZATION OF A Y-MATRIX OF ORDER n WITH MORE THAN n+1 NODES

For the sake of simplicity, let us initially consider resistive n-port networks. If a symmetric Y-matrix of order n with all real elements is given, we can always determine if this matrix is realizable with a resistive network of $n+1$ nodes. If this matrix is not realizable with the minimum number of nodes, namely $(n+1)$, it may be realizable with more than $n+1$ nodes, that is, with $n+k$ nodes and $1 < k \leqslant n$. The following sufficient condition has been given by Slepian and Weinberg for realization with $2n$ nodes.

Theorem 11. A sufficient condition for the realization of a real, symmetric matrix **Y** of order n as the short-circuit admittance of a resistive n-port with $2n$ nodes is that the matrix be dominant.

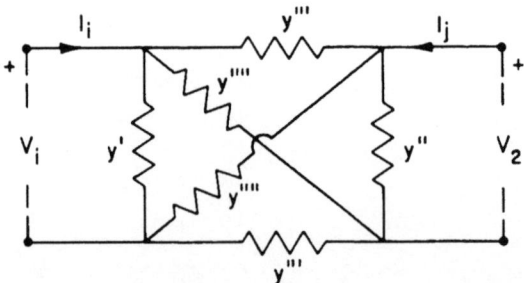

Fig. 4.1. Realization of a dominant matrix

Proof: The proof of the theorem is given in terms of a synthesis procedure. The scheme for developing the network is given in Fig. 4.1, where

$$
\begin{aligned}
y' &= -|y_{i1}| - |y_{i2}| - \ldots + y_{ii} - \ldots - |y_{in}| \\
y'' &= -|y_{j1}| - |y_{j2}| - \ldots + y_{jj} - \ldots - |y_{jn}| \\
y''' &= |y_{ij}| - y_{ij} \\
y'''' &= |y_{ij}| + y_{ij}
\end{aligned}
\tag{4.1}
$$

and the values of y's are in mhos. Examination of the network of Fig. 4.1 shows that the short-circuit admittance matrix of the n-port indeed satisfies the properties of $\mathbf{Y}$. This completes the proof of the sufficiency of the theorem.

Example 14: Realize the Slepian–Weinberg matrix with eight nodes:

$$
\mathbf{Y} = \begin{bmatrix}
7 & 1 & 2 & 3 \\
1 & 12 & 4 & 5 \\
2 & 4 & 15 & 6 \\
3 & 5 & 6 & 18
\end{bmatrix}.
\tag{4.2}
$$

Before we try to realize the given Y-matrix, one should note that the matrix was considered previously and found *not* to be realizable with five nodes. However, it is a dominant matrix and therefore can be realized with eight nodes. The realization is shown in Fig. 4.2.

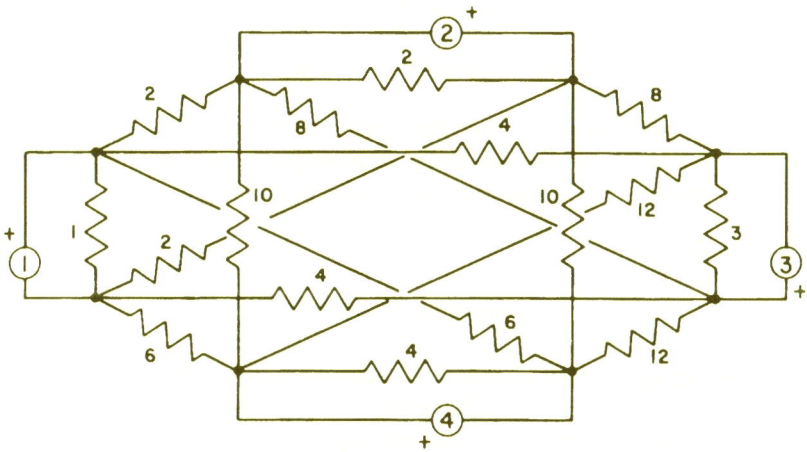

Fig. 4.2. *Realization of the* Y-*matrix of Eq.* (4.2)

The previous theorem proves that the property of dominance is a sufficient condition for the realization of a Y-matrix with $2n$ nodes. This condition is, of course, not necessary because there exist matrices which are not dominant but which nevertheless are realizable with $2n$ nodes. Let us consider the following example.

Example 15: The realization of a nondominant matrix

$$\mathbf{Y} = \begin{bmatrix} \frac{4}{3} & -1 & -1 \\ -1 & \frac{5}{3} & 1 \\ -1 & -1 & 2 \end{bmatrix} \tag{4.3}$$

with $2n = 6$ nodes is shown in Fig. 4.3.

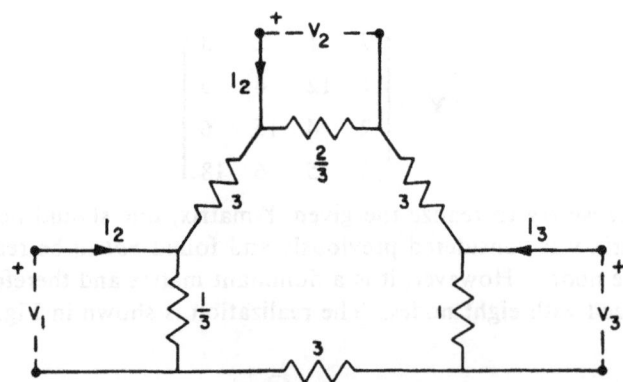

Fig. 4.3. Realization of the Y-matrix of Eq. (4.2)

Let us now explore further the general problem of realization with more than $n+1$ nodes. As a first step we give a number of physical interpretations of certain operations on a Y-matrix:

Eliminating both kth row and column implies that port k is short-circuited;

Applying pivotal condensation with respect to y_{kk} implies that port k is open-circuited;

Changing signs of all elements of kth row and column implies that the polarity of port k is changed;

Interchanging ith and jth rows and columns implies an interchange in the labeling of ports i and k.

If a Y-matrix is realizable with more than $n+1$ nodes, then the port-structure does not form a tree-structure, instead it consists of a number of separate parts, each part forming a sub-stree, The problem of determining which ports belong to each one of the sub-trees has not yet been solved. A necessary condition that q ports belong in a sub-tree T_p is that upon short-circuiting and open-circuiting all other ports, the q-port network is realizable with $q+1$ nodes and with the port-structure T_p. Let us assume that the over-all port-structure, $T_1, T_2, \ldots, T_p$ is known. Even with this information available the problem of finding the graph and element values to realize a Y-matrix is still a formidable one. A possible approach to this problem is the following:

Insert ports $p_1, p_2, \ldots, p_{p-1}$ between some of the sub-trees so as to form a tree T. (See Fig. 4.4.) Now augment matrix $\mathbf{Y}$ by inserting

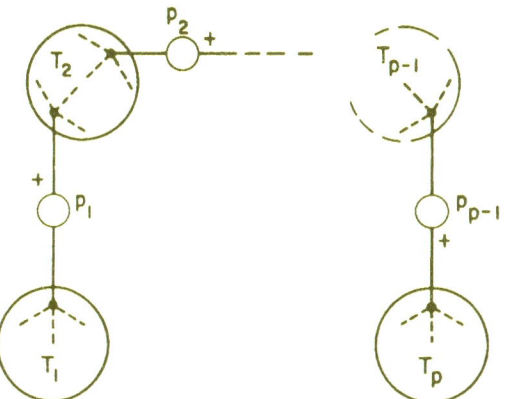

Fig. 4.4. Port-structure of a $(p-1)$-port with more than p nodes

$(p-1)$ rows and columns in the appropriate places resulting in a matrix $\mathbf{Y}$. The matrix $\bar{\mathbf{Y}}$ must satisfy the following properties:

$\bar{\mathbf{Y}}$ is a paramount matrix of order $(n+p-1)$;

$\bar{\mathbf{Y}}$ is realizable with $n+p$ nodes and with a tree-port-structure as shown in Fig. 4.4;

$\bar{\mathbf{Y}}$ yields the matrix $\mathbf{Y}$ upon applying pivotal condensation with respect to elements corresponding to the ports $p_1, p_2, \ldots, p_{p-1}$.

If a matrix $\bar{\mathbf{Y}}$ exists, satisfying the above properties, then this matrix is realized and ports $p_1, p_2, \ldots, p_{p-1}$ are left open-circuited, thus resulting in the realization of matrix $\mathbf{Y}$.

We have discussed a general approach to the problem of realizing a Y-matrix of order n with a resistive n-port with more than $n+1$ nodes. In order to be more specific let us consider again the realization of the matrix of Eq. (4.2).

Let us attempt to realize it with six nodes. An interchange of the second and fourth rows and columns results in the matrix

$$\mathbf{Y}_{1432} = \begin{bmatrix} 7 & 3 & 2 & 1 \\ 3 & 18 & 6 & 5 \\ 2 & 6 & 15 & 4 \\ 1 & 5 & 4 & 2 \end{bmatrix}. \tag{4.4}$$

If port 2 is short-circuited, the submatrix corresponding to ports 1, 4, and 3 is uniformly tapered; the same is true if port 2 is open-circuited. Thus, a possible port-structure for realization of this matrix is shown in Fig. 4.5.

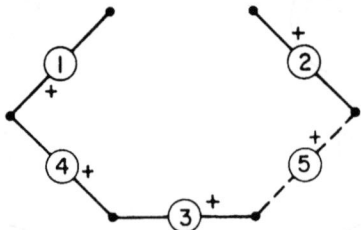

Fig. 4.5. A possible port-structure corresponding to the Y-matrix of Eq. (4.4)

Let us now augment the matrix of Eq. (4.4) by introducing a fifth port connecting ports 3 and 2. The new matrix $\overline{\mathbf{Y}}$ must be uniformly tapered, therefore realizable with ports 1, 4, 3, 5, and 2, in this order. Thus, the matrix $\overline{\mathbf{Y}}$ should have the following form:

$$\overline{\mathbf{Y}} = \begin{bmatrix} x & x & x & \bar{y}_{14} & x \\ x & x & x & \bar{y}_{24} & x \\ x & x & x & \bar{y}_{34} & x \\ \bar{y}_{41} & \bar{y}_{42} & \bar{y}_{43} & \bar{y}_{44} & \bar{y}_{45} \\ x & x & x & \bar{y}_{54} & x \end{bmatrix}, \tag{4.5}$$

where the elements of the fourth column and row can be selected arbitrarily. The remaining elements are obtained from the known elements of $\overline{\mathbf{Y}}$ and the elements of the original matrix $\mathbf{Y}$. For example, element y_{ii} of $\mathbf{Y}$ is given by

$$y_{ii} = \bar{y}_{ii} - (\bar{y}_{i4}\bar{y}_{i4}/\bar{y}_{44}). \tag{4.6}$$

For every choice of elements for the fourth column and row of the augmented matrix $\overline{\mathbf{Y}}$ there corresponds a completely specified matrix $\mathbf{Y}$. Since the number of choices is infinite, so is the number of the augmented matrices which, upon pivotal condensation, result in the matrix $\mathbf{Y}_{1432}$. It is, of course, possible that the set of all augmented matrices contains no matrix which is realizable with the linear-tree port-structure; ports 1, 4, 3, 5, and 2. In such a case the original Y-matrix would not be realizable with the tree port-structure of Fig. 4.4a, thus another port-structure would have to be investigated. This points out the great need for further work in this area. For our example concerning the matrix of Eq. (4.2), an augmented matrix has been proposed by Brown and Tokad[14] which is realizable with the linear-tree, consisting of ports 1, 4, 3, 5, and 2, in that order. For the following choice of arbitrary elements:

$$\bar{y}_{14} = 2$$
$$\bar{y}_{24} = 10$$
$$\bar{y}_{34} = 15 \tag{4.7}$$
$$\bar{y}_{44} = 30$$
$$\bar{y}_{54} = 15$$

and then the augmented matrix is found to be:

$$\overline{\mathbf{Y}} = \begin{bmatrix} \dfrac{214}{30} & \dfrac{11}{3} & 3 & 2 & 2 \\[2mm] \dfrac{11}{3} & \dfrac{64}{3} & 11 & 10 & 10 \\[2mm] 3 & 11 & \dfrac{45}{2} & 15 & \dfrac{23}{2} \\[2mm] 2 & 10 & 15 & 30 & 15 \\[2mm] 2 & 10 & \dfrac{23}{2} & 15 & \dfrac{39}{2} \end{bmatrix}. \tag{4.8}$$

The matrix of Eq. (4.8) is uniformly tapered, and thus is realizable with a linear-tree. If port 5 is left open-circuited, the resulting 4-port represents a realization of the Slepian–Weinberg matrix. This realization is shown in Fig. 4.6.

Example 15 points out that if a Y-matrix of order n is not realizable with $n+1$ nodes, then the search for a realization with more than $n+1$ nodes becomes very tedious if not impossible. Cederbaum has offered a theorem which makes our task easier. For a certain class

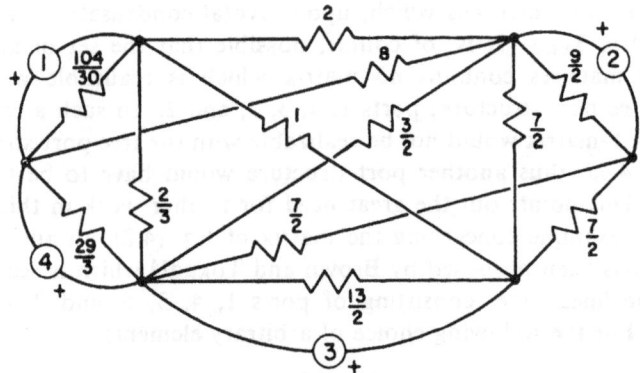

Fig. 4.6. Realization of the Y-matrix of Eq. (4.2) with a four-port of six nodes

of Y-matrices no realization exists with any number of nodes if the matrix is not realizable with $n+1$ nodes. This class of matrices is defined in the following theorem.

Theorem 12. A Y-matrix of order n, whose elements for some index j satisfy the conditions $|y_{kj}| = y_{kk}$ for $k = 1, 2, \ldots, n$, is realizable only if it can be realized with a resistive n-port with $n+1$ nodes.

The proof of this theorem can be found in Reference 6f. An important consequence of Theorem 12 is the conclusion that the paramount character of a Y-matrix is not a sufficient condition for realization as a short-circuit admittance matrix. This is shown in the following example.

Example 16. Consider the matrix proposed by Foster,[15]

$$Y = \begin{bmatrix} 3 & 2 & 1 & 3 \\ 2 & 3 & 2 & 3 \\ 1 & 2 & 3 & 3 \\ 3 & 3 & 3 & 5 \end{bmatrix}. \tag{4.9}$$

This matrix is paramount and satisfies the conditions of Theorem 12. A permutation of the rows and columns 1, 2, 3, 4 into 1, 3, 4, 2 results in

$$
\mathbf{Y}_{1342} =
\begin{bmatrix}
3 & 3 & 2 & 1 \\
3 & 5 & 3 & 3 \\
2 & 3 & 3 & 2 \\
1 & 3 & 2 & 3
\end{bmatrix},
\tag{4.10}
$$

which is not uniformly tapered and hence not realizable with five nodes. We conclude that the paramount matrix of Eq. (4.9) is not at all realizable.

Thus far we have considered resistive *n*-port networks and exposed the difficulties involved for realization with more than $n+1$ nodes. We now extend our study to the *RLC* case. Theorem 11 on dominant matrices can be generalized to apply to *LC* networks.

Theorem 13. Let $\mathbf{Y}(s)$ of order *n* be a matrix of rational functions, and assume that it is realizable as the short-circuit admittance matrix of an *LC n*-port with ideal transformers. If $\mathbf{Y}(s)$ possesses dominant "residue matrices," it is realizable as the short-circuit admittance matrix of an *LC n*-port without transformers.[10]

Proof: The proof provides a synthesis procedure for the realization of $\mathbf{Y}(s)$. Each of the elements of the matrix is expanded in a partial fraction and conjugate terms are combined. Then each of the resulting terms is of the form $k^{(\infty)}s$ or $k(0)/s$ or $2k^{(i)}s/(s^2+\omega_i^2)$, where the *k*'s are residues and $\omega_i^2 > 0$ for $s = j\omega$. Next we write $\mathbf{Y}(s)$ as a sum of matrices and that each matrix corresponds to a particular pole. The general expansion is of the form:

$$
\mathbf{Y}(s) = s
\begin{bmatrix}
k_{11}^{(\infty)} k_{12}^{(\infty)} \dots & k_{1n}^{(\infty)} \\
k_{21}^{(\infty)} k_{22}^{(\infty)} \dots & k_{2n}^{(\infty)} \\
\cdot \quad \cdot \quad \cdot \quad \cdot \quad \cdot \quad \cdot \quad \cdot \\
k_{n1}^{(\infty)} k_{n2}^{(\infty)} \dots & k_{nn}^{(\infty)}
\end{bmatrix}
$$

$$
+ \frac{1}{s}
\begin{bmatrix}
k_{11}^{(0)} k_{12}^{(0)} \dots & k_{1n}^{(0)} \\
k_{21}^{(0)} k_{22}^{(0)} \dots & k_{2n}^{(0)} \\
\cdot \quad \cdot \quad \cdot \quad \cdot \quad \cdot \quad \cdot \quad \cdot \\
k_{n2}^{(0)} k_{n2}^{(0)} \dots k_{nn}^{(0)}
\end{bmatrix}
$$

$$+ \frac{2s}{s^2 + \omega_1^2} \begin{bmatrix} k_{11}^{(1)}k_{12}^{(1)} & & k_{1n}^{(1)} \\ k_{21}^{(1)}k_{22}^{(1)} \ldots & k_{2n}^{(1)} \\ \cdot & \cdot & \cdot & \cdot & \cdot & \cdot & \cdot \\ k_{n1}^{(1)}k_{n2}^{(1)} \cdot \cdot & k_{nn}^{(i)} \end{bmatrix}$$

$$+ \ldots + \frac{2s}{s^2 + \omega_i^2} \begin{bmatrix} k_{11}^{(i)}k_{12}^{(i)} \ldots & k_{1n}^{(i)} \\ k_{21}^{(i)}k_{22}^{(i)} \ldots & k_{2n}^{(i)} \\ \cdot & \cdot & \cdot & \cdot & \cdot & \cdot & \cdot \\ k_{n1}^{(i)}k_{n2}^{(i)} \ldots & k_{nn}^{(i)} \end{bmatrix}, \tag{4.11}$$

where each matrix in the right-hand expression is called a "residue matrix." Each dominant residue matrix in Eq. (4.11) is realized separately in a manner similar to the realization of a dominant real matrix. For the matrix with the factor $1/s$, each edge is an inductance; for the matrix with the factor s, each edge is a capacitance; and, for the matrices with the factor $2s/(s^2 + \omega_i^2)$, each edge is a series-connection of a capacitance and an inductance. At each port the edges corresponding to the component matrices are connected in parallel.

Finally, before concluding this section, it is worthwhile to mention that the results of the previous chapter on the synthesis of an n-port with $n+1$ nodes can be extended to the realization of a Y-matrix with more than $n+1$ nodes.

Let us consider a graph with $n+1$ nodes and b edges corresponding to an n-port. Now it will be assumed that each edge of the graph is a composite edge, consisting of more than one ordinary edge. Therefore, the weight associated with a composite edge is in general the driving-point admittance function in the form of a rational function of s, i.e., a positive-real function, containing more than a single term. In such a case, we still may be able to decompose the short-circuit admittance matrix of the network, Y, into a form of

$$Y = C_s Y_e C_s{}^t, \tag{4.12}$$

where Y_e is a diagonal matrix with positive-real functions as main diagonal elements.

It is then clear that the decomposition technique of a Y-matrix, which has been investigated thoroughly in the previous chapter, may still be useful for this case. Once C_s and Y_e are found, we may realize C_s with a connected graph (a composite graph) if it is realizable. The graph realized will have a positive-real function specified by each diagonal element of $\mathbf{Y}_e$ as the weight of each edge. We therefore now realize the positive-real function assigned to each edge by known one-port network synthesis methods, such as the Brune and Bott–Duffin methods. This completes the realization of an n-port with more than $n+1$ nodes. However, in this generalized approach, there still remain many unsolved problems and the reader should refer to Reference 6g.

REALIZATION OF A Z-MATRIX

For the realization of a Z-matrix we will face problems similar to those we have encountered in realizing a Y-matrix. Suppose that a matrix $\mathbf{Z}(s)$ of order n is decomposed into the form

$$\mathbf{Z}(s) = \mathbf{B}_s\mathbf{Z}_e(s)\mathbf{B}_s^t, \qquad (4.13)$$

where $\mathbf{B}_s$ is a real matrix of order n by b and of rank n, and each element is 1, -1, or zero; $\mathbf{Z}_e(s)$ is a diagonal matrix of order b in which each diagonal element is a positive-real function of s.

Then, in order to realize $\mathbf{Z}(s)$ as an open-circuit impedance matrix of an ordinary *RLC* n-port network, $\mathbf{B}_s$ must be realizable as a special loop matrix or a modified basic loop matrix or a basic loop matrix of a connected graph. We shall assume without loss in generality that $\mathbf{B}_s$ is a special loop matrix, i.e. a submatrix of a modified basic loop matrix, so that the subsequent realization technique becomes quite general. We therefore let $\mathbf{B}_s$ be a submatrix of a modified basic loop matrix $\mathbf{B}_{fm}$. It is then obvious that $\mathbf{B}_{fm}$ is always decomposed into two submatrices with respect to a tree (a modified tree) as:

$$\mathbf{B}_{fm} = [\mathbf{B}_{fm_{11}} \quad \mathbf{U}_n], \qquad (4.14)$$

where $\mathbf{U}_n$ is a unity matrix of order n in which each column corresponds to a chord of the modified tree, and $\mathbf{B}_{fm}$ is of order n by $b+d$ for $d = 0, 1, 2, \ldots$.

For the same tree on which $\mathbf{B}_{fm}$ was formed, the modified basic cut-set matrix, $\mathbf{C}_{fm}$ is then found to be (see Part I)

$$\mathbf{C}_{fm} = [\mathbf{U}_k \quad -\mathbf{B}_{fm_{11}}^t], \qquad (4.15a)$$

where

$$\mathbf{C}_{fm} = [\mathbf{U}_k \quad \mathbf{C}_{fm_{11}}] \qquad (4.15b)$$

and $k = b+d-n$.

Thus the problem of the realization of a loop matrix is reduced to realizing a cut-set matrix. One must, however, note that matrix $\mathbf{B}_s$ should be augmented with columns, each of which contains only one 1 such that the matrix is converted into a form of Eq. (4.14). We therefore now try to realize C_{fm}. The realization of a cut-set matrix is again always reduced to the realization of a *Y*-matrix. That is,

$$\mathbf{Y} = \mathbf{C}_{fm}\mathbf{U}_k\mathbf{C}_{fm}{}^t = \mathbf{C}_{fm}\mathbf{C}_{fm}{}^t. \tag{4.16}$$

The decomposition of a *Z*-matrix into the form of Eq. (4.13) can be accomplished by letting s take a positive and real value, $s = k > 0$, and then applying Cederbaum's algorithm. That is,

$$\mathbf{Z}(k) = \mathbf{B}_s\mathbf{Z}_e(k)\mathbf{B}_s{}^t. \tag{4.17}$$

Let us illustrate the procedure by considering the following example.

Example 17. Realize the Foster matrix as an open-circuit impedance matrix,

$$\mathbf{Z} = \begin{bmatrix} 3 & 3 & 2 & 1 \\ 3 & 5 & 3 & 3 \\ 2 & 3 & 3 & 2 \\ 1 & 3 & 2 & 3 \end{bmatrix}. \tag{4.18}$$

We then decompose the matrix into the form

$$\mathbf{Z} = \begin{bmatrix} 1 & 1 & 1 & 0 & 0 \\ 1 & 1 & 1 & 1 & 1 \\ 1 & 0 & 0 & 1 & 1 \\ 1 & 1 & 0 & 1 & 0 \end{bmatrix} \begin{bmatrix} 1 & 0 & 0 & 0 & 0 \\ 0 & 1 & 0 & 0 & 0 \\ 0 & 0 & 1 & 0 & 0 \\ 0 & 0 & 0 & 1 & 0 \\ 0 & 0 & 0 & 0 & 1 \end{bmatrix} \begin{bmatrix} 1 & 1 & 1 & 1 \\ 1 & 1 & 0 & 1 \\ 1 & 1 & 0 & 0 \\ 0 & 1 & 1 & 1 \\ 0 & 1 & 1 & 0 \end{bmatrix}$$

$$= \mathbf{B}_s\mathbf{Z}_e\mathbf{B}_s{}^t. \tag{4.19}$$

Let us augment the matrix of Eq. (4.19) by inserting the missing columns until the matrix is converted into a modified basic loop matrix $\mathbf{B}_{fm}$ as:

$$\begin{matrix} \quad\; 1 \quad 2 \quad 3 \quad 4 \quad 5 \quad 6 \quad 7 \quad 8 \quad 9 \\ \mathbf{B}_{fm} = \begin{bmatrix} 1 & 1 & 1 & 0 & 0 & 1 & 0 & 0 & 0 \\ 1 & 1 & 1 & 1 & 1 & 0 & 1 & 0 & 0 \\ 1 & 1 & 0 & 1 & 0 & 0 & 0 & 1 & 0 \\ 1 & 0 & 0 & 1 & 1 & 0 & 0 & 0 & 1 \end{bmatrix}. \end{matrix} \tag{4.20}$$

And the corresponding modified basic cut-set matrix is

$$
C_{fm} =
\begin{bmatrix}
1 & 0 & 0 & 0 & 0 & -1 & -1 & -1 & -1 \\
0 & 1 & 0 & 0 & 0 & -1 & -1 & -1 & 0 \\
0 & 0 & 1 & 0 & 0 & -1 & -1 & 0 & 0 \\
0 & 0 & 0 & 1 & 0 & 0 & -1 & -1 & -1 \\
0 & 0 & 0 & 0 & 1 & 0 & -1 & 0 & -1
\end{bmatrix}.
\qquad (4.21)
$$

The short-circuit admittance matrix $Y = C_f C_f{}^t$ (here we assume all elements having a resistance of 1 ohm, elements 6, 7, 8, 9 are of course fictitious) is

$$
Y = C_{fm} C_{fm}{}^t =
\begin{bmatrix}
5 & 3 & 2 & 3 & 2 \\
3 & 4 & 2 & 2 & 1 \\
2 & 2 & 3 & 1 & 1 \\
3 & 2 & 1 & 4 & 2 \\
2 & 1 & 1 & 2 & 3
\end{bmatrix}.
\qquad (4.22)
$$

A permutation of the rows and columns 1, 2, 3, 4, 5, into 3, 2, 1, 4, 5 results in the matrix

$$
Y_{32145} =
\begin{bmatrix}
3 & 2 & 2 & 1 & 1 \\
2 & 4 & 3 & 2 & 1 \\
2 & 3 & 5 & 3 & 2 \\
1 & 2 & 3 & 4 & 2 \\
1 & 1 & 2 & 2 & 3
\end{bmatrix}.
\qquad (4.23)
$$

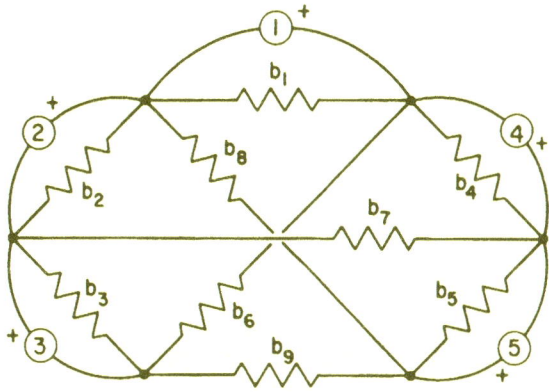

Fig. 4.7. Realization of the Y-matrix of Eq. (4.24) where all resistors are one ohm

This matrix is uniformly tapered and the realization is shown in Fig. 4.7. Ports 1, 2, 3, and 4 of the impedance matrix correspond to the impedanceless chords 6, 7, 8, and 9. If the five ports of Fig. 4.7 are open-circuited and ports 1, 2, 3, and 4 are inserted in series with the impedanceless chords 6, 7, 8, and 9 then the four-port network realizing the Foster matrix is obtained as shown in Fig. 4.8.

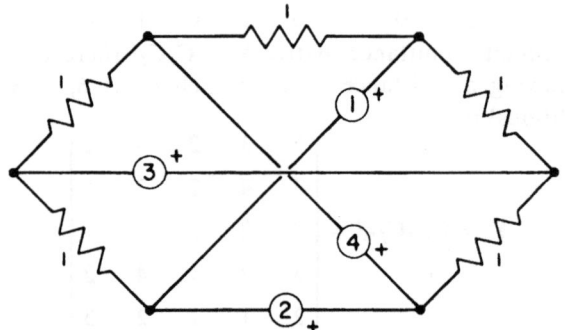

Fig. 4.8.   Final realization of the Z-matrix of Eq. (4.18)

Note here that the matrix considered in this example was also examined in Example 16. It was shown there that this matrix is not realizable as a short-circuit admittance matrix with any number of nodes. We thus conclude that the conditions on a matrix for realization as a short-circuit admittance matrix or as an open-circuit impedance matrix are not the same in each case.

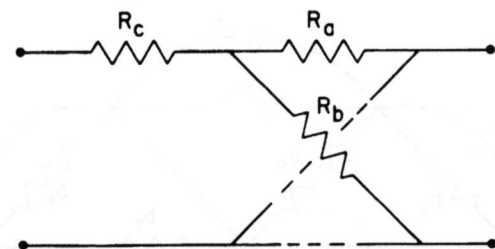

Fig. 4.9.   Realization of a 2 by 2 paramount matrix

The problem of realizing a Z-matrix by a network with more than n independent loops is similar to the problem of realizing a Y-matrix with more than $n+1$ nodes. The process of augmenting a Z-matrix can be used, and the procedure is similar to the one discussed in the previous section.

In the previous section, the dominance condition was given as sufficient for the realization of a short-circuit admittance matrix; in the case of an open-circuit impedance matrix no procedure exists for realizing a dominant matrix. Tellegen has shown that all 2×2 and 3×3 paramount matrices are realizable as open-circuit impedance matrices. The realization is shown in Figs. 4.9 and 4.10.

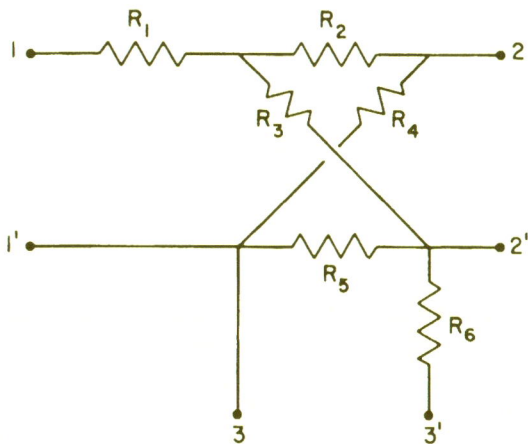

Fig. 4.10. *General three-port corresponding to any third-order para-mount matrix*

The element values of Fig. 4.10 can be obtained by solving for the resistances in terms of the matrix elements, and the voltage polarities and numbering of the ports can be selected so that each one of the resistances is non-negative. Thus we shall let $R_{11} \geqslant R_{22}$ with no loss in generality, then

$$R_c = R_{11} - R_{22}$$
$$R_a = R_{22} - R_{21} \qquad (4.24)$$
$$R_b = R_{22} + R_{21}.$$

Finally, before concluding this section, we give a theorem which is very useful in the investigation of the realization of a special class of Z-matrices.

Theorem 14. A Z-matrix of order n, whose elements for some value of k, which satisfying the condition $|z_{kj}| = z_{jj}, j = 1, 2, \ldots, n$, is realizable only if it can be realized as the impedance matrix corresponding to a system of basic loops of some resistive network.[6f]

Since the proof of the theorem is given in Reference 6f. the following example is given as an illustration of the theorem.

Example 18. Let us consider the matrix

$$
\mathbf{Z} = \begin{bmatrix} 3 & 1 & 2 & 3 \\ 1 & 5 & 4 & 5 \\ 2 & 4 & 6 & 6 \\ 3 & 5 & 6 & 57/7 \end{bmatrix}. \tag{4.25}
$$

This is a paramount matrix satisfying Theorem 14. Since $z_{k4} = z_k$ for $k = 1, 2, 3, 4$. If we try to decompose $\mathbf{Z}$ into the unimodular congruence using Cederbaum's algorithm, we find after five steps that the procedure diverges. The failure of the decomposition procedure shows that this matrix satisfying Theorems 12 and 14 can be neither an admittance nor an impedance matrix of a resistive 4-port.

Thus far we have investigated the various known realization methods for a short-circuit admittance matrix as well as an open-circuit impedance matrix. The synthesis of a multiport is actually accomplished by the realization of a basic cut-set matrix or a basic loop matrix, i.e., unimodular matrices, in terms of a connected graph. There exists, however, an interesting and important contribution by So[16] to the realization of a loop impedance matrix of a network. We shall briefly review the work of So, since it is directly related to the subject of our discussion in this part.

Let us consider a real matrix $\mathbf{M} = [m_{ij}]$ of order n by n.

Definition 5: $\mathbf{M}$ is "definite dominant" if

$$
m_{ii} > \sum_{j=1}^{n} |m_{ij}|, \quad i \neq j.
$$

Definition 6: A real matrix $\mathbf{M}$ of order $n(n \geqslant 4)$ is "definite slant-dominant" if the rows of $\mathbf{M}$ may be arranged in such an order that (a) the first four rows are definite dominant; (b) any row r_i, $4 < i \leqslant n$ is definite $(n - i + 3)$ dominant. That is, m_{ii} is greater than the sum of the absolute values of any $(n - i + 3)$ of the $(n - 1)$ elements of the r_ith row.

Definition 7: A realization of $\mathbf{M}$ is coherent if there exist no two loops l_i and l_j such that l_i and l_j agree in orientation in one resistor R_a but oppose in orientations in another resistor R_b.

So has obtained a synthesis procedure which, when applied to a definite slant-dominant matrix, either leads to a coherent realization or furnishes proof that no such realization exists. If $\mathbf{M}$ is not definite slant-dominant, application of the procedure may or may not lead to a network, even if **M** is coherently realizable.

In order to use So's results for the realization of a loop matrix **B** of order $n \times b$ and rank n with elements ± 1 and 0's, where **B** is not necessarily unimodular, we must limit ourselves to the sub-class of **B** matrices which could be *coherently realizable*. That is, if $\mathbf{B} = [b_{ij}]$, then

$$b_{ik}b_{jk} = +1, 0 \qquad (4.26a)$$

or

$$b_{ik}b_{jk} = -1, 0, \qquad (4.26b)$$

for all i, j, k and $i \neq j$.

If **B** satisfies Eq. (4.26), then we may form

$$\mathbf{M} = \mathbf{BB}^t. \qquad (4.27)$$

If in Eq. (4.27) **M** is definite slant-dominant, then So's procedure can be used. Since this procedure is complicated and requires the construction of a number of tables, interested readers should refer to Reference 16.

REFERENCES

1. Y. Oono, "Synthesis of a finite $2n$-terminal network as the extension of Brane's two-terminal network theory," *J. Elec. Commun. Eng., Japan*, (1948).

2. B. MacMillan, "Introduction to formal realizability theory," *Bell System Tech. J.*, 31 (1952), 217–79, 541–600.

3. I. W. Sandberg, "Synthesis of active N-port networks," *Bell System Tech. J.*, 40 (1961), 761–83.

4. H. J. Carlin and D. C. Youla, "Network synthesis with negative resistors," *Proc. IRE*, 49 (1961), 907–20.

5a. E. A. Guillemin, *Introductory Circuit Theory*. New York, John Wiley & Sons, 1953.

5b. E. A. Guillemin, "On the analysis and synthesis of single element kind networks," *IRE Trans.* CT-7 (1960), 303–12.

6a. I. Cederbaum, "On networks without ideal transformers," *IRE Trans.* CT-3 (1956), 179–82.

6b. I. Cederbaum, "Matrices all of whose elements and subdeterminants are 1, −1, or 0," *J. Math. Phys.*, 36 (1958), 351–61.

6c. I. Cederbaum, "Conditions for the impedance and admittance matrices of N-ports without ideal transformers," *IEE (London)*, Monograph 276R, January, 1958.

6d. I. Cederbaum, "Applications of matrix algebra to network theory," *IRE Trans.* CT-6, Special Supplement (1959), 127–37.

6e. I. Cederbaum, "On matrices with some form of dominance of the main diagonal," *The Matrix and Tensor Quart.*, 9 (1959), 29–39.

6f. I. Cederbaum, "Topological considerations in the realization of resistive N-port networks," *IRE Trans.* CT-9 (1961), 324–29.

7. M. B. Reed, "The seg: a new class of subgraphs," *IRE Trans.* CT-8 (1961), 17–22.

8a. G. Biorci and P. P. Civalleri, "Alcune considerazioni sulla sintesi dei multipoli resistivi," *Atti accad. sci. Torino*, 94 (1959–1960).

8b. G. Biorci and P. P. Civalleri, "On the conductance matrices with all all-positive elements," *IRE Trans.* CT-8 (1961), 76–77.

8c. G. Biorci and P. P. Civalleri, "On the synthesis of resistive N-port-networks," *IRE Trans.* CT-8 (1961), 22–28.

8d. G. Biorci, "Sign matrices and the realizability of conductance matrices," *IEE (London)*, Monograph No. 424E, December, 1960.

9. B. D. H. Tellegen, "Theorie der Wisselstromen," Deel IV, Theorie der Electrische Netwerken, P. Noordhoff N.V., Gronigen, Djakarta, 1952, 166–68.

10. P. Slepian and L. Weinberg, "Synthesis applications of paramount and dominant matrices," *Proc. Natl. Electronics Conf.*, 14 (1958), 611–30.

11a. P. R. Bryant, Ph.D. thesis, Cambridge University, Cambridge, England, 1959.

11b. P. R. Bryant, "Discussion on conditions for the impedance and admittance matrices of N-ports without ideal transformers," Proc. IEE (London), Part C, 106 (1959), 116.

12. P. P. Civalleri, "A direct procedure for the synthesis of resistive $(n+1)$-poles," *IEE (London)*, Monograph No. 464E, August, 1961.

13a. C. C. Halkias, I. Cederbaum, and W. H. Kim, "Synthesis of resistive N-ports with $(n+1)$ nodes," *IRE Trans.* CT-9 (1962), 69–73.

13b. C. C. Halkias, "Synthesis of N-port networks," Ph.D. thesis, Columbia University, New York, 1962.

14. D. P. Brown and Y. Tokad, "On the synthesis of R networks," *IRE Trans.* CT-8 (1961), 31–39.

15. R. M. Foster, see L. Weinberg, "Circuit theory progress report," *J. Research Natl. Bur. Standards, Part D, Radio Propagation*, 64D (1960). 687–706.

16. Hing Cheong So, "Realization of loop-resistance matrices," Ph.D, thesis, University of Illinois, Urbana, Illinois, 1960; Int. Tech. Rept. No. 17, contract No. DA-11-022-ORD-1983, University of Illinois, 1960.

Synthesis of Single-Contact Switching

Networks and Realization of

Loop and Cut-Set Matrices

PART IV

Synthesis of Single-Contact Switching

Networks and Realization of

Loop and Cut-Set Matrices

Introduction

FROM OUR PREVIOUS STUDY of the synthesis of an ordinary *RLC* multiport the reader must have noted that the realization of a short-circuit admittance matrix or open-circuit impedance matrix is accomplished by synthesizing a connected graph based on a cut-set or loop matrix. The network synthesis problems, whether they are *RLC* networks, contact networks, or flow-nets, are actually reduced to the realization of a graph. In particular, the synthesis of a class of unate switching network (the so-called "single contact network") in minimal form is directly related to the realization of a non-oriented connected graph. We shall therefore discuss the synthesis techniques of a single-contact network in terms of the realizability of a real matrix of integer mod 2 as a loop or cut-set matrix of a connected graph.

Various authors have proposed a number of algorithms for the realization of a loop or cut-set matrix, both with and without relation to the applications of the techniques to network synthesis problems. The method presented here is based on development of the synthesis techniques of a resistive multiport network described previously. This method will be briefly compared to some other known techniques.

We shall first review some fundamental concepts of the theory of switching networks. Then, we shall develop a simple synthesis method for a single-contact switching network in terms of the realizability of a path, loop, or cut-set matrix. It may therefore be necessary for the reader to have a clear understanding of some concepts of the theory of linear graphs defined in Part I and the realization techniques of a multiport network discussed in Part III.

CHAPTER 1

Fundamentals of the Theory

of Contact Networks

THE SIMPLEST and the most basic types of switching networks are the so-called "contact networks." Each element of a contact network is a contact switch which has two states—open and closed. If a contact is in the open state there exists no conduction of current through the element, while if it is closed there will be a conductive path in either direction through the switch. It is further assumed that the terminals of a contact can be reversed without affecting the operation of a contact network. It should be understood that although each state of the contact is determined by some means, whether electrical, magnetical, or mechanical, the means will be considered external to the network. In other words, the conduction of currents through the various paths of a contact network depends only on the combination of the state of each contact of the network. Such a network is called a "combinational switching network."

We assign to a contact a binary variable x taking the value 0 when the contact is open and the value 1 when the contact is closed. With each variable x, we associate a symbol x' to denote the corresponding contact which is open while x is closed, and closed when x is open. This symbol x' is called the "complement" of variable x. We also denote the output of a contact network, or the output of the current detector,† by F which is a binary function of the binary variables associated with the contacts of the network. A function F is usually called a "switching function." A switching function is normally expressed in Boolean algebra which has three fundamental operations defined by:

---

† It is assumed that a current detector is connected between the input and the output of a network in order to measure the conduction of currents through the network.

sum:
$$x_i + x_j = 0, \qquad \text{if } x_i = x_j = 0$$
$$x_i + x_j = 1, \qquad \text{otherwise;} \tag{1.1a}$$

product:
$$x_i x_j = 1, \qquad \text{if } x_i = x_j = 1$$
$$x_i x_j = 0, \qquad \text{otherwise;} \tag{1.1b}$$

complementation:
$$(x)' = 0, \qquad \text{if } x = 1$$
$$(x)' = 1, \qquad \text{if } x = 0. \tag{1.1c}$$

The contact network diagrams corresponding to the addition and multiplication of Boolean operations are shown in Fig. 1.1. The

(a)   (b)

Fig. 1.1. Parallel connection for $x_i + x_j$ (a) and series connection for $x_i x_j$ (b)

parallel connection of x and its complement x' is equivalent to a direct connection, i.e., a shorted wire, since $x + x' = 1$. The series connection of x and x' is equivalent to an open-wire since their

(a)   (b)

Fig. 1.2. Physical implications for the sum (a) and the product (b) of a binary variable and its complement

product is always zero. They are illustrated in Fig. 1.2, where the symbol $\sim$ is used to indicate an "equivalence class."

If more than one path exists between the input and output terminals of a contact network, the output of the network may be expressed by the sum of the currents flowing through the various paths. Since each path for current flow should result only when every contact comprising the path is in the closed state, we therefore set the switching function describing the output equal to the sum of products of the variables associated with each path. When a switching functiion is expressed by the sum of products of variables, it is said to be in "normal form."

As an illustration, let us consider a graphical representation of a

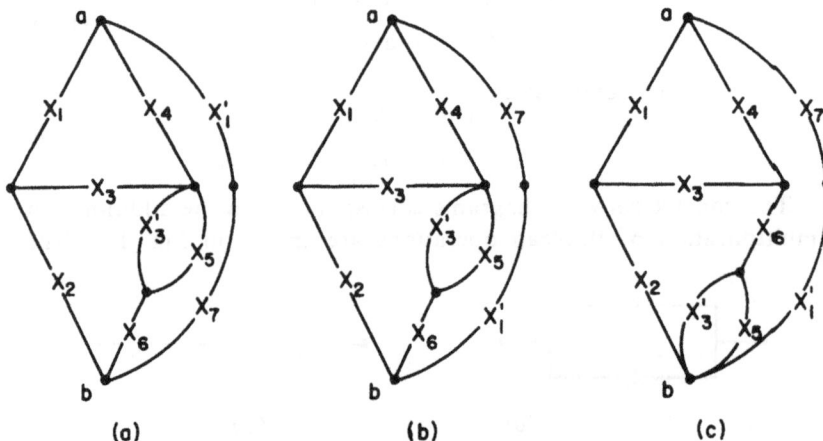

Fig. 1.3. Graphically 2-isomorphic and electrically equivalent contact networks

contact network shown in Fig. 1.3a, in which each edge represents a contact with a binary variable assigned.

Then the switching function F which describes all the conducting paths between terminals a and b of the network is:

$$F = x_1x_2 + x_1x_3x'_3x_6 + x_1x_3x_5x_6 + x'_3x_4x_6 + x_4x_5x_6 + x'_1x_7 + x_2x_2x_4$$

$$= x_1x_2 + x_1x_3x_5x_6 + x'_3x_4x_6 + x_4x_5x_6 + x'_1x_7 + x_2x_2x_4, \qquad (1.2)$$

where $x_1x_3x'_3x_6 = 0$, since $x_ix'_i = 0$ for all i.

In Fig. 1.3a, if we operate on edges x_7 and x'_1 such that the resulting graph becomes 2-isomorphic to the original one as shown in Fig. 1.3b, then the new network will have the same switching function with respect to terminals a and b. The two networks are therefore "electrically equivalent" or simply "equivalent" with each other. In Fig. 1.3b, we again reverse the connection of the subnetwork consisting of edges x'_3, x_5, and x_6 and obtain another equivalent network as shown in Fig. 1.3c. Thus the networks shown in Fig. 1.3 are equivalent contact networks and are topologically 2-isomorphic with each other. In order to generate a class of 2-isomorphic graphs with respect to a given graph, the graph must be isolated into two subgraphs by cutting it at two nodes. In other words, the graph should contain a subgraph(s) which has only two common nodes with its complement of the graph. For this reason, Gould (Reference 15 of Part I) named such a type of subgraph a "two-terminal subgraph" and it plays an important role in his realization techniques."

When a binary variable x takes the value of one, its complement x' takes zero. The complement of a function of binary variables, F, is defined in the same way, such that if $F = 1$, then $F' = 0$ or vice versa. The complement function F' of a switching function F is obtained by the so-called "dual operations":

$$(x_i x_j)' = x'_i + x'_j \tag{1.3a}$$

$$(x_i + x_j)' = x'_i x'_j \tag{1.3b}$$

$$(x')' = x. \tag{1.3c}$$

As an illustration, let us consider the switching function of the network of Fig. 1.3a, i.e., Eq. (1.2). Then the complement of F is readily obtained as:

$$F' = (x' + x'_2)(x_1 + x'_7)(x'_1 + x'_3 + x'_5 + x'_6)(x_3 + x'_4 + x'_6)(x'_4 x'_5 + x'_6)$$
$$\times (x'_2 + x'_3 + x'_4). \tag{1.4}$$

One more fundamental concept in the theory of switching network is the so-called "canonical form" of a switching function. The canonical form of a switching function F which is in a normal form can be obtained by the following procedure: (a) multiply each product in F which contains neither x_i nor x'_i as a factor by the sum $(x_i + x'_i)$ for every product term and for every variable in F; (b) remove every redundant term. These steps are obviously valid because $x_i + x_i = 1$ for all i in F. It should also be clear that every product term resulting in the canonical form of F contains either x_i or x'_i for all i, and that no product appears twice.

Given a switching function F in a normal form as:

$$F = x_1 x_2 + x_2 x'_3 + x'_1 x_2 x_4. \tag{1.5}$$

Then the canonical form of F is found to be

$$F = (x_3 + x'_3)(x_4 + x'_4)x_1 x_2 + (x_1 + x'_1)(x_4 + x'_4)x_2 x'_3 + (x_3 + x'_3)x'_1 x_2 x_4$$
$$= x_1 x_2 x_3 x_4 + x'_1 x_2 x_3 x_4 + x_1 x_2 x'_3 x_4 + x_1 x_2 x'_3 x'_4 +$$
$$+ x'_1 x_2 x'_3 x_4 + x'_1 x_2 x'_3 x'_4. \tag{1.6}$$

Now let us consider the converse process, that is, a switching function F is given in the conical form, and we try to simplify the expression in order to reduce it to a normal form in which no product term is included in any other product terms of the function with fewer literals. Such a product term in a normal form is called a "prime implicant" by Quine.[1] Moreover, if some canonical term is included in only one prime implicant, then that prime implicant is called "essential." The prime implicant of a function can be obtained by algebraic manipulation or by the so-called "Karnaugh maps" or by a chart method.[2-6]

For instance, let us assume that we are given a switching function F of Eq. (1.6) and wish to obtain the simplified form of Eq. (1.5) from it. Then we can factor F to give

$$F = x_1 x_2 (x_3 x_4 + x'_3 x_4 + x_3 x'_4 + x'_3 x'_4) + x_2 x'_3 (x_1 x_4 + x'_1 x_4 +$$
$$+ x_1 x'_4 + x'_1 x'_4) + x'_1 x_2 x_4 (x_3 + x'_3). \tag{1.7}$$

The terms in parentheses contain all the possible combinations of the variables, and they are therefore each equal to 1. Note in Eq. (1.7) that in order to make the factorization more easy to carry out, terms such as $x_1 x_2 x'_3 x_4$, $x_1 x_2 x'_3 x'_4$, and $x'_1 x_2 x'_3 x_4$ are introduced twice. These extra terms will not affect the given switching function because the Boolean addition of two identical binary functions (or variables) is equal to the function (or the variable) itself. Equation (1.7) is reduced to

$$F = x_1 x_2 + x_2 x'_3 + x'_1 x_2 x_4. \tag{1.8}$$

For more detailed discussions of the reduction of a switching function, see References 1–6.

Since the aim of our study of contact networks is to develop a topological realization method of switching functions, the function which will be considered is a "proper switching function," i.e., a switching function in which none of the variables occurs vacuously. As an illustration, let us consider the function given in Eq. (1.9a):

$$F = x'_1 x'_2 x'_3 + x'_1 x'_2 x_3 + x_1 x'_2 x'_3 + x'_1 x_2 x_3 + x_1 x'_2 x_3 + x_1 x_2 x_3. \tag{1.9a}$$

If we factor on variable x_1, we obtain

$$F = x'_1 (x'_2 x'_3 + x'_2 x_3 + x_2 x_3) + x_1 (x'_2 x'_3 + x'_2 x_3 + x_2 x_3) \tag{1.9b}$$

and it is found that the coefficient of x'_1 is identical to that of x_1. The variable x_1 is therefore a "vacuous variable" and F is *not* a proper function.

Let us again rearrange F of Eq. (1.9a) to be factored on variable x_2. Thus we have

$$F = x'_2 (x'_1 x'_3 + x'_1 x_3 + x_1 x'_3 + x_1 x_3) + x_2 (x'_1 x_3 + x_1 x_3). \tag{1.9c}$$

Then the coefficient of x'_2 includes that of x_2, and Eq. (1.9c) is therefore reduced to

$$F = x'_2(x'_1x'_3 + x_1x'_3) + x_1x'_3 + x_1x_3; \qquad (1.9d)$$

it is clear that variable x_2 is not included in F. Then F is said to be "isotone in x'_2" and "monotone in x."

SINGLE-CONTACT NETWORKS

We shall restrict our discussion to the analysis and synthesis of single-contact networks. Furthermore, we assume that the state of each contact is independent of the states of all the other contacts in the network. Thus we guarantee that the switching function of a single-contact network, called a "single-contact function," or a "network function," is a proper function in which none of the variables is *vacuous*, and in which no variable appears both negated and unnegated, i.e., the variable and its complement.† We can therefore evaluate the switching function of a two-terminal single-contact network as the *Boolean sum* of all the "path-products" between the terminal nodes. (A path product is the product of the variables associated with all the edges included in a path.) In other words, a two-terminal single-contact network and its corresponding graph have the same function. For this reason, the switching function of a single-contact network is sometimes referred to as a "graph

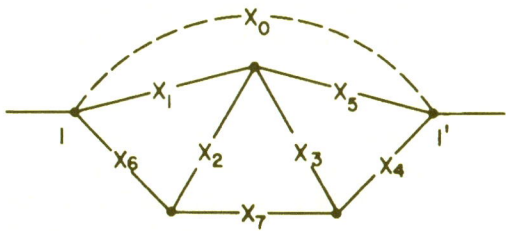

Fig. 1.4. A single-contact network

function." Moreover, a single-contact function is a completely monotone function of all the variables associated with the contacts of the network.

† It should, therefore, be clear that a single-contact network is a special case of a so-called "unate switching network" in which no same contact (or the variable associated with a contact) appears more than once. From now on, although no negated variable will be used, the following discussions are valid for a switching function with some negated variables as long as unnegated forms of the same variables are not contained in the function or the network.

Let us consider a single-contact network of Fig. 1.4. Then the path matrix of the network, $\mathbf{P}$ for all possible paths between terminals 1 and 1′ is given by

$$
\mathbf{P} = \begin{array}{c} \begin{array}{ccccccc} x_1 & x_2 & x_3 & x_4 & x_5 & x_6 & x_7 \end{array} \\ \begin{bmatrix} 1 & 0 & 0 & 0 & 1 & 0 & 0 \\ 0 & 1 & 0 & 0 & 1 & 1 & 0 \\ 0 & 0 & 1 & 0 & 1 & 1 & 1 \\ 0 & 0 & 0 & 1 & 0 & 1 & 1 \\ 1 & 0 & 1 & 1 & 0 & 0 & 0 \\ 1 & 1 & 0 & 1 & 0 & 0 & 1 \\ 0 & 1 & 1 & 1 & 0 & 1 & 0 \end{bmatrix} \end{array} \qquad (1.10)
$$

and the switching function of the network with respect to terminals 1 and 1′, F, is then found as the Boolean sum of the path-products of the variables corresponding to the rows of the path matrix of Eq. (1.10). We therefore have

$$F = x_1x_5 + x_2x_5x_6 + x_3x_5x_6x_7 + x_4x_6x_7 + x_1x_3x_4 + x_1x_2x_4x_7 + x_2x_3x_4x_6.$$

$$(1.11)$$

We should notice here that every product term which appears in Eq. (1.11) is an *essential implicant* or simply *essential* for the function F, and thus the switching function of Eq. (1.11) is in the simplest normal form. In other words, there is no way of further reducing the function of Eq. (1.11) by the Boolean operations.

However, if we consider the nonoriented graph of the network of Fig. 1.4, then there should exist only four independent paths, since the rank of the path matrix of the graph is four (mod 2). That is, all the paths represented by the product terms of the function of Eq. (1.11), or by the rows of the path matrix of Eq. (1.10), are not independent of each other. For instance, modulo 2 sum of the last four rows yields a row of all zeros. It is thus clear that for the topological realization of a single-contact network we should consider only the set of independent paths between a pair of terminal nodes of the network. We shall therefore concentrate our study on the realization of a graph for a given path matrix.

Now let us introduce an extra element with variable x_0 between the terminals of the network as indicated in Fig. 1.4 by a dotted line. This newly introduced element may have a current detector as its physical counterpart. However, the existence of the physical implication is immaterial, because all we are trying to do is to convert every path into a loop with that edge. The path matrix of the network will then be augmented with an additional column corresponding to the new edge x_0 such that we obtain a loop matrix B as given in Eq. (1.12):

$$B = \begin{array}{c} \begin{array}{cccccccc} x_1 & x_2 & x_3 & x_4 & x_5 & x_6 & x_7 & x_0 \end{array} \\ \begin{bmatrix} 1 & 0 & 0 & 0 & 1 & 0 & 0 & 1 \\ 0 & 1 & 0 & 0 & 1 & 1 & 0 & 1 \\ 0 & 0 & 1 & 0 & 1 & 1 & 1 & 1 \\ 0 & 0 & 0 & 1 & 0 & 1 & 1 & 1 \\ 1 & 0 & 1 & 1 & 0 & 0 & 0 & 1 \\ 1 & 1 & 0 & 1 & 0 & 0 & 1 & 1 \\ 0 & 1 & 1 & 1 & 0 & 1 & 0 & 1 \end{bmatrix} \end{array}. \qquad (1.12)$$

This loop matrix obtained from a path matrix appending a column of all 1's will be referred to as a "converted loop matrix."

We now eliminate the dependent rows in the converted loop matrix B and then transform the remainder into the form of $[U_4 \quad B_{12}]$, where U_4 is a unit matrix of order four, since the rank of B is four. We shall refer to this form of loop matrix as a "basic form."†

For the transformation of a loop matrix into a basic form we shall use Jordan's method of elimination. For instance, in B of Eq. (1.12), a row having a 1 in the first column is added to every other row which also contains a 1 in the first column. That is, the first row is added to the fifth and sixth rows mod 2, and the resulting matrix will be denoted by B_1. We therefore have

$$B_1 = \begin{bmatrix} 1 & 0 & 0 & 0 & 1 & 0 & 0 & 1 \\ 0 & 1 & 0 & 0 & 1 & 1 & 0 & 1 \\ 0 & 0 & 1 & 0 & 1 & 1 & 1 & 1 \\ 0 & 0 & 0 & 1 & 0 & 1 & 1 & 1 \\ 0 & 0 & 1 & 1 & 1 & 0 & 0 & 0 \\ 0 & 1 & 0 & 1 & 1 & 0 & 1 & 0 \\ 0 & 1 & 1 & 1 & 0 & 1 & 0 & 1 \end{bmatrix}. \qquad (1.13a)$$

† This form of a loop matrix is called by some authors the "normal form." However, we are not using this definition because it would be confused with a "normal form" of a switching function.

Then a row containing a 1 in the second column is again added to every other row having a 1 in the same column. The second row of B_1 of Eq. (1.13a) will be added to the sixth and seventh rows. Then the resulting matrix, B_2, will be

$$B_2 = \begin{bmatrix} 1 & 0 & 0 & 0 & 1 & 0 & 0 & 1 \\ 0 & 1 & 0 & 0 & 1 & 1 & 0 & 1 \\ 0 & 0 & 1 & 0 & 1 & 1 & 1 & 1 \\ 0 & 0 & 0 & 1 & 0 & 1 & 1 & 1 \\ 0 & 0 & 1 & 1 & 1 & 0 & 0 & 0 \\ 0 & 0 & 0 & 1 & 0 & 1 & 1 & 1 \\ 0 & 0 & 1 & 1 & 1 & 0 & 0 & 0 \end{bmatrix}. \qquad (1.13b)$$

In B_2 of Eq. (1.13b), the fifth and seventh rows are identical, and the fourth and sixth rows are also identical. We therefore remove the seventh and sixth rows from B_2. The remainder is then denoted by B_3 and is given by

$$B_3 = \begin{bmatrix} 1 & 0 & 0 & 0 & 1 & 0 & 0 & 1 \\ 0 & 1 & 0 & 0 & 1 & 1 & 0 & 1 \\ 0 & 0 & 1 & 0 & 1 & 1 & 1 & 1 \\ 0 & 0 & 0 & 1 & 0 & 1 & 0 & 1 \\ 0 & 0 & 1 & 1 & 1 & 0 & 0 & 0 \end{bmatrix}. \qquad (1.13c)$$

In B_3, we add the third row to the fifth row. We therefore have B_4 as

$$B_4 = \begin{bmatrix} 1 & 0 & 0 & 0 & 1 & 0 & 0 & 1 \\ 0 & 1 & 0 & 0 & 1 & 1 & 0 & 1 \\ 0 & 0 & 1 & 0 & 1 & 1 & 1 & 1 \\ 0 & 0 & 0 & 1 & 0 & 1 & 1 & 1 \\ 0 & 0 & 0 & 1 & 0 & 1 & 1 & 1 \end{bmatrix}. \qquad (1.13d)$$

We again find in B_4 that the last two rows are identical, and the last row is therefore eliminated. Thus we have

$$B_5 = \begin{array}{c} \begin{array}{cccccccc} x_1 & x_2 & x_3 & x_4 & x_5 & x_6 & x_7 & x_0 \end{array} \\ \begin{bmatrix} 1 & 0 & 0 & 0 & 1 & 0 & 0 & 1 \\ 0 & 1 & 0 & 0 & 1 & 1 & 0 & 1 \\ 0 & 0 & 1 & 0 & 1 & 1 & 1 & 1 \\ 0 & 0 & 0 & 1 & 0 & 1 & 1 & 1 \end{bmatrix} \end{array}. \qquad (1.13e)$$

The matrix $\mathbf{B}_5$ which we have thus obtained corresponds to the basic loop matrix of the network of Fig. 1.4 with respect to the tree, consisting of edges x_5, x_6, x_7, and x_0. The independent paths corresponding to the rows of $\mathbf{B}_5$, except the last column, denoted by P_f and called a "basic path set," will then be

$$P_f = \{x_1x_5, \quad x_2x_5x_6, \quad x_3x_5x_6x_7, \quad x_4x_6x_7\}. \tag{1.14}$$

We now face the question of whether the elimination process which we have used, the so-called "Jordan elimination procedure," always reduces a converted loop matrix into a basic form. However, this is assured by the first property of Property 17 of Part I. That is, the ring sum of two paths is a loop or an edge-disjoint union of loops and the switching function of a single-contact network includes all possible paths of the network between a pair of nodes, so does the path matrix formed from the switching function. It is therefore clear that the converted loop matrix, generated from the path matrix appending a column of all 1's, should contain the sufficient number of independent loops, at least $(b-n+1)$ independent loops if the corresponding augmented network has b edges and n nodes (see Reference 15 of Part I for additional discussion and examples). We have thus reduced the realization of a single-contact function to the realizability of a loop matrix, and this will be investigated thoroughly in the following chapter.

CHAPTER 2

Realization of Loop
and Cut-Set Matrices

REALIZABILITY OF A MATRIX OF INTEGERS MOD 2 AS A LOOP MATRIX

FOR A LONG TIME it has been a challenge for both network theorists and mathematicians to find the necessary and sufficient conditions for the realizability of a loop matrix as well as a cut-set matrix.

Let us consider a matrix $\mathbf{B}$, of order $(b-n+1)$ by b for $n = 2, 3, \ldots$, and $n \leqslant b$ in a basic form

$$\mathbf{B} = [\mathbf{U}_2 \quad \mathbf{B}_{12}], \tag{2.1}$$

where $\mathbf{U}_2$ is a unit matrix of order $(b-n+1)$. Then, in order for $\mathbf{B}$ to be a basic loop matrix of a connected nonoriented graph, the matrix must satisfy the following properties:

1. The elements of $\mathbf{B}$ are 1's and 0's, i.e., $\mathbf{B}$ is a matrix of integers mod 2;

2. Each row of $\mathbf{B}$ corresponds to a loop, and modulo 2 addition of any number of rows yields only a single loop or an edge-disjoint union of loops;

3. The columns of $\mathbf{B}_{12}$ correspond to a set of the branches of a tree of a connected graph, while the columns of the unit matrix correspond to the chords of the tree.

The first two properties are the necessary conditions and the last one is sufficient for a matrix $\mathbf{B}$ to be realizable as a basic loop matrix of a connected graph.

Let us consider the following matrix:

$$\mathbf{B} = \begin{array}{c} \begin{array}{cc} abcd & efg \end{array} \\ \begin{bmatrix} 1000 & 110 \\ 0100 & 101 \\ 0010 & 011 \\ 0001 & 111 \end{bmatrix} \end{array}. \tag{2.2}$$

Some modulo 2 sums of rows of **B** are listed in Table 1.

Table 1

Rows added	Result
1, 2, and 3	*abc*
1 and 2	*abfg*
2 and 3	*bcef*
4	*defg*
1, 2, 3, and 4	*abcdefg*

Let us examine the set of edges in the fifth combination of Table 1. This set contains the set (*defg*), the fourth combination of Table 1, as a proper subset. Thus, if **B** were realizable, the set (*abcdefg*) could not form a single loop.

If **B** were realizable, the set (*efg*) would be a tree of the graph corresponding to **B**, and therefore (*defg*) would be a Hamilton loop of the graph which corresponds to the last row of **B**. It is however clear that the set (*abc*) would be a single loop, since it is impossible for a set containing three edges to form an edge-disjoint union of loops. Thus, if **B** were realizable, (*abcdefg*) would be the disjoint union of loops (*abc*) and (*defg*).

However, since (*defg*) is a Hamilton loop, the subgraph consisting of edges *d*, *e*, *f*, and *g* contains all the nodes of the graph. The subgraph would appear as shown in Fig. 2.1. (The relative position of the edges is irrelevant to the argument to follow.)

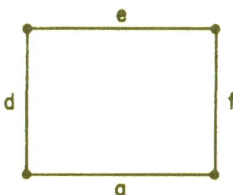

Fig. 2.1. Subgraph consisting of edges d, e, f, and g

To the subgraph in Fig. 2.1 we must add edges *a*, *b*, and *c* in such a way that the following conditions are satisfied:

No additional nodes may be introduced;

Edges *a*, *b*, and *c* must form a loop;

No loop containing two edges may be created, since no modulo two linear combination of rows of **B** yields a set consisting of only two edges.

It is impossible to satisfy all three of the above conditions. Therefore, the set (abcdefg) cannot correspond either to a single loop or to an edge-disjoint union of loops. Hence, due to Property 2, **B** is not realizable.

The second property is, however, not sufficient. To see this, consider the matrix

$$
\mathbf{B} = \begin{array}{cc} abc & defg \\ \begin{bmatrix} 100 & 1101 \\ 010 & 1011 \\ 001 & 0111 \end{bmatrix} \end{array} . \qquad \cdot (2.3)
$$

The linear combinations of rows of **B** mod 2 are:

adeg	abef	abcg
bdfg	acdf	
cefg	bcde	

If **B** were realizable, all combinations of rows would correspond to single loops. Thus we cannot use Property 2 to conclude that **B** is not realizable. Nevertheless, there is no graph corresponding to **B**.

We therefore find some sufficient conditions[6] for a matrix **B** to be realizable, i.e., showing the existence of a topological graph.

Theorem 1. Given a matrix **B** in a basic form of Eq. (2.1) which satisfies Properties 1 and 2. If there exists an arrangement of the columns of $\mathbf{B}_{12}$ such that in any row of $\mathbf{B}_{12}$ no two 1's are separated by one or more 0's, then **B** is always realizable as a basic loop matrix of a connected graph with b edges and n nodes.

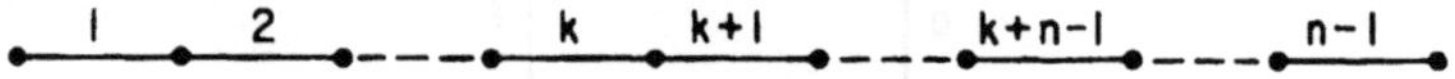

Fig. 2.2. Linear-tree

In particular, it is possible to realize **B** with a graph G, which has the following properties:

a. The tree of G corresponding to the columns of $\mathbf{B}_{12}$ is a linear-tree;

b. The edges of the linear-tree are ordered in the same way as the columns of $\mathbf{B}_{12}$. In other words, if the columns of $\mathbf{B}_{12}$ (after re-arrangement) are labeled $1, 2, \ldots, k, k+1, \ldots, k+n-1, \ldots, n-1$, then the linear-tree appears as shown in Fig. 2.2.

Proof: Let the columns of $\mathbf{B}_{12}$ be arranged so that no two 1's are separated by one or more 0's. Suppose a particular row of $\mathbf{B}_{12}$

contains m 1's, where, of course, $m \leqslant (n-1)$. By hypothesis, the 1's must occur in columns k, $k+1$, $k+2$, . . ., $k+m-1$ for some k. In Fig. 2.2, if we connect the left node of edge k to the right node of edge $k+n-1$, we have formed a loop which corresponds to the given row of $\mathbf{B}_{12}$, the edge which completes the loop is the chord associated with the given row. It is clear that the above procedure can be repeated for each row of $\mathbf{B}_{12}$. Hence the theorem.

Example 1. Let

$$
\mathbf{B} = \begin{array}{c} \\ \end{array}
\begin{array}{cc} abcdef & ghij \\
\left[\begin{array}{cc} 100000 & 1111 \\ 010000 & 1010 \\ 001000 & 1011 \\ 000100 & 1101 \\ 000010 & 1001 \\ 000001 & 0101 \end{array}\right]. \end{array}
\tag{2.4}
$$

Rearrange the columns of $\mathbf{B}$ so that we obtain $\mathbf{B}_1$ as follows:

$$
\mathbf{B}_1 = \begin{array}{cc} abcdef & igjh \\
\left[\begin{array}{cc} 100000 & 1111 \\ 010000 & 1100 \\ 001000 & 1110 \\ 000100 & 0111 \\ 000010 & 0110 \\ 000001 & 0011 \end{array}\right]. \end{array}
\tag{2.5}
$$

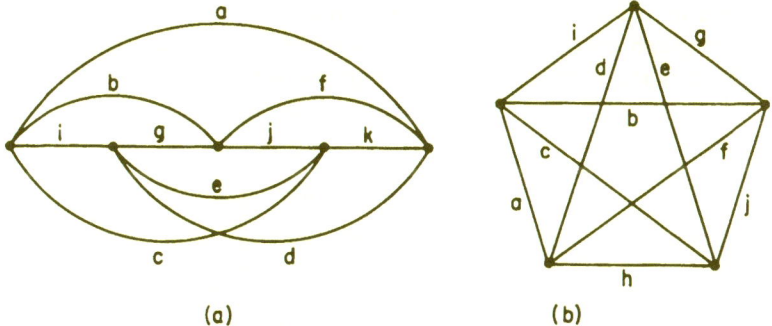

(a)   (b)

Fig. 2.3. Final realization of the loop matrix of Example 1

The realization of $\mathbf{B}_1$ is then shown in Fig. 2.3a which is redrawn in Fig. 2.3b. Notice that if a row of $\mathbf{B}_{12}$ contains all 1's, the loop corresponding to this row is a Hamilton loop of the graph. That is, the first row of $\mathbf{B}$ of Eq. (2.4) corresponds to a Hamilton loop of edges a, g, h, i, and j.

Even if a matrix does not satisfy all hypotheses of Theorem 1, it may still be possible to realize the matrix with a graph satisfying (a) and (b) of Theorem 1. For example, consider the following matrix:

$$\mathbf{B} = \begin{array}{c} \begin{array}{cc} efghij & abcd \end{array} \\ \left[\begin{array}{cc} 100000 & 1011 \\ 010000 & 1100 \\ 001000 & 1101 \\ 000100 & 1010 \\ 000010 & 1001 \\ 000001 & 0110 \end{array} \right] \end{array}. \qquad (2.6)$$

There is no arrangement of the columns of $\mathbf{B}_{12}$ such that in any row of $\mathbf{B}_{12}$ no two 1's are separated by a 0. However, let us form modulo 2 sum of the last four rows of $\mathbf{B}$. The result, ($aghij$), must be a loop or an edge-disjoint union of loops in the graph corresponding to $\mathbf{B}$, provided that graph exists. If we wrote out all modulo 2 sums of rows of $\mathbf{B}$, we would find that there is no set of combinations whose union would yield ($aghij$). But if $\mathbf{B}$ is realizable, all linear combinations, mod 2, of rows of $\mathbf{B}$ generate *all* loops of the graph as well as all *disjoint* unions of loops. Therefore, *if* $\mathbf{B}$ *is realizable*, ($aghij$) forms a single loop. The graph corresponding to $\mathbf{B}$, if it exists, has five nodes. Thus, if $\mathbf{B}$ is realizable, ($aghij$) is a Hamilton loop.

It is clear that the removal of one edge from a Hamilton loop results in a linear-tree. Thus, if $\mathbf{B}$ is realizable, ($ghij$) is a linear-tree of the graph corresponding to $\mathbf{B}$. Now let us rearrange the columns of $\mathbf{B}$ so that columns g, h, i, and j appear on the right,

$$\mathbf{B} = \begin{array}{c} \begin{array}{cc} abcdef & ghij \end{array} \\ \left[\begin{array}{cc} 101110 & 0000 \\ 110001 & 0000 \\ 110100 & 1000 \\ 101000 & 0100 \\ 100100 & 0010 \\ 011000 & 0001 \end{array} \right] \end{array}. \qquad (2.7)$$

Since modulo 2 sum of two loops is a loop or an edge-disjoint union of loops, any elementary row transformation on a loop matrix results in a loop matrix. If we diagonalize the submatrix of **B** formed by the first six columns, using Jordan's elimination method, we get the matrix $\mathbf{B}_1$ shown in Eq. (2.8):

$$
\mathbf{B}_1 = \begin{array}{c} \phantom{\mathbf{B}_1 = } \begin{array}{cc} abcdef & ghij \end{array} \\ \left[\begin{array}{cc} 100000 & 1111 \\ 010000 & 1010 \\ 001000 & 1011 \\ 000100 & 1101 \\ 000010 & 1001 \\ 000001 & 0101 \end{array} \right]. \end{array} \tag{2.8}
$$

Then $\mathbf{B}_1$ is identical with the matrix given in Example 1, Eq. (2.4), and its graph is the complete pentagon of Fig. 2.3, which is related to $\mathbf{B}_1$ by a nonsingular row transformation, is a loop matrix of the graph.

Corollary 1. If **B** in a basic form satisfies Properties 1 and 2, and $\mathbf{B}_{12}$ has two columns or less, **B** is always realizable.

Proof: If $\mathbf{B}_{12}$ has two columns or less, it is impossible for two 1's to be separated by a 0. Thus **B** satisfies the hypothesis of Theorem 1 and is thus realizable. The following corollary also follows immediately from Theorem 1.

Corollary 2. If **B** in a basic form satisfies Properties 1 and 2 and also has only one or two rows, **B** is always realizable.

If we restrict our attention to planar graphs, we have the following theorem.[6]

Theorem 2. $\mathbf{B}_1 = [\mathbf{U}_2 \quad \mathbf{B}_{12}]$ is a basic loop matrix of a planar graph G if and only if $\mathbf{B}_2 = [\mathbf{U}_1 \quad \mathbf{B}_{12}{}^t]$ is a basic loop matrix of a planar graph G', where G' is the dual of G, and $\mathbf{U}_2$ and $\mathbf{U}_1$ are unit matrices of order $(b-n+1)$ and $(n-1)$, respectively.

Proof: Suppose $\mathbf{B}_1$ is a basic loop matrix of a planar graph G. The dual of a loop is a basic cut-set, and therefore $\mathbf{B}_1$ is a basic cut-set matrix of the dual graph G'. By rearranging columns, we may write $\mathbf{B}_1$ as $[\mathbf{B}_{12} \quad \mathbf{U}_2]$; $\mathbf{B}_1$ contains a unit matrix of order $(b-n+1)$ and is therefore a basic cut-set matrix of G'. The basic loop matrix $\mathbf{B}_f$ and the basic cut-set matrix $\mathbf{C}_f$ of a graph with respect to the same tree are related by

$$\mathbf{C}_f = [\mathbf{C}_{f_{11}} \quad \mathbf{U}_1] \tag{2.9a}$$

$$\mathbf{B}_f = [\mathbf{U}_2 \quad \mathbf{C}_{f_{11}}{}^t]. \tag{2.9b}$$

Thus the basic loop matrix of G' is $[\mathbf{U}_1 \quad \mathbf{B}_{12}{}^t] = \mathbf{B}_2$, and $\mathbf{B}_2$ is realizable.

Now suppose that $\mathbf{B}_2 = [\mathbf{U}_1 \quad \mathbf{B}_{12}{}^t]$ is realizable as a planar graph G'. The above argument shows that $[\mathbf{U}_2 \quad (\mathbf{B}_{12}{}^t)^t]$ is realizable as a planar graph $(G')'$. But $(\mathbf{B}_{12}{}^t)^t = \mathbf{B}_{12}$ and $(G')' = G$, i.e., the dual of a dual is the original graph. Therefore $\mathbf{B}_1 = [\mathbf{U}_2 \quad \mathbf{B}_{12}]$ is realizable and the theorem is proved.

Note here that Theorem 2 shows that the transpose of any invalid submatrix of $\mathbf{B}_{12}$ is also invalid. We now come to an important theorem for the realization of a B-matrix.[6]

Theorem 3. If $\mathbf{B}$ in a basic form satisfies the two required properties 1 and 2, and, in addition, (a) $\mathbf{B}_{12}$ has three or more columns, and (b) every row and column of $\mathbf{B}_{12}$ contains exactly two 1's, with no two rows or columns of $\mathbf{B}_{12}$ being identical, then $\mathbf{B}$ is realizable as a basic loop matrix of a connected graph, and this graph is unique, i.e., no 2-isomorphisms are possible.

Proof: If $\mathbf{B}$ and $\mathbf{B}_{12}$ satisfy the hypothesis of the theorem, $\mathbf{B}_{12}$ must be a square matrix. The rows and columns of $\mathbf{B}_{12}$ may always be rearranged so that $\mathbf{B}_{12}$ appears in the form shown below:

$$\mathbf{B}_{12} = \begin{array}{c} \\ \\ 1 \\ 2 \\ 3 \\ \vdots \\ k+1 \\ k \end{array} \overset{\displaystyle 123 \quad \dots \quad k}{\begin{bmatrix} 110 & \dots & 0 \\ 011 & \dots & 0 \\ 0011 & \dots & 0 \\ \vdots & & \vdots \\ 000 & \dots & 11 \\ 100 & \dots & 01 \end{bmatrix}}. \tag{2.10}$$

A graph corresponding to $\mathbf{B} = [\mathbf{U}_k \quad \mathbf{B}_{12}]$ is shown in Fig. 2.4 where $\mathbf{U}_k$ is a unit matrix of order k.

By inspection of the graph, it is seen that no 2-isomorphisms are possible, in other words, it is impossible to construct a graph with the same loop matrix and a different incidence matrix. Hence the theorem.

Corollary 3. If $\mathbf{B}$ in a basic form satisfies Properties 1 and 2, and if $\mathbf{B}_{12}$ has exactly three rows and three columns, $\mathbf{B} = [\mathbf{U}_2 \quad \mathbf{B}_{12}]$ is always realizable.

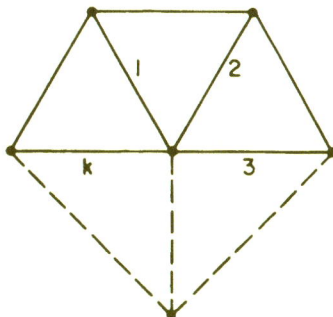

Fig. 2.4. Graph corresponding to $B = [U_k \quad B_{12}]$ where U_k is a unit matrix of order k

Proof: We may exclude the trival cases where two rows or columns of B_{12} are identical and also the cases where a row or column of B_{12} contains only a single 1. The only remaining possibilities are shown below.

$$B_{12} = \begin{bmatrix} 111 \\ 110 \\ 011 \end{bmatrix} \quad \text{(Case 1)} \qquad\qquad B_{12} = \begin{bmatrix} 110 \\ 011 \\ 101 \end{bmatrix} \quad \text{(Case 2)}.$$

In Case 1, **B** is realizable by Theorem 1. In Case 2, **B** is realizable by Theorem 3.

Notice that in the graph corresponding to the matrix of Theorem 3 all sets of three or more edges in the tree $(12 \ldots k)$ are star-connected. If a *square submatrix* of B_{12} satisfies the conditions of the theorem, namely, that each row and each column of the submatrix contains exactly two 1's, with no rows or columns of the submatrix being identical, then the same conclusion may be reached concerning the edges corresponding to the columns of the submatrix; if **B** is realizable, and the edges corresponding to the columns of the submatrix are called $1, 2, \ldots, k$, then all sets of three or more edges in the set $(12 \ldots k)$ are star-connected.

Thus Theorem 3 suggests that if B_{12} contains any of the following submatrices given in Table 2, $B = [U_2 \quad B_{12}]$ is not realizable. Due to Theorem 2, the transpose of any invalid submatrix is also invalid.

Table 2. *Some invalid submatrices for* $\mathbf{B}_{12}$, *the transpose of the invalid matrices are also invalid*

$$
\begin{bmatrix} 110 \\ 101 \\ 011 \\ 111 \end{bmatrix}
\quad
\begin{bmatrix} 1001 \\ 0101 \\ 0011 \\ 1110 \end{bmatrix}
\quad
\begin{bmatrix} 1010 \\ 0110 \\ 0101 \\ 1001 \\ 1110 \end{bmatrix}
\quad
\begin{bmatrix} 1010 \\ 0110 \\ 0101 \\ 1001 \\ 1011 \end{bmatrix}
\quad
\begin{bmatrix} 10100 \\ 00101 \\ 00011 \\ 01010 \\ 11000 \\ 01110 \end{bmatrix}
\quad
\begin{bmatrix} 11000 \\ 00101 \\ 01001 \\ 00110 \\ 10010 \\ 10101 \end{bmatrix}
$$

One should note in Table 2 that the first matrix or its transpose, both of which are invalid, are essentially included as a submatrix in all other invalid matrices in the table. For instance, in the second matrix in the table let us add the last row to all other rows mod 2; then it will be seen clearly that the resulting matrix contains the first matrix as its submatrix. We therefore know that the first matrix in Table 2 and the transpose of the matrix are bases of all invalid matrices. Moreover, a matrix which contains the pattern of nonzero elements of the first matrix or its transpose is also nonunimodular.

Following Theorem 3 and Table 2 one can easily construct or detect a single-contact function which is not realizable with the minimal network. For example, a function given in Eq. (2.11),

$$ F = x_1x_5x_6 + x_2x_5x_7 + x_3x_6x_7 + x_4x_5x_6x_7 \tag{2.11} $$

is not a single-contact function because the path matrix corresponding to the function as well as the converted loop matrix will contain the first invalid matrix listed in Table 2 as a submatrix.

We have thus proved that if a basic form of a matrix of integer mod 2, $\mathbf{B}$, contains the first matrix in Table 2 or the transpose of the matrix, then there exists no topological graph which can be drawn for $\mathbf{B}$. These particular matrices, the first matrix in Table 2 and its transpose, have actually been studied by many other authors including Whitney (Reference 9f of Part I), Gould, Cederbaum (Reference 6b of Part III), and Tutte.[7] Their investigations were, however, mainly based on abstract graphs and were not, except for that of Gould, based on topological construction of graphs. We shall now need to introduce a definition in order to discuss the work of Tutte.

Definition 1: A matrix of integers mod 2 is "valid" if no basic form of the matrix contains the first matrix of Table 2 or the transpose of the matrix. Some authors refer to such a matrix as a "regular matrix."

We thus have the following theorem due to Tutte, and also translated into familiar terminology by Seshu (Reference 7 of Part I), incorporating the work of Cederbaum.

Theorem 4. A matrix of integers mod 2 is a loop matrix (or cut-set matrix) of a graph if and only if it is valid and no basic form of the matrix contains a cut-set matrix (or loop matrix) of any of Kuratowski's nonplanar graphs.

The necessity parts of the theorem are obvious, because (a) due to Theorem 3 and the foregoing arguments the matrix must be valid; and (b) there exists no dual of a nonplanar graph, and Kuratowski's nonplanar graphs are the basis of all nonplanar graphs.

It is also clear, however, that the proof of sufficiency as well as a possible algorithm based on the theorem will not be a simple one. That is, one must consider all possible permutations and elementary transformations on the rows and columns of a matrix to be realized in order to determine if the matrix is valid. This is a main drawback. However, the theorem proposes the complete answer to the problem of realization of a loop matrix (or a cut-set matrix). It also states that it is immaterial which basic form is chosen; that is, one should consider any basic form of the given matrix, and, if one basic form is not realizable, then the original matrix is also not realizable. We therefore investigate a simple realization procedure of a basic loop and cut-set matrix.

REALIZATION OF A BASIC CUT-SET MATRIX

Let us write the basic loop matrix, $\mathbf{B}_f$, of a connected graph with n nodes and b edges in a basic form:

$$\mathbf{B}_f = [\mathbf{U}_2 \quad \mathbf{B}_{f_{12}}], \tag{2.12}$$

where $\mathbf{U}_2$ is a unit matrix of order $(b-n+1)$. Then, as found previously, the basic cut-set matrix C_f, with respect to the same tree, is found by (Property 15 of Part I)

$$\mathbf{C}_f = [\mathbf{B}_{f_{12}} \quad \mathbf{U}_1] \tag{2.13}$$

and $\mathbf{U}_1$ is a unit matrix of order $(n-1)$ and the columns of the unit matrix correspond to the branches of the tree.

From Eqs. (2.12) and (2.13) it should be clear that if $\mathbf{B}_{f_{12}}$ contains one of the invalid matrices in Table 2, the corresponding basic cut-set matrix is also not realizable because the transpose of an invalid matrix is also invalid.

We therefore try to realize a graph for a basic cut-set matrix instead of a basic loop matrix. It is, however, necessary for us to know more about a basic cut-set matrix of a connected graph. Let C_f be in a basic form, i.e., $C_f = [C_{f_{11}} \quad U_1]$. Then we have the following properties.[8]

Lemma 1: If C_{f11} of a basic cut-set matrix C_f contains a column of all 1's, then the tree on which C_f is based is a linear-tree.

The proof of this lemma is obvious because a column of all 1's in $C_{f_{11}}$ corresponds to the row of $B_{f_{12}}$ of all 1's [from Eqs. (2.12) and (2.13)], and because a row of all 1's of $B_{f_{12}}$, together with a column of U_2, is a Hamilton loop of the graph corresponding to C_f if it exists.

Lemma 2: If $C_{f_{11}}$ contains a column with 1's in the rows $i, j, \ldots, k$, then the edges $i, j, \ldots, k$ are path-connected; i.e., the edges are included in a path, if all other edges are shorted. The proof of this lemma is similar to the proof of Lemma 1.

With the background of Lemmas 1 and 2, we shall proceed to realize a basic cut-set matrix. To find an algorithm for the realization of a basic cut-set matrix, we shall make use of the synthesis techniques which were developed in Part III. We therefore briefly review the synthesis of the multiport network.

If we form an $(n-1)$-port of n nodes and b edges based on a linear-tree port-structure such that the orientation of the ith port is taken in the same direction as the orientation of the ith branch of the tree, then the short-circuit admittance matrix of the $(n-1)$-port, Y, is given by

$$Y = C_f Y_e C_f{}^t, \tag{2.14a}$$

where Y_e is a diagonal matrix in which each diagonal element is the admittance of an edge of the network. Since the tree is a linear-tree, all nonzero elements of C_f can be considered as $+1$ so that we may consider C_f as a matrix of integer mod 2.

If we assume that the network consists entirely of unit conductances, the matrix Y_e is reduced to a unit matrix. Thus we rewrite Y as:

$$Y = C_f C_f{}^t. \tag{2.14b}$$

It is therefore clear that if Y is realizable as the short-circuit admittance matrix of an $(n-1)$-port with n nodes and b edges of unit conductance, then C_f should be realizable. And if C_f contains a column of all 1's, due to Lemma 1 the network should have a linear-tree port-structure. If C_f is not realizable, then Y is also not realizable.

We now propose the following theorem.[8]

Theorem 5. If a basic cut-set matrix C_f has a column containing 1's in the rows $i, j, \ldots, k$, then

$$Y_{ij\ldots k} = (C_{f_{ij\ldots k}})(C_{f_{ij\ldots k}}{}^t) \tag{2.15}$$

is realizable as a short-circuit admittance matrix of a multiport resistive network of unit resistors and with a linear-tree port-structure formed from ports $i, j, \ldots, k$. The subscripts indicate the rows included in Y. The proof of this theorem follows from Lemma 2 and the foregoing arguments.

We now summarize the realization procedure of a basic cut-set matrix.

Case 1: C_f contains a column of all 1's:

a. Form $Y = C_f C_f{}^t$
b. Place Y in a uniformly tapered form (if this is impossible, then C_f is not realizable)
c. Realize Y with the specified number of nodes and unit conductances
d. Open-circuit the ports of Y, and the resulting graph realizes C_f.

Case 2: C_f does *not* contain a column with all 1's:

a. Form $Y = C_f C_f{}^t$
b. Examine the columns with the largest numbers of 1's in the positions $i, j, \ldots, k$, and obtain the matrix $Y_{ij\ldots k}$
c. Place $Y_{ij\ldots k}$ in a uniformly tapered form and thus derive the order of edges $i, j, \ldots, k$ corresponding to the ports $i, j, \ldots, k$
d. Repeat steps b and c for the other columns and obtain the tree on which C_f is based. The realization of C_f can now be deduced by inspection of C_f.

The procedures will be best illustrated in the following example.

Example 2: This example was considered by Gould. Let us consider a single-contact function given in a normal form,

$$F = abceg + acdf + aefh + bceh + bd + cdfgh + efg. \tag{2.16}$$

The converted loop matrix B corresponding to the switching function is then

$$\mathbf{B} = \begin{bmatrix} a & b & c & d & e & f & g & h & x_0 \\ 1 & 1 & 1 & 0 & 1 & 0 & 1 & 0 & 1 \\ 1 & 0 & 1 & 1 & 0 & 1 & 0 & 0 & 1 \\ 1 & 0 & 0 & 0 & 1 & 1 & 0 & 1 & 1 \\ 0 & 1 & 1 & 0 & 1 & 0 & 0 & 1 & 1 \\ 0 & 1 & 0 & 1 & 0 & 0 & 0 & 0 & 1 \\ 0 & 0 & 1 & 1 & 0 & 1 & 1 & 1 & 1 \\ 0 & 0 & 0 & 0 & 1 & 1 & 1 & 0 & 1 \end{bmatrix} \qquad (2.17)$$

Using Jordan's elimination procedure, we found $\mathbf{B}$ in a basic form, $\mathbf{B}_1$, as:

$$\mathbf{B}_1 = \begin{bmatrix} a & b & c & e & d & f & g & h & x_0 \\ 1 & 0 & 0 & 0 & 0 & 0 & 1 & 1 & 0 \\ 0 & 1 & 0 & 0 & 1 & 0 & 0 & 0 & 1 \\ 0 & 0 & 1 & 0 & 1 & 1 & 1 & 1 & 1 \\ 0 & 0 & 0 & 1 & 0 & 1 & 1 & 0 & 1 \end{bmatrix} \qquad (2.18)$$

and the corresponding basic cut-set matrix to be realized, $\mathbf{C}_f$, is then found to be:

$$\mathbf{C}_f = \begin{bmatrix} a & b & c & e & d & f & g & h & x_b \\ 0 & 1 & 1 & 0 & 1 & 0 & 0 & 0 & 0 \\ 0 & 0 & 1 & 1 & 0 & 1 & 0 & 0 & 0 \\ 1 & 0 & 1 & 1 & 0 & 0 & 1 & 0 & 0 \\ 1 & 0 & 1 & 0 & 0 & 0 & 0 & 1 & 0 \\ 0 & 1 & 1 & 1 & 0 & 0 & 0 & 0 & 1 \end{bmatrix} . \qquad (2.19)$$

At this point, one may attempt to find a linear transformation which transforms $\mathbf{C}_f$ into an incidence matrix with at most two 1's in every column. The search for such a transformation can become very laborious (See References 9–11 and also Reference 15 of Part I). Using the procedure summarized previously, we note that the matrix $\mathbf{C}_f$ contains a column with all 1's. Hence the following matrix:

$$\mathbf{Y} = \mathbf{C}_f \mathbf{C}_f = \begin{bmatrix} 3 & 1 & 1 & 1 & 2 \\ 1 & 3 & 2 & 1 & 2 \\ 1 & 2 & 4 & 2 & 2 \\ 1 & 1 & 2 & 3 & 1 \\ 2 & 2 & 2 & 1 & 4 \end{bmatrix} \qquad (2.20)$$

must be realized if $\mathbf{C}_f$ is realizable by a five-port resistive network of unit resistors with a linear-tree port-structure, where ports 1, 2, 3, 4, 5 correspond to the branches d, f, g, h, x_0.

If we transpose the rows and columns of $\mathbf{Y}$ in the order of the positions 4, 3, 2, 5, 1, we obtain the uniformly tapered form:

$$\mathbf{Y}_{43251} = \begin{bmatrix} 3 & 2 & 1 & 1 & 1 \\ 2 & 4 & 2 & 2 & 1 \\ 1 & 2 & 3 & 2 & 1 \\ 1 & 2 & 2 & 4 & 2 \\ 1 & 1 & 1 & 2 & 3 \end{bmatrix}. \tag{2.21}$$

The realization of the five-port is then obtained as shown in Fig. 2.5. If we open-circuit the ports and eliminate edge x_0, then we obtain the minimal realization of the switching function, which is shown in Fig. 2.6.

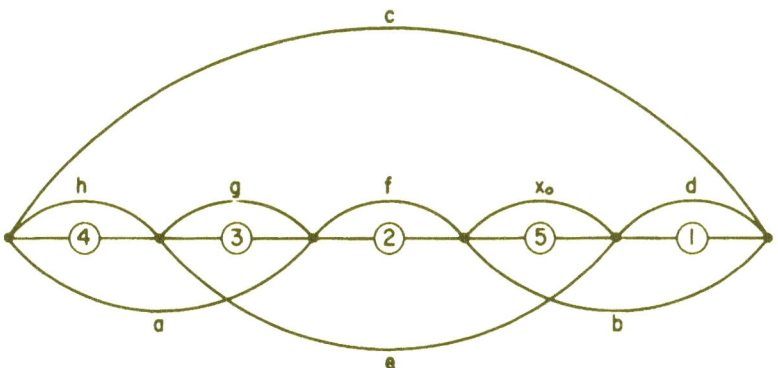

Fig. 2.5. Five-port realization of the Y-matrix of Eq. (2.20)

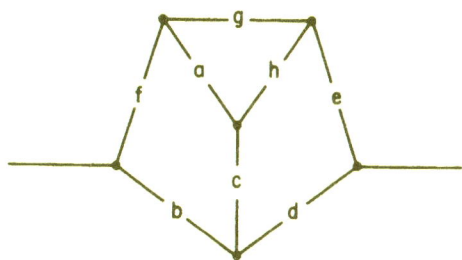

Fig. 2.6. Final realization of the loop matrix of Example 2

Example 3. Let us consider the realization of the following matrix as a basic cut-set matrix of a connected graph (this matrix was considered by Mayeda[12]),

$$
C_f = \begin{array}{c} \begin{array}{ccccccccc} a & b & c & d & e & f & g & h & i \end{array} \\ \begin{bmatrix} 1 & 1 & 1 & 0 & 0 & 1 & 0 & 0 & 0 \\ 1 & 0 & 1 & 0 & 0 & 0 & 1 & 0 & 0 \\ 0 & 0 & 1 & 1 & 1 & 0 & 0 & 1 & 0 \\ 0 & 1 & 1 & 0 & 1 & 0 & 0 & 0 & 1 \end{bmatrix} \end{array}. \qquad (2.22)
$$

We now form Y as:

$$
Y = C_f C_f^t = \begin{bmatrix} 4 & 2 & 1 & 2 \\ 2 & 3 & 1 & 1 \\ 1 & 1 & 4 & 2 \\ 2 & 1 & 2 & 4 \end{bmatrix}, \qquad (2.23)
$$

which should be realized (if C_f is realizable) as the short-circuit admittance matrix of four-ports with five nodes and a linear-tree port-structure (because C_f has a column of all 1's).

We therefore rearrange the rows and columns of Y so that it becomes a uniformly tapered matrix as shown in Eq. (2.24):

$$
Y_{2143} = \begin{bmatrix} 3 & 2 & 1 & 1 \\ 2 & 4 & 2 & 1 \\ 1 & 2 & 4 & 2 \\ 1 & 1 & 2 & 4 \end{bmatrix}. \qquad (2.24)
$$

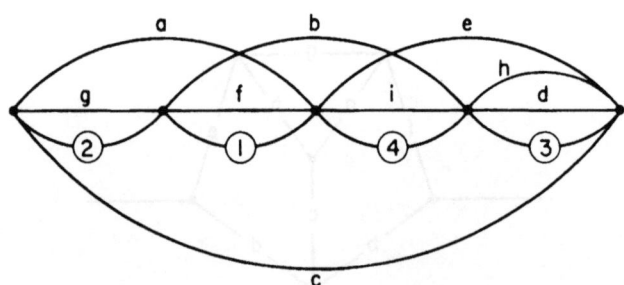

Fig. 2.7. Four-port realization of the Y-matrix of Eq. (2.24)

The four-port resistive network of unit resistors corresponding to the Y-matrix is given in Fig. 2.7. If all the ports of the network are opened, the resulting graph is the final realization of the matrix of Eq. (2.22), which is shown in Fig. 2.8.

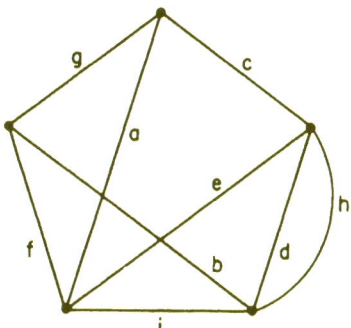

Fig. 2.8. Final realization of the matrix of Example 3

Example 4. Let us now consider the following loop matrix of integers mod 2 in a basic form which was also considered by Guillemin[13]:

$$
\mathbf{B} =
\begin{array}{c}
 \\
1 \\
2 \\
3 \\
4 \\
5 \\
6 \\
7 \\
8 \\
9 \\
10
\end{array}
\begin{array}{cccccccccccccccc}
1 & 2 & 3 & 4 & 5 & 6 & 7 & 8 & 9 & 10 & 11 & 12 & 13 & 14 & 15 \\
\left[\begin{array}{ccccccccccccccc}
1 & 1 & 0 & 0 & 0 & 1 & 0 & 0 & 0 & 0 & 0 & 0 & 0 & 0 & 0 \\
1 & 0 & 1 & 0 & 0 & 0 & 1 & 0 & 0 & 0 & 0 & 0 & 0 & 0 & 0 \\
1 & 0 & 0 & 1 & 0 & 0 & 0 & 1 & 0 & 0 & 0 & 0 & 0 & 0 & 0 \\
1 & 0 & 0 & 0 & 1 & 0 & 0 & 0 & 1 & 0 & 0 & 0 & 0 & 0 & 0 \\
0 & 1 & 0 & 1 & 0 & 0 & 0 & 0 & 0 & 1 & 0 & 0 & 0 & 0 & 0 \\
0 & 0 & 1 & 0 & 1 & 0 & 0 & 0 & 0 & 0 & 1 & 0 & 0 & 0 & 0 \\
1 & 1 & 0 & 0 & 1 & 0 & 0 & 0 & 0 & 0 & 0 & 1 & 0 & 0 & 0 \\
1 & 0 & 1 & 1 & 0 & 0 & 0 & 0 & 0 & 0 & 0 & 0 & 1 & 0 & 0 \\
1 & 1 & 1 & 0 & 0 & 0 & 0 & 0 & 0 & 0 & 0 & 0 & 0 & 1 & 0 \\
1 & 0 & 0 & 1 & 1 & 0 & 0 & 0 & 0 & 0 & 0 & 0 & 0 & 0 & 1
\end{array}\right]
\end{array}
\cdot \quad (2.25)
$$

Then the basic cut-set matrix, $\mathbf{C}_f$, corresponding to the matrix to be realized is found to be

$$
C_f = \begin{array}{c} \\ 1 \\ 2 \\ 3 \\ 4 \\ 5 \end{array}
\begin{array}{cccccccccccccccc}
1 & 2 & 3 & 4 & 5 & 6 & 7 & 8 & 9 & 10 & 11 & 12 & 13 & 14 & 15 \\
\begin{bmatrix} 1 & 0 & 0 & 0 & 0 & 1 & 1 & 1 & 1 & 0 & 0 & 1 & 1 & 1 & 1 \\
0 & 1 & 0 & 0 & 0 & 1 & 0 & 0 & 0 & 1 & 0 & 1 & 0 & 1 & 0 \\
0 & 0 & 1 & 0 & 0 & 0 & 1 & 0 & 0 & 0 & 1 & 0 & 1 & 1 & 0 \\
0 & 0 & 0 & 1 & 0 & 0 & 0 & 1 & 0 & 1 & 0 & 0 & 1 & 0 & 1 \\
0 & 0 & 0 & 0 & 1 & 0 & 0 & 0 & 1 & 0 & 1 & 1 & 0 & 0 & 1 \end{bmatrix}
\end{array} . \quad (2.26)
$$

We now form **Y** as:

$$
\mathbf{Y} = \mathbf{C}_f \mathbf{C}_f{}^t = \begin{bmatrix}
9 & 3 & 3 & 3 & 3 \\
3 & 5 & 1 & 1 & 1 \\
3 & 1 & 5 & 1 & 1 \\
3 & 1 & 1 & 5 & 1 \\
3 & 1 & 1 & 1 & 5
\end{bmatrix} . \quad (2.27)
$$

Since the twelfth, thirteenth, fourteenth, and fifteenth columns of $\mathbf{C}_f$ have the largest number of ones, we therefore form submatrices of **Y** containing the rows which have ones in the twelfth, thirteenth, fourteenth, and fifteenth columns, respectively. In other words, ports $(1, 2, 5)$ are path-connected if we short ports 3 and 4, ports $(1, 3, 4)$ are path-connected if ports 2 and 5 are shorted, ports $(1, 2, 3)$ are path-connected if ports 4 and 5 are shorted, and ports $(1, 4, 5)$ are path-connected if ports 2 and 3 are shorted because of Lemma 2. The corresponding submatrices are

$$
\mathbf{Y}_{125} = \begin{bmatrix}
9 & 3 & 3 \\
3 & 5 & 1 \\
3 & 1 & 5
\end{bmatrix} \quad (2.28a)
$$

$$
\mathbf{Y}_{134} = \begin{bmatrix}
9 & 3 & 3 \\
3 & 5 & 1 \\
3 & 1 & 5
\end{bmatrix} \quad (2.28b)
$$

$$
\mathbf{Y}_{123} = \begin{bmatrix}
9 & 3 & 3 \\
3 & 5 & 1 \\
3 & 1 & 5
\end{bmatrix} \quad (2.28c)
$$

$$
\mathbf{Y}_{145} = \begin{bmatrix}
9 & 3 & 3 \\
3 & 5 & 1 \\
3 & 1 & 5
\end{bmatrix} . \quad (2.28d)
$$

In order to find the ordering of the ports, each submatrix of Eq. (2.28) should be converted into a uniformly tapered form; these are shown in Eq. (2.29):

$$Y_{215} = \begin{bmatrix} 5 & 3 & 1 \\ 3 & 9 & 3 \\ 1 & 3 & 5 \end{bmatrix} \qquad (2.29a)$$

$$Y_{314} = \begin{bmatrix} 5 & 3 & 1 \\ 3 & 9 & 3 \\ 1 & 3 & 5 \end{bmatrix} \qquad (2.29b)$$

$$Y_{213} = \begin{bmatrix} 5 & 3 & 1 \\ 3 & 9 & 3 \\ 1 & 3 & 5 \end{bmatrix} \qquad (2.29c)$$

$$Y_{415} = \begin{bmatrix} 5 & 3 & 1 \\ 3 & 9 & 3 \\ 1 & 3 & 5 \end{bmatrix}. \qquad (2.29d)$$

It is now readily recognized that the order of ports should be (2, 1, 5), (3, 1, 4), (2, 1, 3), and (4, 1, 5).

From the first two submatrices the over-all port-structure is necessarily of the two types shown in Fig. 2.9. However, to satisfy

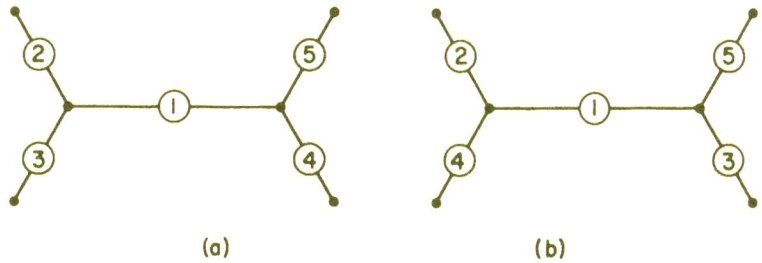

(a)   (b)

Fig. 2.9. Possible port-structures which satisfy the first two matrices of Eq. (2.28)

the last two submatrices, ports (2, 1, 3) and also ports (4, 1, 5) should be path-connected. We therefore eliminate the possible port-structure of Fig. 2.9a and know that the over-all port-structure should be the one shown in Fig. 2.9b. The final realization of the loop matrix of Eq. (2.25), alternatively the cut-set matrix of Eq. (2.26), is then obtained by opening all the ports of the five-port resistive network of unit resistors corresponding to the Y-matrix of Eq. (2.27). This is shown in Fig. 2.10.

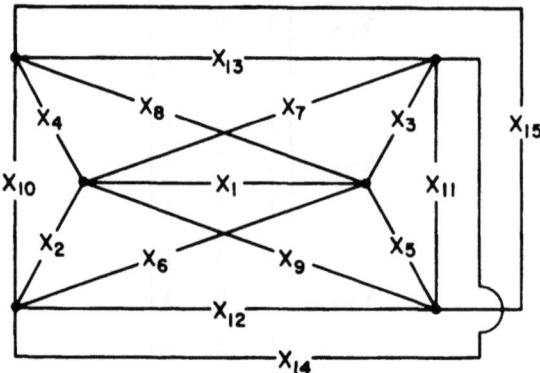

Fig. 2.10. Final realization of the loop matrix of Example 4

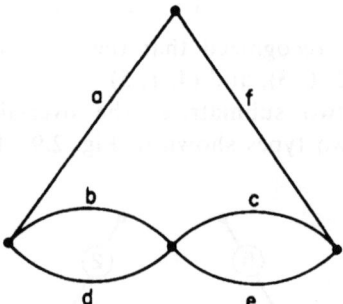

Fig. 2.11. A connected graph

The realization methods for a basic loop and cut-set matrix thus far described are due to Halkias and Kim. The methods proposed respectively by Guillemin and Mayeda shall be illustrated in a little more detail so as to compare them with the method we have used here.

The work of Guillemin and the method proposed here both use the short-circuit admittance matrix corresponding to a given basic loop or cut-set matrix. The methods of Guillemin and Mayeda are related such that both methods propose the process of "how to grow a tree." The former is, however, searching for the end-branches (or tips) of a tree, while the latter is looking for the branches which are not end-branches, the so-called "inside-branches." If we delete a row which contains an end-branch of a tree from a basic cut-set matrix of a connected graph, the resulting graph corresponding to the submatrix remains connected. We repeat this "sorting process" until we identify all the branches of the tree. This idea is used by Guillemin in developing his method. Mayeda, however, utilizes the fact that if one removes a row of a basic cut-set matrix such that the basic cut-set edges corresponding to the row contain no end-branch, then the resulting cut-set matrix may be reduced to two disjoint submatrices. The search for end-branches may meet some difficulty (see Reference 13), while the latter concept may need a modification in some cases. For instance, let us consider a graph given in Fig. 2.11 in which a tree is chosen such that it consists of edges a, b, and c. Then the basic cut-set matrix, C_f, with respect to the tree, will be

$$\mathbf{C}_f = \begin{array}{c} \begin{array}{cccccc} a & b & c & d & e & f \end{array} \\ \begin{bmatrix} 1 & 0 & 0 & 0 & 0 & 1 \\ 0 & 1 & 0 & 1 & 0 & 1 \\ 0 & 0 & 1 & 0 & 1 & 1 \end{bmatrix} \end{array}. \tag{2.30}$$

If we delete the first row from C_f and the columns containing 1's in the first row, then remainder C_{f_1} is given by

$$\mathbf{C}_{f_1} = \begin{array}{c} \begin{array}{cccc} b & c & d & e \end{array} \\ \begin{bmatrix} 1 & 0 & 1 & 0 \\ 0 & 1 & 0 & 1 \end{bmatrix} \end{array}. \tag{2.31}$$

Now in C_{f_1} we interchange the second and third columns. We then have C_{f_2},

$$\mathbf{C}_{f_2} = \begin{array}{c} \begin{array}{cccc} b & d & c & e \end{array} \\ \begin{bmatrix} 1 & 1 & 0 & 0 \\ 0 & 0 & 1 & 1 \end{bmatrix} \end{array}. \tag{2.32}$$

The matrix thus obtained contains two disjoint submatrices, even though a row corresponding to a cut-set containing an end-branch was removed. In other words, it may not always be so easy to judge whether the branch under consideration is an end-branch or an inside-branch (see Reference 12 for detailed discussion).

Before concluding this part, we should mention the possible extension of the realization technique of a nonoriented graph into the realization of an oriented graph. The realization of an oriented graph for a given loop or cut-set matrix is accomplished by following the procedures described in Part III for the synthesis of multiport networks. That is, one should form a Y-matrix and realize it with a resistive multiport of unit resistors (see Reference 8). The realization procedures which have been discussed here are, for the most part, directly applicable for the case of oriented graphs.

REFERENCES

1a. W. V. Quine, "The problems of simplifying truth functions," *Am. Math. Monthly*, 59 (1952), 521–31.

1b. W. V. Quine, "A way to simplify truth functions," *Am. Math. Monthly*, 628 (1955), 627–31.

2. M. Karnaugh, "The map method for synthesis of combinational logic circuits," *Trans. AIEE*, 72, Part I (1953), 593–99.

3. Staff of the Computation Laboratory, *Synthesis of Electronic Computing and Control Circuits*. Cambridge, Massachusetts, Harvard University Press, 1951.

4. E. J. McCluskey, Jr., "Algebraic minimization and design of the terminal contact networks," *Bell System Tech. J.*, 35 (1956), 1417–44.

5. S. H. Caldwell, *Switching Circuits and Logical Design*. New York, John Wiley & Sons, 1958.

6. R. B. Ash and W. H. Kim, "On the realizability of a circuit matrix," *IRE Trans.* CT-6 (1959), 219–23.

7a. W. T. Tutte, "A homotopy theorem for matroids, I, II," *Trans. Am. Math. Soc.*, 88 (1958), 144–74.

7b. W. T. Tutte, "Matroids and graphs," *Trans. Am. Math. Soc.*, 90 (1959), 527–52.

8. C. C. Halkias and W. H. Kim, "Realization of circuit and cut-set matrices," in 1962 *IRE International Convention Record*.

9a. S. Okada, "Topology applied to switching circuits," in *Proceedings of a Symposium on Information Networks*, pp. 267–90. New York, Poly-Technic Institute of Brooklyn, 1954.

9b. S. Okada and K. Young, "Ambit realization of cut-set matrices into graphs," Tech. Report, Contract AF-19(604)-6620, Polytechnic Institute of Brooklyn, New York, 1961.

10. S. Seshu, "On electric circuits and switching circuits," *IRE Trans.* CT-3 (1956), 172–78.

11. L. Lofgren, "Irredundant and redundant Boolean branch networks," *IRE Trans.*, *Special Supplement*, CT-6 (1959), 158–75.

12. W. Mayeda, "Necessary and sufficient conditions for realizability of cut-set matrices," *IRE Trans.* CT-7 (1960), 79–81.

13. E. A. Guillemin, "How to grow your own trees from cut-set or tie-set matrices," *IRE Trans.*, *Special Supplement*, CT-6 (1959), 110–26.

PART V

Analysis and Synthesis

of Communication Nets

Introduction

IN THIS PART we are concerned with a topic of current research interest, namely, the analysis and synthesis of communication nets. A communication net is a mathematical model of a large class of traffic systems such as railroad complexes, highway nets, communication networks, flow networks for industrial plants, power distribution systems, and many others. A number of important features of these systems are common to all. These common features justify the use of a common mathematical model.

First, all systems under consideration have the same topological properties. They are, topologically, a network of interconnected links and terminals. All links are of finite capacity. Second, the main purpose of the systems is to be able to transport some "commodity" from one set of terminals to another set through the network, and the network facilities are shared between all sources and destinations. Third, systems requirements are usually of a statistical nature. Costs for network elements are in general a nonlinear function of capacity and location.

To encompass all topological properties within the model, we construct a communication net which is a linear graph with weights on edges. The weight of an edge corresponds to the capacity of that edge.

Given a communication net together with all the detailed structures, one would then want to know the systems capabilities of this net. Such questions could be phrased in many ways. In the simplest case, one could ask, "What is the maximum flow possible from node i to node j?" A possibly more advanced question would be, "What is a good way of describing mathematically the systems capability of a communication net?" Such questions concerning the analysis of communication nets have been much studied, by mathematicians as well as engineers.

From an engineering point of view, another class of questions might seem more pressing and therefore more interesting, that is, the question of synthesis with minimum cost, given systems requirements. However, as it has often turned out, many basic analytical properties of a communication net must be known before a good treatment of synthesis can begin. As the present state of art in communication net theory stands, only a restricted class of synthesis problems has been solved. This, nevertheless, is significant, especially when one realizes that this field has been explored only within the last decade.

For pure convenience and clarity of technical presentation, the material is organized as follows. Chapters 1 and 2 treat, respectively, the theory of oriented and nonoriented communication nets. Although the basic theorem for analysis of communication nets is first given for oriented nets, a great deal more is known in the case of nonoriented nets, as they have more symmetry and are therefore a comparatively simple problem. Much is known concerning the synthesis of nonoriented nets; for oriented nets, however, even the realizability condition is known only for special cases. Chapter 3 treats a number of disjointed but interesting topics, such as maximizing total communication capacity for given cost and simultaneous-flow multicommodity problems. Each selected topic is interesting in its own right, and all are discussed in great detail. Many of the "still open" questions are mentioned and are discussed either in the text or in the last section of Chapter 3.

CHAPTER 1

Oriented Communication Nets

ORIENTED COMMUNICATION NETS

An oriented communication net is an oriented graph with weights on edges. The weight of an edge is identified with its capacity and denoted by C. Thus the capacity of an edge is equal to C in the direction of its orientation. It is zero in the reverse direction. Furthermore, we shall assume C to be a non-negative real number.

A convenient matrix representation of the graph, complete with the capacities, is possible. This is done through the use of the edge matrix.[†] An entry e_{ij} of the edge matrix E is simply the capacity of the direct link from i to j. If there is no such link in the graph, e_{ij} is zero. The diagonal elements, e_{ii}, are not defined; we shall simply write the numbering of the node i in ith place along the diagonal.

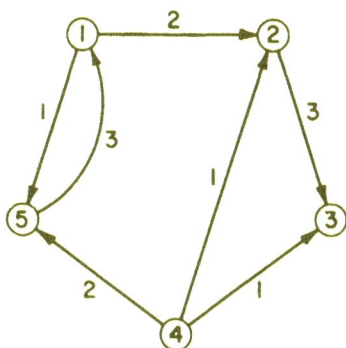

Fig. 1.1. Graph of an oriented communication net

In Fig. 1.1 is shown the graph of an oriented communication net whose edge matrix may be written down easily as

† Sometimes referred to in literature as "branch capacity matrix."

$$\mathbf{E} = \begin{bmatrix} ① & 2 & 0 & 0 & 1 \\ 0 & ② & 3 & 0 & 0 \\ 0 & 0 & ③ & 0 & 0 \\ 0 & 1 & 1 & ④ & 2 \\ 3 & 0 & 0 & 0 & ⑤ \end{bmatrix}. \tag{1.1}$$

It is seen that the edge matrix has all the information concerning the graph, thus giving either the graph or the edge matrix is sufficient.

Definition 1. The flow f_{ij} or $f(n_i, n_j)$ of an edge (n_i, n_j) is defined as a non-negative real number such that[1]:

$$f(n_i, n_j) \leqslant C(n_i, n_j). \tag{1.2}$$

It satisfies Kirchhoff's current law, that is, for some quantity F,

$$\sum_{\substack{j \\ (j \neq i)}} \{f(n_i, n_j) - f(n_j, n_i)\} = F, \qquad n_i \text{ is the input;}$$

$$\sum_{\substack{j \\ (j \neq i)}} \{f(n_i, n_j) - f(n_j, n_i)\} = -F, \qquad n_i \text{ is the output;}$$

$$\sum_{\substack{j \\ (j \neq i)}} \{f(n_i, n_j) - f(n_j, n_i)\} = 0, \qquad \text{otherwise.} \tag{1.3}$$

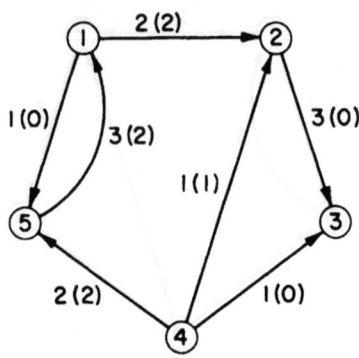

Fig. 1.2. A possible flow pattern of the net of Fig. 1.1

In what follows the flow of any edge shall be written in parentheses following the capacity of the edge. For instance, a possible flow pattern for the graph in Fig. 1.1 is shown in Fig. 1.2.

THE MAX-FLOW MIN-CUT THEOREM

A fundamental property of the oriented communication net is contained in a theorem formulated by Ford, Fulkerson, and Dantzig.[1a,b,2] It is called the "Max-Flow Min-Cut Theorem."[13] This theorem answers the question: "Suppose a flow of magnitude F is supplied at node i and taken away at node j. What is the maximum value of F obtainable through the net?" This question is a basic one and is analogous to the driving-point problem in electrical network theory.

Definition 2: Let $n_1, \ldots, n_k$ be a set of distinct nodes of a graph G.

If for each i $(i = 1, 2, \ldots, k - 1)$, (n_i, n_{i+1}) is an edge of G, then the sequence of nodes and edges is called a "directed path from n_1 to n_k." A directed path is called a "cycle" (or directed loop) if $n_1 = n_k$.

If for each i $(i = 1, 2, \ldots, k - 1)$, either (n_i, n_{i+1}) or (n_{i+1}, n_i) is an edge, then the sequence of nodes and edges is called a "path." We note that in nonoriented graphs the concept of a directed path and the concept of a path coincide. (Cf. Reference 5, Part I.)

Definition 3: A "cut" of a connected oriented graph G is a minimal set of edges the removal of which partitions G into two subgraphs G_1 and its complement, $\bar{G}_1$, such that for any node n_1 in G_1 or n_2 in $\bar{G}_1$ all directed paths from n_1 to n_2 are destroyed.†

Definition 4: The "capacity of a cut" is the sum of the capacities of all edges of the cut.

Theorem 1. For an oriented communication net G the maximal flow from a node n_1 to another node n_2 is equal to the capacity of the min-cut (cut of minimal capacity) which cuts all directed paths from n_1 to n_2.

Proof: It is clear that the maximal flow from n_1 to n_2 will not exceed the capacity of any cut which cuts all directed paths from n_1 to n_2, for the removal of edges belonging to such a cut would reduce the flow to zero.

Let there be a maximal flow F in G. We partition the nodes of G into two subgraphs G_1 and its complement $\bar{G}_1$, such that:

n_1 is in G_1

For any n_o in G_1, n'_o also is in G_1 if either

$$f(n_o, n'_o) < C(n_o, n'_o)$$

or

$$f(n'_o, n_o) > 0.$$

† Note here that a "cut" is not in general a "cut-set" as defined in Part I. After the removal of a cut the remaining graph may still be connected.

First, we shall show that n_2 is in $\overline{G}_1$. Suppose n_2 is in G_1; then there is a path from n_1 to n_2 such that all forward edges (where the edge orientation agrees with the path orientation) have flows lower than the edge capacity, and all reverse edges (where the edge orientation disagrees with the path orientation) have a positive flow. Thus there exists a positive quantity ϵ such that if flows in forward edges are increased by an amount equal to ϵ they are still under the edge capacity. Also, reducing flows in reverse edges by an amount equal to ϵ will not make them negative. However, by doing this we have increased the flow by an amount equal to $\epsilon > 0$. This contradicts our original assumption that the flow was maximal. Therefore n_2 belongs to $\overline{G}_1$.

Let us now focus our attention to the set of edges, S, that lead from G_1 to $\overline{G}_1$. If S is not empty, for each edge e_i in S,

$$f(e_i) = C(e_i). \tag{1.4}$$

However, for any node n_i in G_1,

$$\sum_{\substack{j \\ (j \neq i)}} \{f(n_i, n_j) - f(n_j, n_i)\} = F, \qquad n_i = \text{input};$$

$$\sum_{\substack{j \\ j \neq i}} \{f(n_i, n_j) - f(n_j, n_i)\} = 0, \qquad \text{otherwise}. \tag{1.5}$$

This implies

$$\sum_{\substack{i \\ i\epsilon G_1}} \sum_{\substack{j \\ (j \neq i)}} \{f(n_i, n_j) - f(n_j, n_i)\} = F \tag{1.6}$$

and therefore

$$\sum_{e_i \epsilon S} f(e_i) = F. \tag{1.7}$$

From Definition 4 the capacity of the cut S is

$$C(S) = \sum_{e_i \epsilon S} f(e_i) = F. \tag{1.8}$$

Thus we have exhibited a cut S whose capacity is equal to the maximal flow F. This, together with the fact that for any cut S' which separates n_1 from n_2

$$F \leqslant C(S'), \tag{1.9}$$

leads to the conclusion that

$$F = \min_{S'} \{C(S')\}. \tag{1.10}$$

From the proof of the last theorem we readily have a way of finding the maximal flow of an oriented communication net. The procedure is as follows:

1. Find a path such that all forward edges are not saturated ($f < C$) and all reverse edges have nonzero flow.

2. Let Δf_1 be the minimum of all the differences $(C - f)$ for forward edges and Δf_2 be the minimum of all flows of reverse edges, and $\Delta f = \min(\Delta f_1, \Delta f_2)$. Increase the flow of all forward edges of the path by Δf and reduce the flow of all reverse edges of this path by Δf.

3. Repeat steps 1 and 2 until no more path as described in step 1 can be found.

Example 1: Suppose the communication net in Fig. 1.3a is given. We wish to find the maximal flow from node 1 to node 3. Applying

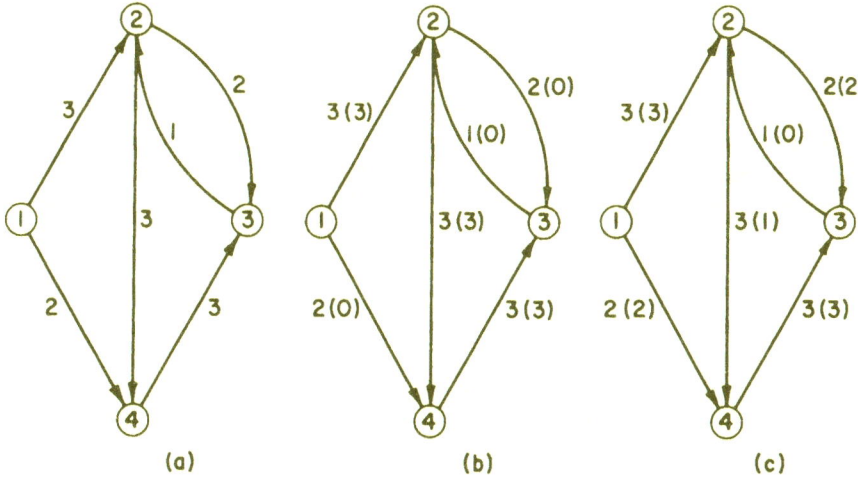

Fig. 1.3. (a) A communication net, (b) an intermediate flow assignment, and (c) the final and maximal flow assignment

step 1 we obtain the path which allows us to assign a flow of 3 units. We show this by inserting 3 after the capacity as shown in Fig. 1.3b.

Repeating 1 we find the path (1, 4) (4, 2) (2, 3) in which two forward edges could have an increase of at most two units and the reverse edge could have a reduction of as much as three units. Thus we increase the flow of this path by two units and obtain the flow assignment as shown in Fig. 1.3c. This is actually the maximal flow, for no more path as described in 1 could be found.

TERMINAL MATRIX AND ITS PROPERTIES

The maximal-flow capacity from node i to node j through a communication net is called the terminal capacity t_{ij}. A "terminal matrix" of a communication net is a matrix whose elements are terminal capacities of the net.[3]† As the edge matrix exhibits the structure of the net, the terminal matrix demonstrates the capacity of the net as a unit. It specifies its systems capability. The terminal matrix for the net in Fig. 1.3 is

$$
T = \begin{bmatrix} ① & 4 & 5 & 5 \\ 0 & ② & 5 & 3 \\ 0 & 1 & ③ & 1 \\ 0 & 1 & 3 & ④ \end{bmatrix}. \tag{1.11}
$$

Since the terminal matrix exhibits the maximal-flow capacity between different terminal pairs of the same communication net, one would expect some sort of relationship among the terminal capacities. This is indeed the case. The known properties of the terminal matrix of an oriented communication net are discussed in detail in this section.[4]

Theorem 2. Upon rearranging the rows and the corresponding columns (if necessary), a terminal matrix T is always partitionable into the form:

$$
T = \begin{bmatrix} A & | & T_1 \\ \hline C & | & B \end{bmatrix}, \tag{1.12}
$$

such that:

Submatrices A and B are square

T_1 is a uniform matrix with value t_1

Every element in A, B, or C is not smaller than the value of t_1

Both A and B are again partitionable in the same fashion, satisfying the same conditions until finally each submatrix becomes a one-by-one matrix.

† The "terminal matrix" defined in this section is sometimes referred to in literature as "the terminal capacity matrix."

Proof: There exists a weighted graph with **T** as its terminal capacity matrix. Among all possible cuts of this graph, there is a minimum-valued cut.

This cut cuts all the possible paths from some subgraph **A** to some subgraph **B**. If we "collect" all the nodes in the subgraph **A** by rearranging rows and columns, we have square matrices **A** and **B**. Suppose the value of this minimum cut is t_1; then the terminal capacity from any node in **A** to any node in **B** must be t_1. Therefore T_1 is uniform with value t_1. Every other element in matrix **T** corresponds to some cut value of the graph and therefore cannot be smaller than t_1. To show that either **A** or **B** is again partitionable in the same fashion we observe that among all the cuts separating any pair of nodes in subgraph **A**, there exists a cut with value t_2 such that every other cut of **A** assumes a value not smaller than t_2. Since this can always be done for any subgraph corresponding to a square matrix located along the diagonal line, a partition is always possible for this matrix unless it contains one element only as was stated in the theorem.

Corollary 1. Any terminal matrix **T** can be written (upon permuting the nodes) in such a form that there are at most $(n-1)$ distinct numbers above the diagonal line, i.e., for all $t_{ij}, j > i$.

Proof: If we carry out the partitions according to Theorem 1, the **T** matrix will be of the form

$$\mathbf{T} = \begin{bmatrix} \text{(see figure)} \end{bmatrix} \tag{1.13}$$

Since all $t_{i,i+1}$ $(i = 1, \ldots, n-1)$ may be distinct, and since each $t_{i,i+1}$ corresponds to exactly one partitioning, the corollary is proved.

Theorem 3. In association with the partitioning of a terminal matrix **T**, if t_1 corresponds to a minimum cut S_1 cutting all directed paths from subgraph **A** to subgraph **B**, and if t_2 corresponds to another minimum cut S_2 cutting all directed paths from A_1 to A_2 (both subgraphs of **A**), then S_2 cannot be a minimum cut of any two subgraphs of **B** unless $t_2 = t_1$. If $t_2 = t_1$, and if S_2 is also a minimum cut cutting all directed paths from some B_3 to B_4 (both nonempty subgraphs of **B**), then there exist at least two more cuts with the same minimum value t (see Fig. 1.4).

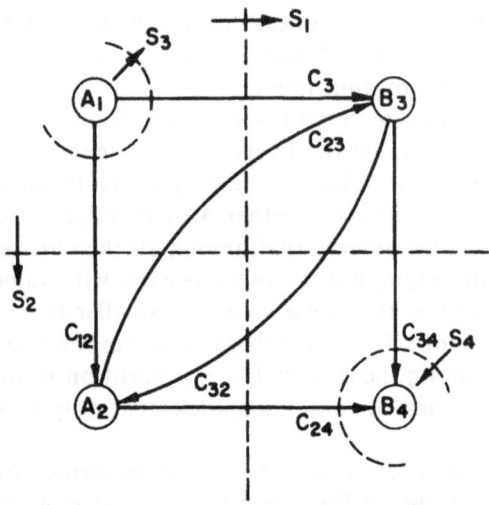

Fig. 1.4. Partitioning of a communication net used for the proof of Theorem 3

Proof: For the first part of the theorem, assume $t_2 > t_1$, and that it cuts **A** into nonempty subgraphs A_1 and A_2, and **B** into B_3 and B_4, respectively. In other words, the cut S_2 is the minimum cut cutting all directed paths from subgraph $A_1 \cup B_3$ to subgraph $A_2 \cup B_4$. Consider all possible directed paths from a node a_1 in A_1 to a node b_4 in B_4; there exists a cut S_1, assuming the value $t_1 < t_2$ which is a contradiction. Therefore $t_2 = t_1$.

To prove the second part of the theorem, let A_1, A_2, B_3, B_4, be the subgraphs obtained by cutting S_1 and S_2 through the graph as described above. Also let C_{13} be the sum of capacities of all edges going from subgraph A_1 to subgraph B_3, C_{12} the sum of capacities of all edges going from A_1 to A_2, etc., as shown in Fig. 1.4. We have the following identities:

$$C_{13} + C_{23} + C_{14} + C_{24} = t_1 \qquad (1.14)$$

$$C_{12} + C_{32} + C_{14} + C_{34} = t_1 \qquad (1.15)$$

and thus

$$C_{12} + C_{13} + C_{24} + C_{34} + 2C_{14} + C_{23} + C_{32} = 2t_1. \qquad (1.16)$$

Let S_3 and S_4 be the cuts around A_1 and B_4, respectively, as shown in Fig. 1.4; then we have:

$$C_{13} + C_{14} + C_{12} = t_{A_1} \geqslant t_1 \qquad (1.17)$$

$$C_{34} + C_{14} + C_{24} = t_{B_4} \geqslant t_1. \qquad (1.18)$$

Equation (1.16) becomes

$$t_{A_1} + t_{B_4} + C_{23} + C_{32} = 2t_1 \qquad (1.19)$$

or

$$t_{A_1} + t_{B_4} \leqslant 2t_1, \qquad (1.20)$$

since C_{23} and C_{32} are non-negative. But from (1.17) and (1.18) we have

$$t_{A_1} + t_{B_4} \geqslant 2t_1. \qquad (1.21)$$

Combining (1.20) and (1.21) we have

$$t_{A_1} + t_{B_4} = 2t_1. \qquad (1.22)$$

Since $t_{A_1} > t_1$ implies $t_{B_4} < t_1$ which is absurd, and since $t_{B_4} > t_1$ is similarly a contradiction, we thus have

$$t_{A_1} = t_B = t_1. \qquad (1.23)$$

Theorem 4. Let t_{ij} $(i, j = 1, \ldots, n, i \neq j)$ be any element of a T-matrix (i.e., terminal matrix); then

$$t_{ij} \geqslant \min(t_{ik}, t_{kj}),$$

and

$$i, j, k = 1, \ldots, n, i \neq j.$$

Proof: We shall consider k distinct from i and j, for otherwise the theorem is trivial.

If we remove the edges in the cut corresponding to t_{ij} and denote the new terminal capacities by t'_{ij}, t'_{ik}, and t'_{kj}, then

$$t'_{ij} = 0 \qquad (1.24)$$

$$t_{ik} - t_{ij} \leqslant t'_{ik} \qquad (1.25)$$

$$t_{kj} - t_{ij} \leqslant t'_{kj}. \qquad (1.26)$$

Assuming the theorem is false, then

$$t_{ij} < \min(t_{ik}, t_{kj}). \qquad (1.27)$$

With the relationships of (1.25) and (1.26) the expression of (1.27) becomes

$$0 = t'_{ij} < \min(t'_{ik}, t'_{kj}). \qquad (1.28)$$

This is absurd, since the union of the directed paths giving positive t'_{ik} and t'_{kj} contains a directed path from i to j which means $t'_{ij} > 0$, contradicting (1.24).

Theorem 5. Suppose (a) $\mathbf{T}'$ is the terminal matrix of G'; (b) $\mathbf{T}''$ is the terminal matrix of G''. Let

$$G = G' + G'' \text{ (in terms of edge matrices; } \mathbf{E} = \mathbf{E}' + \mathbf{E}'') \qquad (1.29)$$

and

$$\mathbf{T} = \mathbf{T}' + \mathbf{T}''. \qquad (1.30)$$

Then $\mathbf{T}$ is the terminal matrix of G if and only if for each ordered node pair i and j there exists a cut which is a minimum cut for all three graphs G, G', and G''.

Proof: If for each ordered node pair i and j there exists a .cut which remains minimum for G, G', and G'', then the terminal capacities in G' and G'' are determined by this cut and are t'_{ij} and t''_{ij}, respectively. However, since any edge capacity in G is just the sum of the corresponding edge capacities in G' and G'', the capacity of this cut in G is simply $(t'_{ij} + t''_{ij})$. But this cut is also a minimum cut in G by assumption; thus we have $t_{ij} = t'_{ij} + t''_{ij}$. On the other hand, if $t_{ij} = t'_{ij} + t''_{ij}$, and if there exists no cut which is a minimum cut for G, G', and G'', then any cut of G giving t_{ij} cannot be a minimum cut for both G' and G''. That is, this cut will assume values C'_{ij} and C''_{ij} in G' and G'' and either $C'_{ij} > t'_{ij}$ or $C''_{ij} > t''_{ij}$, or both. But since

$$t_{ij} = C'_{ij} + C''_{ij}, \qquad (1.31)$$

now

$$C'_{ij} + C''_{ij} > t'_{ij} + t''_{ij} \qquad (1.32)$$

or

$$t_{ij} > t'_{ij} + t''_{ij}, \qquad (1.33)$$

which is a contradiction.

SYNTHESIS OF A TERMINAL MATRIX

In this section we show that for a graph containing three nodes or less, the conditions stated in Theorems 2 and 4 are necessary and sufficient for the realizability of a terminal matrix. For a graph containing four nodes or more, these conditions are no longer sufficient. A counter example of order four has been found.† The realization techniques of low-order cases have their applications in the realization of higher-order matrices, if they can be partitioned in some special way as will be shown later in this section.

If a graph contains two nodes only, then each of the two cuts contains exactly one edge. The edge matrix is thus the same as the terminal matrix and the synthesis is trivial.

The terminal matrix of a graph containing three nodes is 3 by 3 and the partition must be one–two as follows:

$$
T = \begin{bmatrix} ① & t_1 \\ t_{21} & ② & t_{23} \\ t_{31} & t_{32} & ③ \end{bmatrix}
\quad \text{or} \quad
\begin{bmatrix} ① & t_{12} \\ & & t_1 \\ t_{21} & ② \\ t_{31} & t_{32} & ③ \end{bmatrix}.
\tag{1.34}
$$

We shall consider the partition on the left. The synthesis methods for these two cases are analogous. First we write **T** as the sum of two component matrices:

$$
T = \begin{bmatrix} ① & O \\ t'_{21} & ② & t'_{23} \\ t'_{31} & t'_{32} & ③ \end{bmatrix}
+ \begin{bmatrix} ① & t_1 & t_1 \\ t_1 & ② & t_1 \\ t_1 & t_1 & ③ \end{bmatrix}
$$

$$
= T' + T_u,
\tag{1.35}
$$

where $t_{ij}' = t_{ij} - t_1 \geqslant 0$.

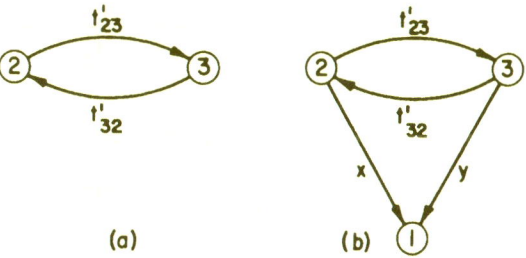

(a)   (b)

Fig. 1.5. *Realization of a terminal matrix of order three*

† Private communication with W. Mayeda.

To realize T', we first realize the 2 by 2 matrix (Fig. 1.5a). Since the first row in T' is zero, the only connections in the realization should be from the subgraph of Fig. 1.5a to node 1. We thus assume edges with capacities x and y as shown in Fig. 1.5b. Note that x and y, if both positive, do not change t'_{32} or t'_{23}. In order that the graph of Fig. 1.5b realize T', the minimum cut requirements must satisfy T' as in equations:

$$\min\{(x+y), (x+t'_{23})\} = t'_{21} \tag{1.36}$$

$$\min\{(y+x), (y+t'_{32})\} = t'_{31} \tag{1.37}$$

or

$$x = \max\{(t'_{21}-t'_{23}), (t'_{21}-y)\} \tag{1.38}$$

$$y = \max\{(t'_{31}-t'_{32}), (t'_{31}-x)\}. \tag{1.39}$$

If we sketch the above two equations on an x, y-plane, we have two "curves" (each of them is piecewise straight) as shown in Fig.

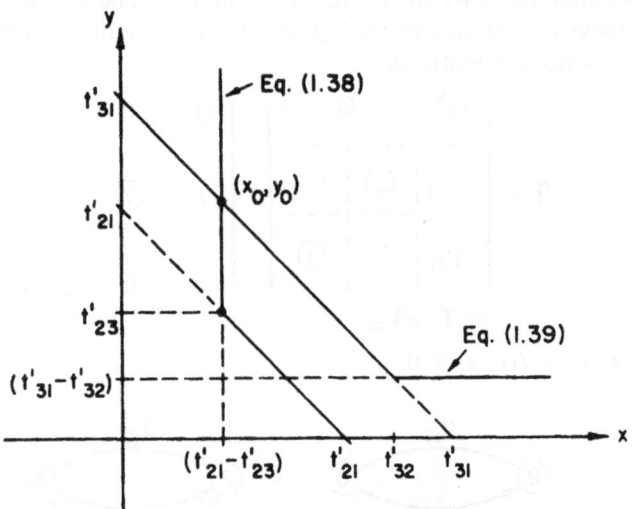

Fig. 1.6. Plots of Eqs. (1.38) and (1.39)

1.6. These "curves" must intersect at a point (x_0, y_c) where if $t'_{31} \geqslant t'_{21}$, we have

$$x_0 = t'_{21} - t'_{23} \tag{1.40}$$

$$y_0 = \max\{(t'_{31}-t'_{32}), (t'_{31}-t'_{21}+t'_{23})\}. \tag{1.41}$$

If $t'_{31} \leqslant t'_{21}$, we have

$$x_0 = \max\{(t'_{21} - t'_{23}), (t'_{21} - t'_{31} + t'_{32})\} \tag{1.42}$$

$$y_0 = t'_{31} - t'_{32}. \tag{1.43}$$

Note that if $t'_{31} = t'_{21} < (t'_{32} + t'_{23})$, the solution is not unique. All the points on the straight line-segment where two curves coincide are solutions.

Furthermore, x_0 and y_0 must assume positive values. To show this, we consider the different cases separately. If $t'_{31} > t'_{21}$, and $x_0 = (t'_{21} - t'_{23}) < 0$, then $t'_{21} < \min(t'_{31}, t'_{23})$, violating Theorem 4. If $t'_{31} > t'_{21}$, and $y_0 = \max\{(t'_{31} - t'_{32}), (t'_{31} - t'_{21} + t'_{23})\} < 0$, then $(t'_{21} - t'_{23}) > t'_{31} > t'_{21}$ which means $t'_{23} < 0$, contradicting the assumption that $t_{ij} \geqslant 0$. One can show similarly that if $t'_{31} < t'_{21}$, both x_0 and y_0 must be non-negative. If $t'_{31} = t'_{21} < (t'_{32} + t'_{23})$, then the line-segment where two curves coincide will pass through the first quadrant since $t'_{31} = t'_{21} \geqslant 0$ by assumption.

The realization of a uniform matrix T_u is a cycle oriented in either direction. The graph may also contain two cycles oriented in different directions (Fig. 1.7). We note that in three-node cases the

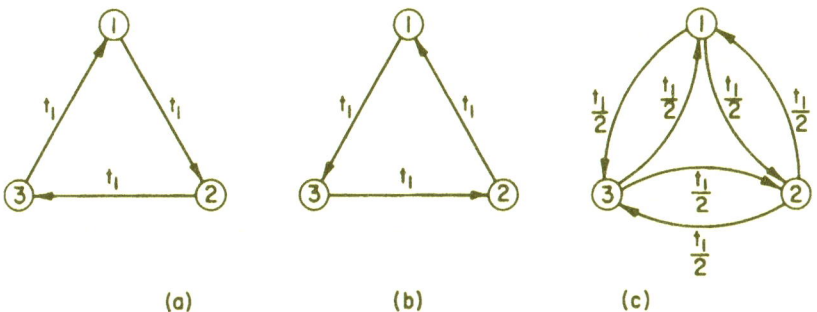

Fig. 1.7. *Realization of a uniform matrix*

minimum cuts of T' always correspond to minimum cuts of T_u since the capacity of all cuts is equal.

When the same approach is used in higher-order cases, a set of simultaneous equations similar to Eqs. (1.36) and (1.37) can be written for the unknown edges connecting two lower-order subgraphs which were already realized. It is found that although the successive approximation method converges to the set of solutions, the triangular relationship of Theorem 4, may not guarantee that the solutions be non-negative. However, in general, the technique of decomposition could be used.

Theorem 6. When a matrix **T** is realizable and

$$\mathbf{T} = \mathbf{T'} + \mathbf{U}, \tag{1.44}$$

where **T'** is realizable as G' and **U** is a uniform matrix of value t_u, then **T** may be realized as

$$G = G' + G_u, \tag{1.45}$$

where G_u is a realization of **U** such that all simple cuts in G have capacity t_u, and minimum cuts of G' correspond to simple cuts in G_u.

Proof: The proof of this theorem follows directly from Theorem 5.

The techniques of realizing a low-order T-matrix can be conveniently applied to a T-matrix of higher order, in general, if it can be partitioned. This is done by rearranging the nodes in such a fashion that the following conditions are satisfied:

1. Each submatrix corresponding to a subcollection of nodes lying along the diagonal line is square and contains elements with values no smaller than the value of any of the elements in an off-diagonal submatrix. We refer to these as "dominant conditions" of partitioning of a terminal matrix.

2. Each off-diagonal submatrix is a constant matrix.

We see that the theorems obtained for nodes apply to submatrices. For instance, $t_{IJ} \geq \min(t_{IK}, t_{KJ})$, as required by Theorem 4. This generalization is immediate since all the minimum cuts corresponding to the off-diagonal matrices must stay outside the subgraphs corresponding to submatrices by Condition 1.

Theorem 7. A matrix **T** satisfying the dominant conditions is realizable as a terminal matrix if: (a) treating these submatrices along the diagonal line as nodes, the matrix **T** is realizable; (b) each submatrix along the diagonal line is realizable.

Proof: The sufficiency of these two conditions to imply realizability is proved by constructing the synthesis procedure demonstrated in the following example.

Example 2: The following matrix is to be realized:

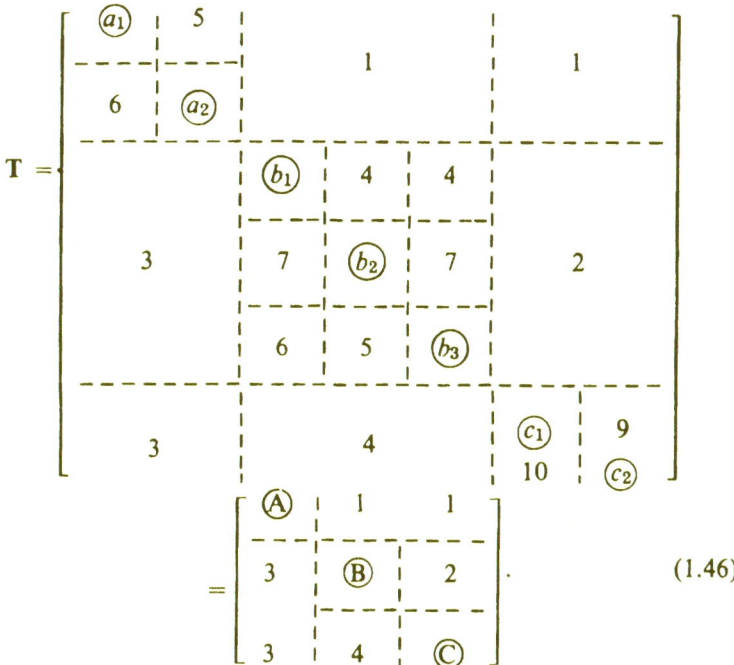

$$
\mathbf{T} =
\begin{bmatrix}
\boxed{a_1} & 5 & & & 1 & & & 1 \\
6 & \boxed{a_2} & & & & & & \\
& & \boxed{b_1} & 4 & 4 & & & \\
3 & & 7 & \boxed{b_2} & 7 & & 2 & \\
& & 6 & 5 & \boxed{b_3} & & & \\
3 & & & 4 & & & \boxed{c_1} & 9 \\
& & & & & & 10 & \boxed{c_2}
\end{bmatrix}
$$

$$
=
\begin{bmatrix}
\circledA & 1 & 1 \\
3 & \circledB & 2 \\
3 & 4 & \copyright
\end{bmatrix}.
\qquad (1.46)
$$

We may write:

$$
\mathbf{T} = \mathbf{T'} + \mathbf{T}_c =
\begin{bmatrix}
\boxed{A'} & 0 & 0 \\
2 & \boxed{B'} & 1 \\
2 & 3 & \boxed{C'}
\end{bmatrix}
+
\begin{bmatrix}
\boxed{A_c} & 1 & 1 \\
1 & \boxed{B_c} & 1 \\
1 & 1 & \boxed{C_c}
\end{bmatrix}.
\qquad (1.47)
$$

The realization of $\mathbf{T'}$ is shown in Fig. 1.8a. Superimposing a cyclic realization of $\mathbf{T}_c$ (Fig. 1.8b), the realization of $\mathbf{T}$ in subgraphs is shown in Fig. 1.8c. Meanwhile, the submatrices $\mathbf{A}$, $\mathbf{B}$, and $\mathbf{C}$ are realized as shown in Figs. 1.9a, b, and c, respectively.

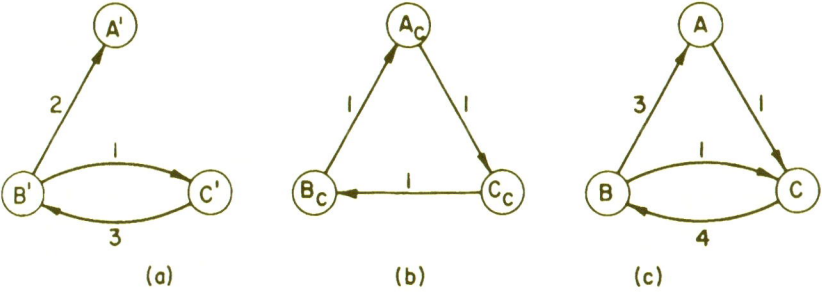

Fig. 1.8. Realization of (a) $\mathbf{T'}$, *(b)* $\mathbf{T}_c$, *and (c)* $\mathbf{T}$ *of Eq. (1.47)*

The easiest way of putting the realizations of the submatrices in place of the nodes of Fig. 1.8c is to choose one node from each subgraph for the outside connections (Fig. 1.10a). However, one

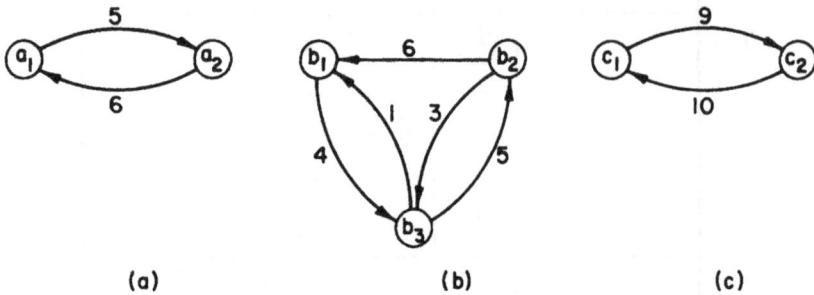

Fig. 1.9. Realization of matrices (a) **A**, *(b)* **B**, *and (c)* **C** *of Eq. (1.46)*

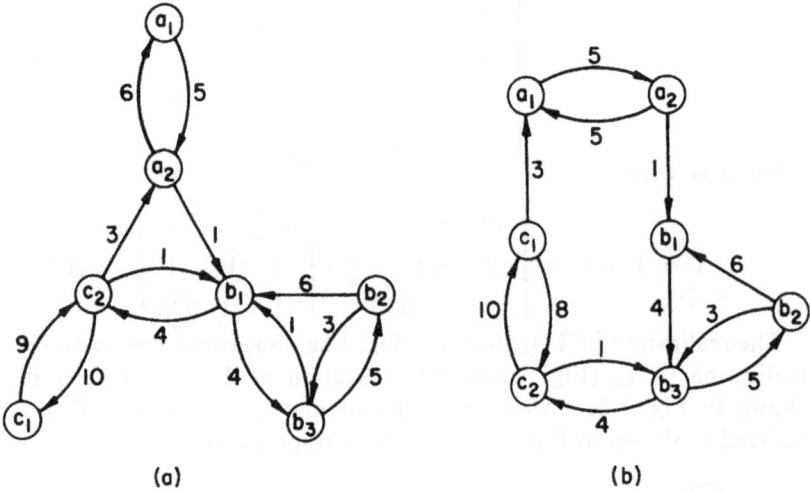

Fig. 1.10. Two possible realizations of the terminal matrix in Example 2

may arbitrarily assign the nodes for outside connections as long as the flow capacity through the outside paths does not exceed the corresponding edge-capacity in the submatrix realization. In such cases, edge-capacity in the final realization of **T** is obtained by subtracting the amount of total outside flow between any ordered pair of nodes of a subgraph from the edge-capacity of the same ordered pair of the subgraph. One such alternative realization is shown in Fig. 1.10b.

CHAPTER 2

Nonoriented
Communication Nets

NONORIENTED COMMUNICATION NETS AND MAXIMUM FLOW

A nonoriented communication net is a nonoriented graph with capacities as weights. The flow in any edge may go in either (but not both) direction as long as its magnitude does not exceed the edge-capacity. The edge matrix for a nonoriented communication net is

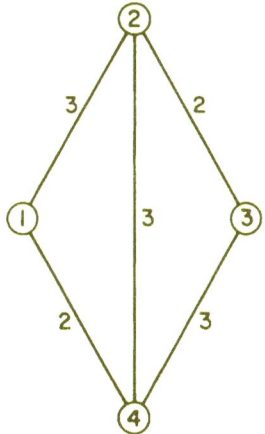

Fig. 2.1. A nonoriented communication net

defined in the same way as in the oriented case. Figure 2.1 shows a nonoriented communication net with the edge matrix

$$\mathbf{E} = \begin{bmatrix} ① & 3 & 0 & 2 \\ 3 & ② & 2 & 3 \\ 0 & 2 & ③ & 3 \\ 2 & 3 & 3 & ④ \end{bmatrix}. \tag{2.1}$$

The edge matrix E is seen to be a symmetrical matrix as a result of the nonoriented character of the graph.

Theorem 8. The maximal flow between nodes i and j is equal to the minimum of the capacities of cut-sets,† the removal of which separates nodes i and j.

This is the max-flow min-cut theorem for nonoriented communication nets. A proof paralleling the reasoning of that in the oriented case is possible; the reader is asked to verify this in detail.

It also follows that the method described in Chapter 1 for finding max-flow of a communication net is applicable in the nonoriented case.

Theorem 9. The set of all cut-sets, $\{S_i\}$, of a graph form a group under the operation ring sum.

Proof: The four properties of a group are:

Closure:

$$S_i, \ S_j \epsilon \ddagger \{S_i\} \text{ imply } (S_i \oplus S_j) \epsilon \{S_i\}; \tag{2.2}$$

Associative law:

$$(S_i \oplus S_j) \oplus S_k = S_i \oplus (S_j \oplus S_k); \tag{2.3}$$

Existence of an identity, I, such that

$$I \oplus S_i = S_i \oplus I = S_i; \tag{2.4}$$

Existence of an inverse S_i^{-1} for every S_i such that

$$S_i \oplus S_i^{-1} = S_i^{-1} \oplus S_i = I. \tag{2.5}$$

Equation (2.3) follows from definition. The identity element of the group is the empty set. The inverse for any S_i is itself for

$$S_i \oplus S_i = I. \tag{2.6}$$

To show that property (2.2) is satisfied we refer to Fig. 2.2. Here a is the set of edges connecting subgraphs G_1 and G_3 and similarly for other sets b, c, d, etc. Thus

$$S_1 = a \oplus c \oplus e \oplus f \tag{2.7}$$
$$S_2 = b \oplus d \oplus e \oplus f \tag{2.8}$$

and

$$S_1 \oplus S_2 = a \oplus b \oplus c \oplus d; \tag{2.9}$$

† A cut-set here implies an edge-disjoint union of one or more simple cut-sets.
‡ Reads as "is contained in."

$S_3 = S_1 \oplus S_2$ is seen to be a cut-set which cuts G into

$$G' = G_1 \cup G_4 \qquad (2.10)$$

and

$$G'' = G_2 \cup G_3. \qquad (2.11)$$

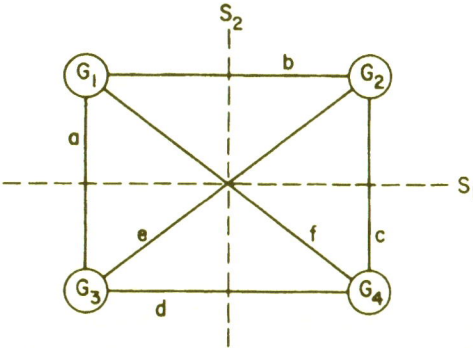

Fig. 2.2. A nonoriented communication net and cut-sets

It is a minimal set, for the addition of any edge will make the graph connected. The group property of $\{S_i\}$ is a useful one. We shall in the following illustrate its usefulness by deriving some properties of the graph in relation to communication nets.

Suppose we now represent the elements of the group by binary sequences of $n(n-1)/2$ digits, such that each position of the sequence corresponds to an edge; if any edge e is in the set we shall write a "one" in its corresponding position; otherwise a zero is entered. We immediately see that these sequences may be treated as vectors and addition of each component is mod 2 ($1 \oplus 1 = 0$). Also, any linear combination of sets with coefficients restricted to zero or one may be obtained. These linear combinations correspond to the ring sum of cut-sets so they are also elements of the group. With these observations we see that the cut-sets of a graph form a vector space with vectors as defined above and scalars zero and one. This observation is very useful in graph theory and will be used later to prove theorems for nonoriented communication nets.[5,6]†

Theorem 10. Given two arbitrary simple cut-sets S_1 and S_2, of a complete graph G, there is at least one edge in common.

† For a more detailed discussion on the relationships between the group of n-tuples with respect to modulo 2 addition and the vector space concept, see any book on modern algebra.

Proof: Refer to Fig. 2.3. Suppose cut-set S_1 cuts the graph into two parts G_1 and G_2. If we delete all edges from the graph except the edges belonging to cut-set S_1, the graph is still connected. That is, there is a path between any two nodes i and j. If nodes i and j

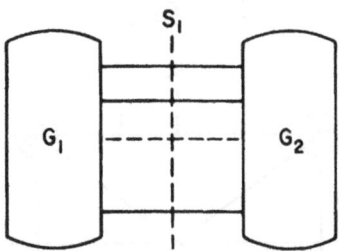

Fig. 2.3. A completely connected graph and simple cut-sets

are both in $G_1(G_2)$, they are both connected to any node in $G_2(G_1)$. If node i is in G_1 and node j is in G_2, they are connected directly. Therefore, with the removal of any subset of the edges one will still leave the graph connected. This means it is not possible to obtain any cut-set which does not contain any element of S_1.

Theorem 11. For a completely connected graph, every cut-set is a simple cut-set.

Proof: From the fact that the cut-sets form a vector space, it is possible to select a base which consists of $(n-1)$ cut-sets. Moreover, it is possible to select $(n-1)$ simple cut-sets as a base. An example of this is the choice of $n-1$ incident sets. Using this set of $(n-1)$ simple cut-sets as a base, every cut-set may be expressed as a linear combination of these basic cut-sets. It is only necessary, then, to prove that for a completely connected graph, the linear combination of any two simple cut-sets is again a simple cut-set.

In the most general case, two cut-sets will cut a completely connected graph into four subgraphs as shown in Fig. 2.2. In particular cases, one of the subgraphs will be empty, but the argument will still apply. It is seen that, graphically, $S_1 \oplus S_2$ will be a cut-set which leaves G_1 and G_4 connected, and similarly for G_2 and G_3. The graph is therefore cut into two parts by the cut-set $(S_1 \oplus S_2)$.

A more rigorous argument is also possible. Suppose S_1 cuts the graph G into two subgraphs G_{1a} and G_{1b}, and S_2 cuts G into G_{2a} and G_{2b}.

Let

G_1 denote the subgraph common to G_{1a} and G_{2a}

G_2 denote the subgraph common to G_{1a} and G_{2b}

G_3 denote the subgraph common to G_{1b} and G_{2a}

G_4 denote the subgraph common to G_{1b} and G_{2b}.

It is seen that G_1, G_2, G_3, and G_4 are connected subgraphs. Also, if any edge is contained in one, it will not be contained in any of the other three subgraphs. Let us focus our attention on G_2 and G_3. After the removal of cut-set S_1 or S_2, G_2 is not connected with G_3. Since we have a completely connected graph, there exists an edge e which connects some node in G_2 and some node in G_3, provided both G_2 and G_3 are nonempty. This edge e is in both S_1 and S_2, consequently not in $(S_1 \oplus S_2)$. Thus G_2 and G_3 are connected after the cut-set $(S_1 \oplus S_2)$ is removed. The same argument may be applied to G_1 and G_4. It then follows that in all cases there will be at most two separated nonempty subgraphs after the cut-set $(S_1 \oplus S_2)$ is removed from the graph.

Definition 5. A function $F_{ij}(S_k)$ is defined for each pair of nodes i, j and each simple cut-set S_k such that:

$F_{ij}(S_k) = -1$, if i and j are not connected with the removal of S_k;

$F_{ij}(S_k) = 1$, if i and j are connected with the removal of S_k. $\hspace{2em}$ (2.12)

Theorem 12. For a completely connected graph,

$$F_{ij}(S_k \oplus S_L) = F_{ij}(S_k)F_{ij}(S_L). \hspace{2em} (2.13)$$

In essence, this theorem says that if both S_k and S_L cut nodes i and j or do not cut i and j, then the set $(S_k \oplus S_L)$ will not cut i, j. If one and only one of S_k or S_L cuts i, j, then the set $(S_k \oplus S_L)$ will also cut i, j.

Proof: We shall first show that

$$F_{ij}(S_k) = F_{ij}(S_L) \hspace{2em} (2.14)$$

implies

$$F_{ij}(S_k \oplus S_L) = 1. \hspace{2em} (2.15)$$

In a completely connected graph, there exists an edge e connecting i and j. If $F_{ij}(S_k) = F_{ij}(S_L)$, either e is contained in both S_k or S_L or e is contained in neither. In both cases, $(S_k \oplus S_L)$ will not contain e. Therefore i and j will be connected with the removal of the cut-set $(S_k \oplus S_L)$.

On the other hand, if

$$F_{ij}(S_k) = -F_{ij}(S_L) \hspace{2em} (2.16)$$

we cannot also have

$$F_{ij}(S_k \oplus S_L) = 1. \tag{2.17}$$

For, without loss of generality, we may assume

$$F_{ij}(S_k) = 1$$
$$F_{ij}(S_L) = -1, \tag{2.18}$$

if, at the same time,

$$F_{ij}(S_k \oplus S_L) = 1. \tag{2.19}$$

It follows from the above that

$$F_{ij}(S_k) = F_{ij}(S_k \oplus S_L), \tag{2.20}$$

which implies

$$F_{ij}(S_k \oplus S_k \oplus S_L) = F_{ij}(S_L) = 1. \tag{2.21}$$

Since it is originally assumed that

$$F_{ij}(S_L) = -1, \tag{2.22}$$

we must conclude that

$$F_{ij}(S_k \oplus S_L) = -1. \tag{2.23}$$

Therefore, in all cases,

$$F_{ij}(S_k \oplus S_L) = F_{ij}(S_k)F_{ij}(S_L). \tag{2.24}$$

Corollary 2. Let S_{G_1,G_2} denote the cut-set which cuts a connected graph G into G_1 and G_2. For a complete graph,

$$S_{G_1,G_2} \oplus S_{G_3,G_4} = S_{G_5,G_6}, \tag{2.25}$$

such that: (a) subgraphs G_1, G_3, G_5 all contain a given reference node; (b) subgraph G_5 contains all nodes which are common to subgraphs G_1 and G_3, or subgraphs G_2 and G_4, and only these nodes. This corollary follows directly from Theorem 12.

Example 3: Figure 2.4 shows a completely connected graph of four nodes. The cut-sets are

$$S_1 = (a,b,f); \quad S_2 = (a,c,d); \quad S_3 = (b,c,e); \quad S_4 = (d,e,f);$$
$$S_5 = (b,c,d,f); \quad S_6 = (a,c,e,f); \quad S_7 = (a,b,d,e). \tag{2.26}$$

Applying the notation in Theorem 12

$$F_{23}(S_1 \oplus S_2) = F_{23}(S_5) = -1 \tag{2.27}$$
$$F_{23}(S_1)F_{23}(S_2) = (1)(-1) = -1. \tag{2.28}$$

Also

$$S_1 = (a, b, f) = S_{1,234} \qquad (2.29)$$

$$S_2 = (a, c, d) = S_{134,2}. \qquad (2.30)$$

It does follow that

$$S_5 = S_1 \oplus S_2 = (b, c, d, f) = S_{12,34}. \qquad (2.31)$$

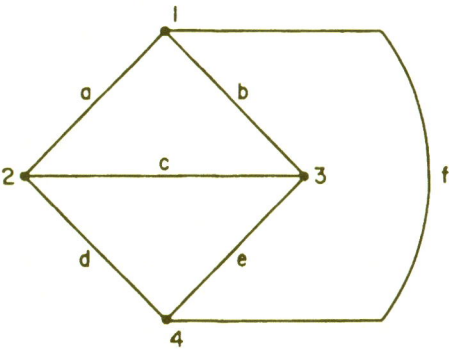

Fig. 2.4. A completely connected graph of four nodes

The terminal-capacity t_{ij}, between a pair of nodes i and j, is the maximum-flow capacity between the two points through the net.

The maximum flow between a pair of nodes is equal to the minimum of the values of cut-sets that cut i and j. The value of a cut-set is simply the sum of the capacities for the edges in the cut-set.

The usual method of computing the maximum flow between a pair of nodes is through the construction of all paths between i and j. Due to the fact that the cut-sets of a graph form a vector space, it is possible to generate cut-sets in a systematic manner and thus make use of the digital computer.

The theorems in this section lead to the following outlined procedure.

Given a net N with n nodes, we shall associate it with a completely connected graph N_M also of n nodes such that the set of edges which belong to N_M but not N has zero capacity. Also, an edge which connects the corresponding nodes in N and N_M has the same capacity in both graphs. The cut-set in the associated graph has properties just discussed. It is readily seen that the corresponding terminal-capacities of N and N_M are equal.

From the foregoing results an alternative process of computing all terminal-capacities of a net is outlined as follows:

Form the augmented graph

Select $n-1$ independent simple cut-sets from N_M

List the node pairs that are separated by each cut-set

For a given pair of nodes i and j, classify the cut-sets according to $F_{ij}(S_k)$ is 1 or -1

Form the linear combination of cut-sets such that the number of cut-sets for which $F_{ij}(S_k) = -1$ is odd

Evaluate the value for the cut-sets. The minimum value is the terminal-capacity.

TERMINAL MATRICES OF NONORIENTED COMMUNICATION NETS AND THEIR REALIZATION

Terminal matrices of nonoriented communication nets are defined as in the case of oriented nets. Any entry t_{ij} denotes the maximal-flow capacity from terminal i to terminal j. Since the graph is non-oriented, it then follows that $t_{ij} = t_{ji}$ for all i and j. As a result of this symmetry, there exist many additional properties for terminal matrices of nonoriented nets.

Theorem 13. Upon rearranging the rows and the corresponding columns (if necessary), a T-matrix for nonoriented communication nets is always partitionable into the form[3]:

$$T = \begin{bmatrix} A & T_1 \\ T_1{}^t & B \end{bmatrix}, \qquad (2.32)$$

Such that

Submatrices A and B are square

T_1 is a uniform matrix with value t_1

Every element in A or B is not smaller than t_1

Both A and B are again partitionable in the same fashion satisfying same conditions until finally each submatrix becomes a one-by-one-matrix.

The proof of this theorem is left to the reader.

Corollary 3. There are at most $(n-1)$ distinct elements, i.e., numbers in a T-matrix of a nonoriented communication net.

Corollary 4. Upon rearranging the rows and the corresponding columns, a T-matrix for nonoriented communication net is partitionable into the form[7]:

$$T = \begin{bmatrix} A_1 & T_{12} & \ldots & T_{1k} \\ T_{12}{}^t & A_2 & \ldots & T_{2k} \\ \cdot & \cdot & \cdots & \cdot \\ T_{1k}{}^t & T_{2k}{}^t & \ldots & A_k \end{bmatrix}, \qquad \text{for } k \leqslant n, \qquad (2.33)$$

such that:

Submatrices $A_1, \ldots, A_k$ are square

T_{ij} is a uniform matrix with value t_1

Every element in $A_1, \ldots, A_k$ is greater than t_1 (except for those one-by-one matrices)

Every square submatrix $A_1, \ldots, A_k$ can be partitioned in the same way and satisfies the same conditions.

Proof: Clearly, if a matrix satisfied conditions in Corollary 4, it satisfied the conditions in Theorem 13. Conversely, if a matrix satisfied conditions in Theorem 13 and matrix A_1 (or A_2) contains an element of value t_1, by applying conditions in Theorem 13 once more A_1 may be partitioned as

$$A_1 = \begin{bmatrix} B_1 & T_1 \\ T_1{}^t & B_2 \end{bmatrix}. \tag{2.34}$$

In general if T can be partitioned as

$$T = \begin{bmatrix} A_1 & T_{12} \ldots & T_{1x} \\ T_{12}{}^t & A_2 \ldots & T_{2x} \\ \cdot & \cdot \quad \cdot \quad \cdot \quad \cdot \\ T_{1x}{}^t & T_{2x}{}^t \ldots A_x \end{bmatrix}, \tag{2.35}$$

then either every element in $A_1 \ldots A_x$ has value greater than t_1 or T can be partitioned as

$$T = \begin{bmatrix} A_1 & T_{12} \ldots & T_{1,x+1} \\ T_{12}{}^t & A_2 \ldots & T_{2,x+1} \\ \cdot & \cdot \quad \cdot \quad \cdot \quad \cdot \quad \cdot \\ T_{1,x+1}{}^t & T_{2,x+1}{}^t \ldots A_{x+1} \end{bmatrix}. \tag{2.36}$$

Since n is finite, there exists a $k \leqslant n$ such that conditions in Corollary 4 are satisfied.

Theorem 14. If a matrix satisfies conditions in Theorem 13, then it is realizable as a T-matrix of a nonoriented communication net.

Proof: We may show the sufficiency by a construction procedure. If T satisfies the conditions in Theorem 13, then

$$T = \begin{bmatrix} A_1 & T_1 \\ T_1{}^t & A_2 \end{bmatrix}. \tag{2.37}$$

We connect an edge of value t_1 between some node of A_1 and some node of A_2. An edge of value equal to the minimum cut of A_1 is connected between some node of B_1 to some node of B_2 where

$$A_1 = \begin{bmatrix} B_1 & T_2 \\ T_2{}^t & B_2 \end{bmatrix}. \tag{2.38}$$

As all partitions are carried out $(n-1)$ edges have been used. It is seen also that the resultant graph must be connected. It is then a tree. To show that the graph is actually a realization of T, consider two subgraphs G_1, G_2 at any partitioning stage with t_x as the capacity between G_1 and G_2. Within G_1 and G_2 we have a connected graph each of whose elements has a value greater than t_x and where there is one and only one edge connecting G_1 and G_2. Since removing this edge separates G_1 and G_2, the terminal-capacity between G_1 and G_2 is actually t_x.

Corollary 5. Any realizable T-matrix of a nonoriented communication net may be realized as a path.

Proof: We may rearrange the row and corresponding columns of the T-matrix such that all partitions may be carried out without further rearrangement of rows and columns. Number the node of the matrix in ascending order as $t_{ii} = i$. Whenever an edge is added between a pair of partitions it is connected between two nodes of successive numbers. The resulting graph is a path.

Example 4. The matrix T given in Eq. (2.39) satisfies the necessary and sufficient conditions outlined in Theorem 13:

$$T = \begin{bmatrix}
① & 4 & 4 & 2 & 2 & 2 & 2 & 2 & 2 \\
4 & ② & 4 & 2 & 2 & 2 & 2 & 2 & 2 \\
4 & 4 & ③ & 2 & 2 & 2 & 2 & 2 & 2 \\
2 & 2 & 2 & ④ & 4 & 4 & 2 & 2 & 2 \\
2 & 2 & 2 & 4 & ⑤ & 4 & 2 & 2 & 2 \\
2 & 2 & 2 & 4 & 4 & ⑥ & 2 & 2 & 2 \\
2 & 2 & 2 & 2 & 2 & 2 & ⑦ & 4 & 4 \\
2 & 2 & 2 & 2 & 2 & 2 & 4 & ⑧ & 4 \\
2 & 2 & 2 & 2 & 2 & 2 & 4 & 4 & ⑨
\end{bmatrix}$$

$$
= \begin{bmatrix}
① & 4 & 4 & 2 & 2 & 2 & 2 & 2 & 2 \\
4 & ② & 4 & 2 & 2 & 2 & 2 & 2 & 2 \\
4 & 4 & ③ & 2 & 2 & 2 & 2 & 2 & 2 \\
2 & 2 & 2 & ④ & 4 & 4 & 2 & 2 & 2 \\
2 & 2 & 2 & 4 & ⑤ & 4 & 2 & 2 & 2 \\
2 & 2 & 2 & 4 & 4 & ⑥ & 2 & 2 & 2 \\
2 & 2 & 2 & 2 & 2 & 2 & ⑦ & 4 & 4 \\
2 & 2 & 2 & 2 & 2 & 2 & 4 & ⑧ & 4 \\
2 & 2 & 2 & 2 & 2 & 2 & 4 & 4 & ⑨
\end{bmatrix}. \tag{2.39}
$$

Two realizations of **T** are given in Fig. 2.5.

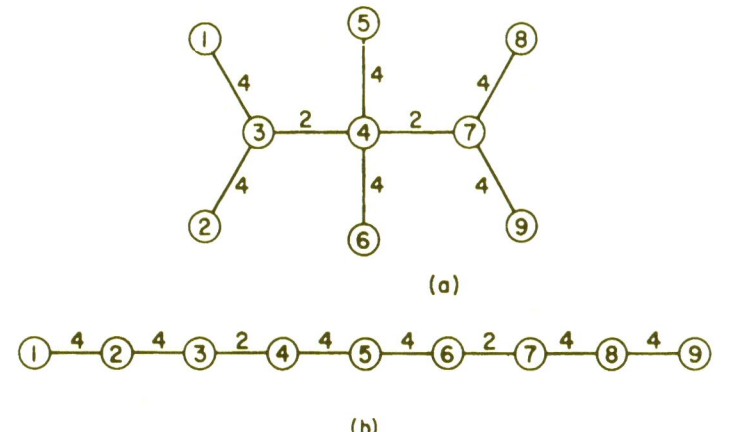

(a)

(b)

Fig. 2.5. Two realizations of the T-matrix of Example 4

Theorem 15. A matrix **T** is realizable as a terminal matrix if and only if for all $i, j, k = 1, 2, \ldots, n$[8],

$$t_{ij} = t_{ji} = \text{a non–negative real number} \tag{2.40a}$$

$$t_{ij} \geq \min\{t_{ik}, t_{kj}\}. \tag{2.40b}$$

Proof: The proof of Theorem 4 can be used here to show that in the nonoriented case a terminal matrix must satisfy the triangular condition. We now show the sufficiency of the conditions in the theorem by construction and induction. The construction is performed by expanding a tree with addition of new nodes. Let us assume that at stage k we have $(k - 1)$ edges in the tree and that the tree realizes all terminal-capacities, t_{ij}, for $i, j = 1, 2, \ldots, k$. Let us also denote the set of k nodes in the tree by $\{x\}$ and the set of $(n-k)$ nodes not in the tree by $\{y\}$. We choose a new node y_0 such that, for some node x_0 in $\{x\}$,

$$t_{x_0 y_0} \geq t_{xy}, \quad \text{for } x = 1, 2, \ldots, k$$
$$\text{and } y = k+1, k+2, \ldots, n. \tag{2.41}$$

Connect $e_{x_0 y_0}$ between nodes x_0 and y_0 with capacity $t_{x_0 y_0}$. The node y_0 and the edge $e_{x_0 y_0}$ thus become a part of the expanded tree, and now

$$\{x\} = \{1, 2, 3, \ldots, k, k+1\}. \tag{2.42}$$

$$\{y\} = \{k+2, k+3, \ldots, n\}. \tag{2.43}$$

We see that the terminal-capacities, t_{ij}, for $i, j = 1, 2, \ldots, k$, have not changed. And $t_{x_0, k+1}$ is precisely what we want. Now we must show that $t_{x,k+1}$ computed from the tree coincides with those as specified in the matrix. For any x we must have

$$t_{x,x_0} \geq \min\{t_{x,k+1}, t_{x_0,k+1}\} \geq t_{x,k+1}. \tag{2.44}$$

Consequently,

$$\min\{t_{x,x}, t_{x_0,k+1}\} \geq t_{x,k+1} \geq \min\{t_{x,x_0}, t_{x_0, k+1}\}, \tag{2.45}$$

which implies

$$t_{x,k+1} = \min\{t_{x_0, x}, t_{x_0, k+1}\}. \tag{2.46}$$

This is precisely the value one would get by computing $t_{x,k+1}$ from the graph.

To begin we choose one of the largest terminal-capacities, say t_{12}, and connect an edge e_{12} of capacity t_{12} between nodes 1 and 2. Thus by induction we have realized the matrix T into a tree.

Example 5. Suppose we again consider the matrix in Example 4. First let us choose $t_{12} = 4$ and, following the procedures outlined in the proof of Theorem 15, we see the realization built up step by step as shown in Fig. 2.6.

When all terminal-capacities of a nonoriented communication net are needed it is not necessary to compute all $n(n-1)/2$ of them. Due to the symmetry and the relationship between them, only $(n-1)$ terminal-capacities need to be computed.

First of all, for any set of three terminal capacities

$$t_{ij} \leq t_{ik} \leq t_{jk}, \tag{2.47}$$

we have

$$t_{ij} \geq \min\{t_{ik}, t_{jk}\}, \tag{2.48}$$

which implies

$$t_{ij} = t_{ik} \leq t_{jk}. \tag{2.49}$$

Also in the calculation of te. minal-capacities, knowledge of other minimum cut-sets often helps to reduce the amount of work involved.

Theorem 16. Let S_k be the minimum cut-set for t_{ij}, also let G' and G'' be the two connected subgraphs of G with the removal of S_k. Let x and y be two nodes in G_1; then there exists a minimum cut-set S_L whose capacity is equal to t_{xy} such that S_L does not cut G_2.

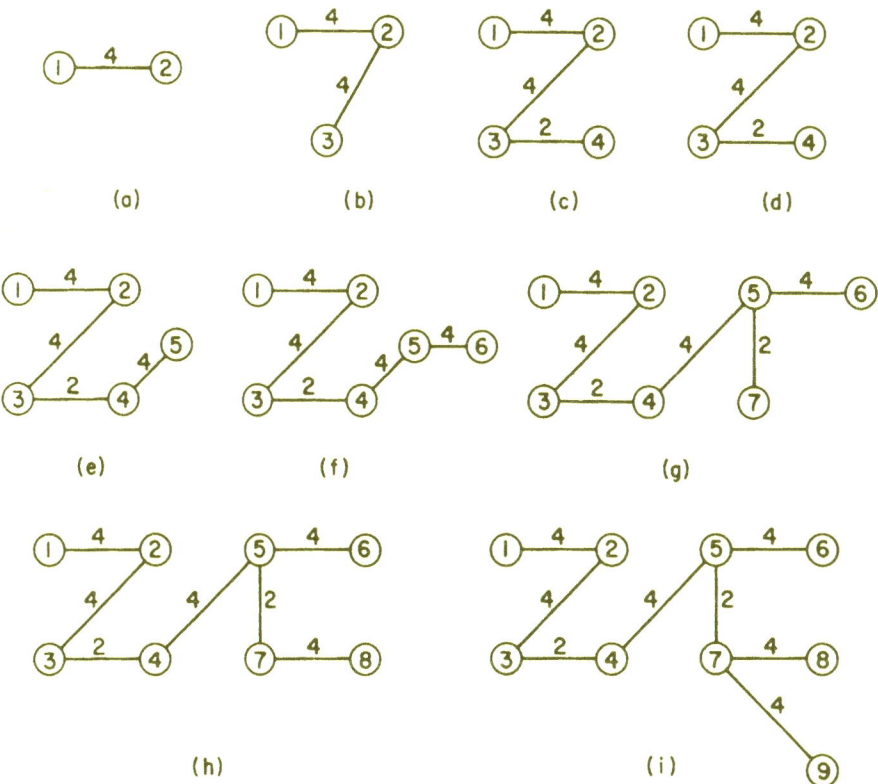

Fig. 2.6. *Realization of the T-matrix of Eq. (2.39) in steps following the proof of Theorem 14*

Proof: We should refer to Fig. 2.2 without any loss of generality. Let

x be in G_1, y in G_3;   (2.50a)

i be in $G' = G_1$, j be in $G'' = G_2$;   (2.50b)

S_1 be the minimum cut-set for x, y, S_2 be the minimum
　　cut-set for i, j.   (2.50c)

We have

$$t_{ij} = C(b) + C(d) + C(e) + C(f)$$ (2.51)

$$t_{xy} = C(a) + C(c) + C(e) + C(f),$$ (2.52)

where $C(a)$ is the capacity of edge a.

Since S_2 is the minimum cut-set, we have

$$[C(b) + C(d) + C(e) + C(f)] \leqslant [C(b) + C(c) + C(e)],$$ (2.53)

which implies

$$[C(d) + C(f)] \leqslant C(c).$$ (2.54)

Thus

$$[C(a) + C(d) + C(e)] \leqslant [C(a) + C(d) + C(e) + 2C(f)]$$

$$\leqslant [C(a) + C(e) + C(f) + C(c)],$$ (2.55)

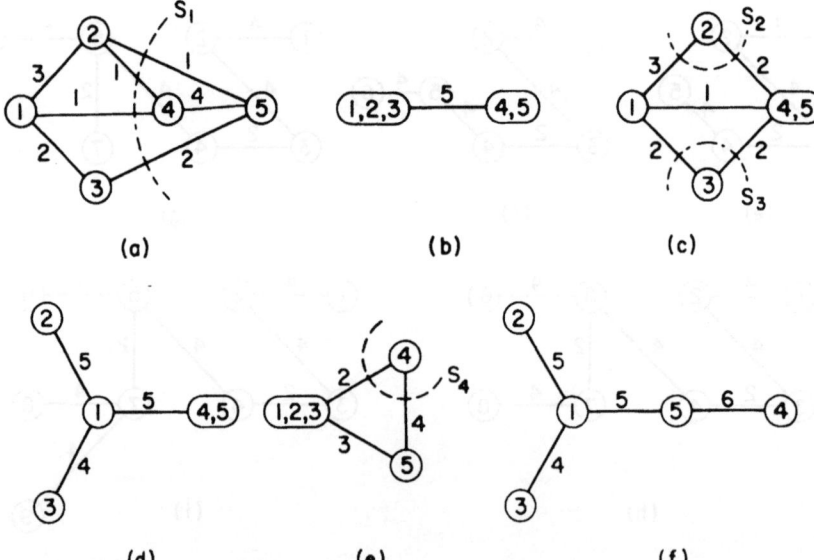

(a)　　　　　　　　　　(b)　　　　　　　　　　(c)

(d)　　　　　　　　　(e)　　　　　　　　　　(f)

Fig. 2.7. *Computation of terminal-capacities for Example 6*

and there is indeed a cut-set, namely, the incident cut-set of G_3, which has the desired property.

Corollary 6.   Let S be the minimum cut-set which separates G into G' and G''; then the terminal-capacity t_{ij} is not changed when all edges in G'' are shorted, provided that i and j are both in G'.

Proof: Since shorting never decreases the flow capacity, this follows directly from Theorem 16.

With successive use of Corollary 6, the computation of all terminal capacities can be done in a rather simple manner. We shall illustrate this with the following example.

Example 6. We are given the nonoriented communication net as shown in Fig. 2.7a, and we find that S_1 is a minimum cut-set for t_{15}. For computing terminal-capacities t_{12}, t_{13}, and t_{23}, the reduced graph in Fig. 2.7c is used. For computing t_{45}, the graph in Fig. 2.7e is used. The terminal capacity for each pair of nodes of the net is given in Fig. 2.7f.

SYNTHESIS OF NONORIENTED COMMUNICATION NETS WITH MINIMUM TOTAL EDGE-CAPACITY

As in most design problems, the realization of a T-matrix is not unique. This has been demonstrated in the last section. In a situation like this, one naturally inquires whether it is possible to impose some further constraints on the realization. In this section we shall discuss realization of matrices with the additional constraint that the total edge-capacity of the realization must be a minimum.[9] Other constraints are discussed in Chapter 3.

First we shall derive a bound for total edge-capacity required for the realization of a given terminal matrix. We observe that for any realization the capacities of the edges, e_{ij}, must satisfy the inequality

$$\sum_j e_{ij} \geq \max_j \{t_{ij}\} = t_{i0}, \qquad i = 1, 2, \ldots, n, \qquad (2.56)$$

where t_{i0} is defined as $\max\{t_{i1}, t_{i2}, \ldots, t_{in}\}$.

Let E_t be the total sum of edge-capacities of the net, called the "total edge-capacity"; then

$$E_t = \tfrac{1}{2}\sum_i \{\sum_j e_{ij}\} = \sum_{i>j} e_{ij} \geq \tfrac{1}{2}\sum_i t_{i0}, \qquad i, j = 1, 2, \ldots, n. \ (2.57)$$

Furthermore, if

$$E_t = \tfrac{1}{2}\sum_{i=1}^{n} t_{i0}, \tag{2.58}$$

then we must have

$$\sum_{j} e_{ij} = t_{i0}, \qquad i = 1, 2, \ldots, n. \tag{2.59}$$

We summarize this in Theorem 17.

Theorem 17. A realization of a terminal matrix has minimum total edge-capacity if and only if for every node i of the realization,

$$\sum_{j} e_{ij} = t_{i0},$$

where

$$t_{i0} = \max\{t_{i1}, t_{i2}, \ldots, t_{in}\}.$$

It is interesting to note that the required total edge-capacity is only a function of the largest terminal-capacity associated with each node. This means that in certain cases one may be able to increase the terminal-capacities of a terminal matrix without necessarily increasing the total edge-capacity. For instance, the following two terminal matrices have the same minimum total edge-capacities, yet they are different in some terminal-capacities:

$$T_1 = \begin{bmatrix} ① & 6 & 4 & 4 \\ 6 & ② & 4 & 4 \\ 4 & 4 & ③ & 5 \\ 4 & 4 & 5 & ④ \end{bmatrix} \tag{2.60}$$

$$T_2 = \begin{bmatrix} ① & 6 & 5 & 5 \\ 6 & ② & 5 & 5 \\ 5 & 5 & ③ & 5 \\ 5 & 5 & 5 & ④ \end{bmatrix}. \tag{2.61}$$

Several methods for the synthesis of nonoriented communication nets with minimum total capacity are known. We shall discuss them in detail.

Method of equal distribution

The method to be described is based on the partitioning process and can best be described along with the partitioning of the graph. Suppose the given matrix is partitioned in the form

$$T = \begin{bmatrix} A & T_1 \\ T_1^t & B \end{bmatrix}, \tag{2.62}$$

where T_1 is a uniform matrix with value t_1. Then its corresponding graph is also partitioned into two subgraphs G_a and G_b, where the terminal capacity between them is t_1. We connect an edge of capacity t_1 between subgraphs G_a and G_b as shown in Fig. 2.8.

Fig. 2.8. *Connection subgraphs G_A and G_B*

Suppose the matrix **A** is partitioned as

$$A = \begin{bmatrix} C & T_2 \\ T_2^t & D \end{bmatrix}, \tag{2.63}$$

where T_2 is a uniform matrix with value t_2 $(t_2 \geqslant t_1)$. Then G_2 is again partitioned into subgraphs G_c and G_d. Each original edge between G_a and G_b is replaced by two new edges, each with half the

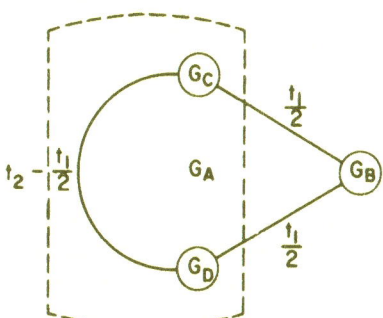

Fig. 2.9. *Connection of subgraphs G_C and G_D*

capacity of that of the original edge, and connected between G_b and G_c, G_b, and G_d, respectively. A new edge is also added between the new subgraphs G_c and G_d with the capacity $(t_2 - t_1/2)$ as shown in Fig. 2.9.

In general, suppose the matrix is already partitioned into k parts and at the next step one of the k subgraphs corresponding to the k

parts, namely G_k, is to be partitioned into two subgraphs G'_k and G''_k. The same process is repeated. Each edge between G_k and G_i ($i = 1, 2, \ldots, k-1$) is replaced by two edges, each with half the capacity of that of the original edge and connected between G_i and G'_k, and G''_k, respectively. The edge $e_{k'k''}$, is determined by the difference,

$$e_{k'k''} = t_{k'k''} - \tfrac{1}{2} \sum_{i=1}^{k-1} e_{ik}. \qquad (2.64)$$

The graphs before and after kth partition are shown in Fig. 2.10. Note that with this process not only is the minimum cut-set equal

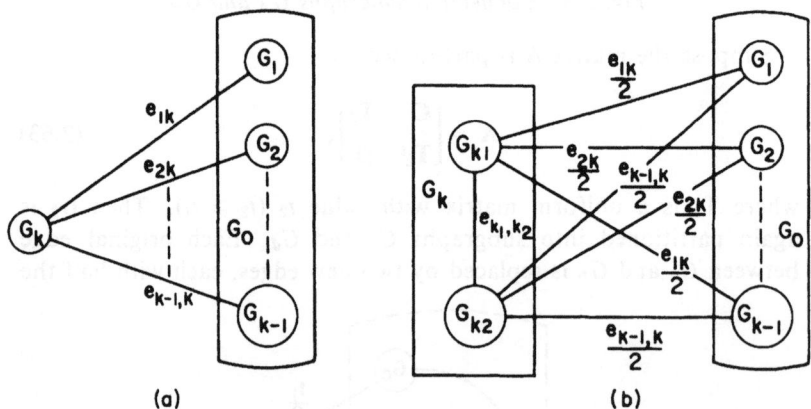

(a)   (b)

Fig. 2.10. Graphs (a) before kth partition and (b) after kth partition

to the desired terminal-capacity, but both cut-sets have the same capacity. Since the necessary and sufficient condition is also based on the partitioning process, the ability to partition is guaranteed.

It is quite clear that this method gives a graph that realizes the given matrix. Furthermore, since the incident cut-set of any node has exactly the same value as the largest terminal-capacity associated with it, this is a realization of minimum total edge-capacity.

Method of elementary matrices[7]

We define "elementary terminal matrix" as a matrix which can be put into the following form:

$$\mathbf{T} = \begin{bmatrix} ① & t_{n-1} & t_{n-2} \dots t_1 \\ t_{n-1} & ② & t_{n-2} \dots t_1 \\ t_{n-2} & t_{n-2} & ③ & \dots t_1 \\ \cdot & \cdot & \cdot & \cdot & \cdot & \cdot & \cdot \\ t_1 & t_1 & t_1 & \dots ⑩ \end{bmatrix},$$ (2.65)

where

$$t_{n-1} \geqslant t_{n-2} \geqslant \dots \geqslant t_2 \geqslant t_1.$$

It should be noted that every elementary terminal matrix is guaranteed to be realizable by the partitioning condition of Theorem 13.

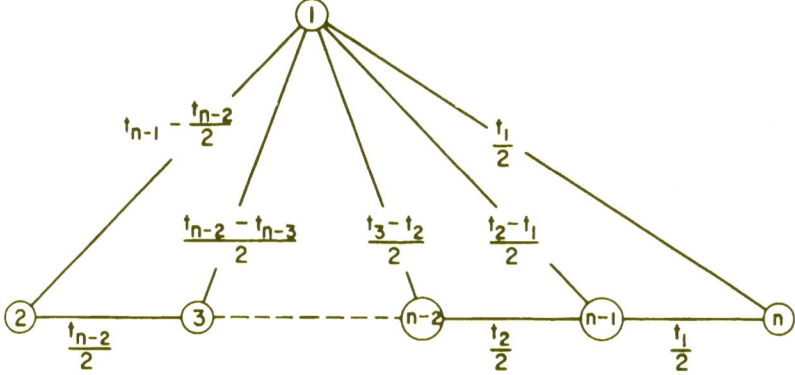

Fig. 2.11. *Minimal realization of an elementary matrix of order n*

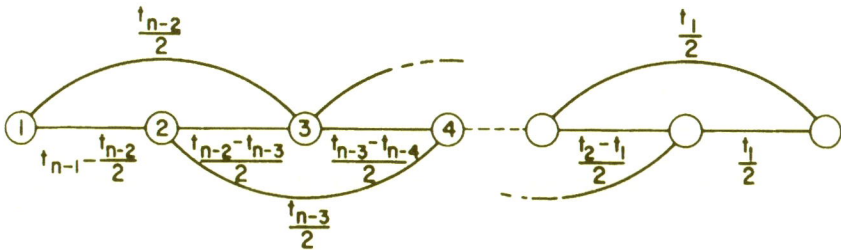

Fig. 2.12. *Alternate realization of an elementary matrix of order n*

We assert that the graph of Fig. 2.11 realizes an elementary terminal matrix of order n at minimum total edge-capacity.

To prove this, we note that at every node the incident cut-set satisfies Eq. (2.59) as an equality; therefore the graph is in fact a minimal realization. For the same reason, the graph in Fig. 2.12 is a minimal realization of an elementary terminal matrix. The two graphs are 2-isomorphic with each other.

If a terminal matrix of order n is partitionable as follows (due to Theorem 13),

$$T = \begin{bmatrix} T_1 & T_o \\ T_o{}^t & T_2 \end{bmatrix}, \tag{2.66}$$

where T_o is uniform with value t_o, where $t_o = \min \{t_{ij}\}$, and T_1 and T_2 are elementary terminal matrices of order k and $(n-k)$, respectively, then the graph of Fig. 2.13 realizes T at minimum total edge-capacity.

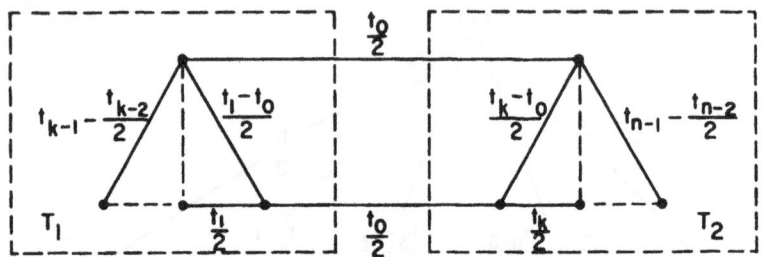

Fig. 2.13. *Minimal realization of the T-matrix of Eq.* (2.66)

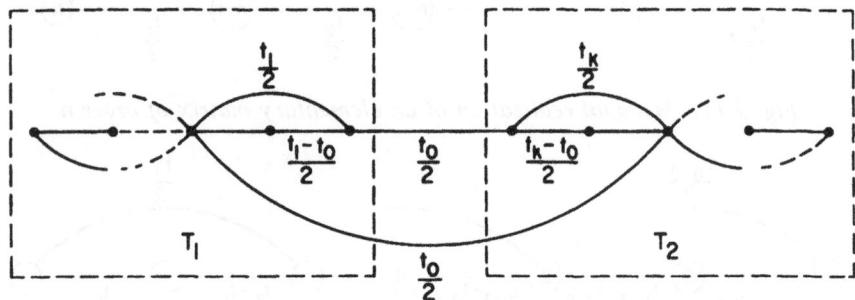

Fig. 2.14. *Alternate minimal realization of a pair of elementary matrices*

It should be noted that the two "linking" edges can be placed between any pair of nodes as long as there is a direct link between this pair of nodes with capacity $t_o/2$ or more. If in T_1 (or T_2), an edge has a capacity equal to $t_o/2$, we may connect the linking edges to the nodes of this edge to make its net capacity zero, consequently reducing the total number of edges. An alternate realization of the above is shown in Fig. 2.14. The generalization of the partitioning shown in **T** of Eq. (2.66) to the case in which the matrix is partition-able into $\mathbf{T}_1, \mathbf{T}_2, \ldots, \mathbf{T}_P$ elementary matrices is obvious. The minimal realization is given in Fig. 2.15 in which every adjacent pair of $\mathbf{T}_1$ $\mathbf{T}_2 \ldots, \mathbf{T}_P$ is realized as in Fig. 2.13 or as its alternate in Fig. 2.14.

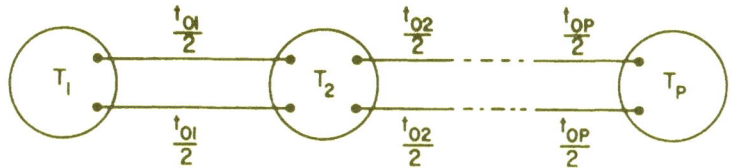

Fig. 2.15. *Minimal realization of a T-matrix having P elementary matrices*

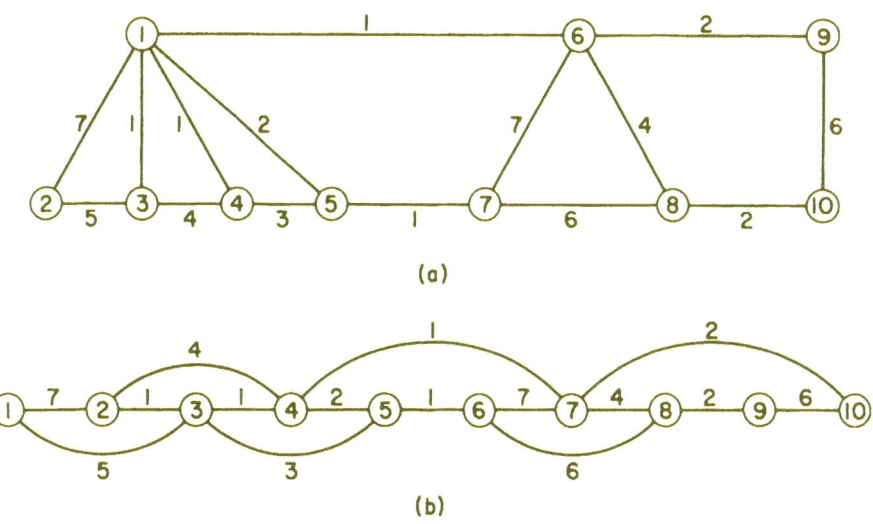

(a)

(b)

Fig. 2.16. *Two minimal realizations of the terminal matrix of Eq. (2.67)*

The number of edges required in either of the two minimal realizations is at most $(2n - P - 2)$, where P is the number of elementary matrices in a given **T**. The actual number will be less if some of the edge-capacities become zero. Figure 2.16 shows the two minimal realizations of the terminal matrix of Eq. (2.67) in which $P = 3$:

$$
T = \begin{bmatrix}
① & 12 & 10 & 8 & 6 & 2 & 2 & 2 & 2 & 2 \\
12 & ② & 10 & 8 & 6 & 2 & 2 & 2 & 2 & 2 \\
10 & 10 & ③ & 8 & 6 & 2 & 2 & 2 & 2 & 2 \\
8 & 8 & 8 & ④ & 6 & 2 & 2 & 2 & 2 & 2 \\
6 & 6 & 6 & 6 & ⑤ & 2 & 2 & 2 & 2 & 2 \\
\hline
2 & 2 & 2 & 2 & 2 & ⑥ & 14 & 12 & 4 & 4 \\
2 & 2 & 2 & 2 & 2 & 14 & ⑦ & 12 & 4 & 4 \\
2 & 2 & 2 & 2 & 2 & 12 & 12 & ⑧ & 4 & 4 \\
2 & 2 & 2 & 2 & 2 & 4 & 4 & 4 & ⑨ & 8 \\
2 & 2 & 2 & 2 & 2 & 4 & 4 & 4 & 8 & ⑩
\end{bmatrix}. \qquad (2.67)
$$

Method of successive expansion

It was shown earlier in this chapter that any realizable matrix may be partitioned according to Eq. (2.33). It should be noted that in this case the partitioning process is unique. The following definition is therefore meaningful.

Definition 6: The index of partitioning I_P of a terminal matrix is the number of operations necessary to partition a T-matrix into a form in which every diagonal submatrix is either of order 2 by 2 or 1 by 1, with the provision that each operation is to be applied to one diagonal submatrix at a time.

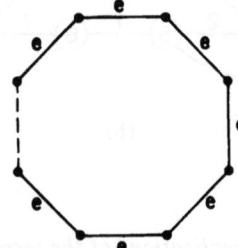

Fig. 2.17. A ring structure

This method of partitioning offers a synthesis method of minimal total edge-capacity and employs relatively few edges. The method is based on the ring structure shown in Fig. 2.17. If the capacity of each element of the ring is e, it is seen that the terminal-capacity between any two points in the ring is $2e$. Using the ring structure, the synthesis of any T-matrix is accomplished as follows.

In the first step, the diagonal submatrices of the first partition are treated as nodes to form a ring, with each ring element equal to $t_0/2$, i.e., half the capacity of the first partition. To realize each diagonal submatrix, a new ring is formed to take the place of the corresponding node and each edge in the new ring will have a capacity of $t_1/2$ except for one edge, which has a capacity of $(t_1 - t_0)/2$ and is the only edge which the new ring shares with the original ring structure as illustrated in Fig. 2.18. Each submatrix is carried out in the same way, but one special case needs to be mentioned. When the

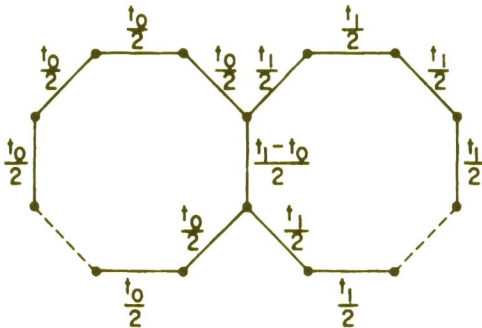

Fig. 2.18. Realization of a terminal matrix of order two

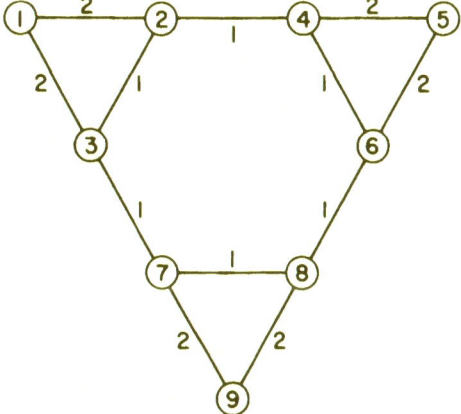

Fig. 2.19. Minimal realization of the terminal matrix of Eq. (2.39)

new partition of the submatrix is of order two, the two edges of the new ring may be combined to form one edge.

The realization thus obtained is minimum total edge-capacity, since every incident cut-set satisfies Eq. (2.59).

The number of edges required by this method is exactly $(I_P + n - 1)$, where I_P is the index of partitioning. This results from the fact that when all entries in T are equal, $I_P = 1$, in which case n edges are needed, and from the fact that an extra edge is added each time a new ring is formed. As an illustration the matrix of Eq. (2.39) is realized as shown in Fig. 2.19.

Method of decomposition of matrices[8]

This method is based on the decomposition of terminal matrices as discussed earlier in this chapter. The given terminal matrix is written as a sum of two terminal matrices of which one is a uniform matrix as:

$$T = T_1 + T_u, \qquad (2.68)$$

where t_u, the uniform element in T_u, is equal to the minimum element of t_{ij}, t_o. Since the zero elements of T_1 will indicate where the minimum cut-set will be in the realization of T_1, one may realize T_u in such a way that the minimum cut-sets of the realization of T_u correspond to the minimum cut-sets of the realization of T_1. The realization of each nonzero part of T_1 is accomplished by further decomposition and realization.

By successive application of the above process, one simply obtains an expansion of a terminal matrix into a sum of uniform matrices as:

$$T = \sum T_{u_i}, \qquad (2.69)$$

where the T_{u_i}'s are realized and combined in such a way that all their minimum cut-sets correspond to each other.

Example 7. Realize the matrix,

$$T = \begin{bmatrix} ① & 5 & 5 & 4 & 4 \\ 5 & ② & 6 & 4 & 4 \\ 5 & 6 & ③ & 4 & 4 \\ 4 & 4 & 4 & ④ & 6 \\ 4 & 4 & 4 & 6 & ⑤ \end{bmatrix}. \qquad (2.70)$$

First we write **T** as

$$\mathbf{T} = \mathbf{T}_{u_1} + \mathbf{T}_{u_2} + \mathbf{T}_{u_3} + \mathbf{T}_{u_4}, \tag{2.71}$$

where

$$\mathbf{T}_{u_1} = \begin{bmatrix} ① & 4 & 4 & 4 & 4 \\ 4 & ② & 4 & 4 & 4 \\ 4 & 4 & ③ & 4 & 4 \\ 4 & 4 & 4 & ④ & 4 \\ 4 & 4 & 4 & 4 & ⑤ \end{bmatrix} \tag{2.72}$$

$$\mathbf{T}_{u_2} = \begin{bmatrix} ① & 0 & 0 & 0 & 0 \\ 0 & ② & 0 & 0 & 0 \\ 0 & 0 & ③ & 0 & 0 \\ 0 & 0 & 0 & ④ & 2 \\ 0 & 0 & 0 & 2 & ⑤ \end{bmatrix} \tag{2.73}$$

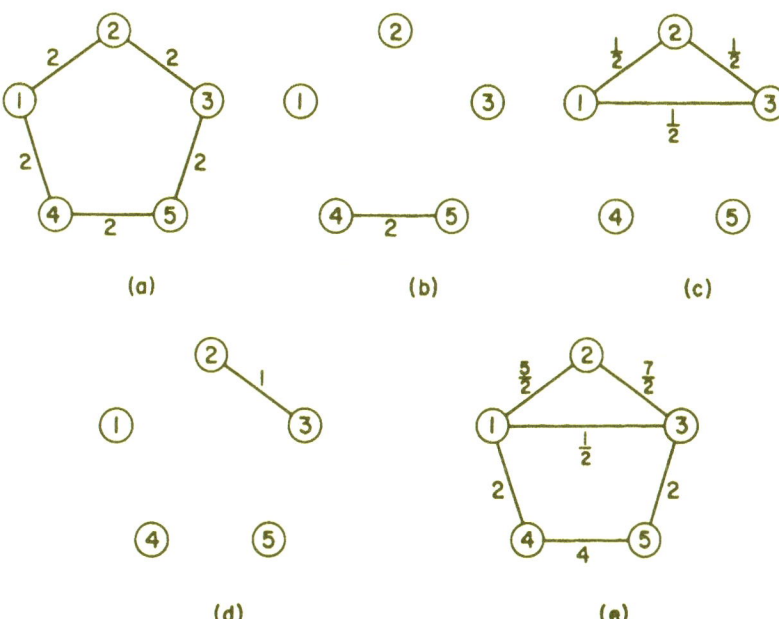

Fig. 2.20. *Minimal realization of the terminal matrix of Example 7*

$$\mathbf{T_{u_3}} = \begin{bmatrix} ① & 1 & 1 & 0 & 0 \\ 1 & ② & 1 & 0 & 0 \\ 1 & 1 & ③ & 0 & 0 \\ 0 & 0 & 0 & ④ & 0 \\ 0 & 0 & 0 & 0 & ⑤ \end{bmatrix} \qquad (2.74)$$

$$\mathbf{T_{u_4}} = \begin{bmatrix} ① & 0 & 0 & 0 & 0 \\ 0 & ② & 1 & 0 & 0 \\ 0 & 1 & ③ & 0 & 0 \\ 0 & 0 & 0 & ④ & 0 \\ 0 & 0 & 0 & 0 & ⑤ \end{bmatrix}. \qquad (2.75)$$

The realization for each uniform matrix $\mathbf{T_{u_i}}$ is shown in Figs. 2.20a, b, c, and d, and the over-all realization is obtained by addition of the individual realizations in proper manner as shown in Fig. 2.20e.

CHAPTER 3

Further Discussion

on Communication Nets

APPLICATION OF LINEAR PROGRAMMING

IN CHAPTER 2 we discussed the synthesis of nonoriented communication nets under some constraints. Several methods for the synthesis of communication nets with minimal total edge-capacity were discussed in detail. Here a more general case will be considered; that is, the case of synthesis with minimum linear cost. A cost c_{ij} is defined as the cost for installing a unit of edge-capacity between the nodes i and j. The total cost of the net is then

$$C_T = \sum_{i,j} c_{ij} e_{ij}, \qquad \text{for } i, j = 1, 2, \ldots, n; \quad i \neq j. \qquad (3.1)$$

Since C_T is seen to be a linear function of the edge-capacities e_{ij}, linear programming techniques turn out to be very useful.[7,10]

The unknowns of the problem are the $[n(n-1)/2]$ e_{ij}'s and $n(n-1)/2$ is the number of edges in a completely connected graph.

Consider a typical cut-set S_k in the set of all cut-sets in a completely connected graph: S_k separates the graph into two subgraphs G_1 and G_2, G_1 may contain $1, 2, \ldots$ or $(n-1)$ nodes. The total number of cut-sets is therefore $(2^{n-1} - 1)$.

Let G_1 and G_2 be separated by S_k so that G_1 contains nodes $v_1, v_2, \ldots, v_p$, and G_2 contains nodes $v_{p+1}, \ldots, v_n$. To satisfy the requirements for the terminal capacity, the value of S_k must be at least as large as the maximum of the set of $p(n-p)$ terminal-capacities t_{ij}, for $i = 1, 2, \ldots, p$ and $j = p+1, \ldots, n$. Applying this principle to all cut-sets, we have a total of $(2^{n-1} - 1)$ inequalities of the form

$$\sum_{i=1}^{p} \sum_{j=p+1}^{n} e_{ij} \geq \max\{t_{1(p+1)}, t_{1(p+2)}, \ldots, t_{1n} \ldots, t_{p(p+1)} \ldots, t_{pn}\}. \qquad (3.2)$$

In addition to the above, we have

$$e_{ij} \geq 0 \qquad i, j = 1, 2, \ldots, n. \tag{3.3}$$

The problem can now be stated in the linear programming form as follows.

"Minimize the total system cost

$$\sum_{i,j} c_{ij} e_{ij}$$

subject to the set of linear constraints of (3.2) and (3.3)." The simplex method or other modified methods can be used to find the optimum solution.

As an example, consider finding a four-node net satisfying a terminal matrix

$$\mathbf{T} = \begin{bmatrix} \textcircled{1} & 8 & 9 & 7 \\ 8 & \textcircled{2} & 8 & 7 \\ 9 & 8 & \textcircled{3} & 7 \\ 7 & 7 & 7 & \textcircled{4} \end{bmatrix}, \tag{3.4}$$

such that the total system cost

$$e_{12} + 2e_{13} + 2e_{14} + 2e_{23} + 5e_{24} + 2e_{34} \tag{3.5}$$

is minimum.

The constraints on the unknown are:

$$\begin{aligned} e_{12} + e_{13} + e_{14} &\geq \max\{t_{12}, t_{13}, t_{14}\} = 9 \\ e_{21} + e_{23} + e_{24} &\geq \max\{t_{21}, t_{23}, t_{24}\} = 8 \\ e_{31} + e_{32} + e_{34} &\geq \max\{t_{31}, t_{32}, t_{34}\} = 9 \\ e_{41} + e_{42} + e_{43} &\geq \max\{t_{41}, t_{42}, t_{43}\} = 7 \\ e_{13} + e_{14} + e_{23} + e_{24} &\geq \max\{t_{13}, t_{14}, t_{23}, t_{24}\} = 9 \\ e_{12} + e_{14} + e_{32} + e_{34} &\geq \max\{t_{12}, t_{14}, t_{32}, t_{34}\} = 8 \\ e_{12} + e_{13} + e_{42} + e_{43} &\geq \max\{t_{12}, t_{13}, t_{42}, t_{43}\} = 9 \end{aligned} \tag{3.6}$$

and

$$e_{ij} \geq 0, \qquad i, j = 1, 2, \ldots, n; \qquad i \neq j. \tag{3.7}$$

The edge-capacity matrix $\mathbf{E}$ which realizes the given $\mathbf{T}$ and at the same time minimizes the system cost is

$$E = \begin{bmatrix} ① & 4 & 1.5 & 3.5 \\ 4 & ② & 4 & 0 \\ 1.5 & 4 & ③ & 3.5 \\ 3.5 & 0 & 3.5 & ④ \end{bmatrix}. \tag{3.8}$$

The corresponding system cost is 29. The network is shown in Fig. 3.1.

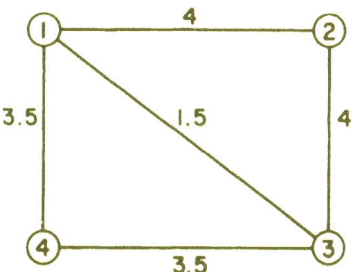

Fig. 3.1. Realization of minimum cost of the terminal matrix of Eq. (3.4)

The number of inequalities is $(2^{n-1}-1)$ and the number of unknowns is $n(n-1)/2$. For any realistic n, the simplex method becomes unmanageable even with a digital computer. The large number of zeros in the set of inequalities suggests that a more efficient method for finding the optimal solution is possible. Some attempts in this direction have been made and more efficient algorithms are now available. However, even with the improved methods, this approach is useful only if it is incorporated with other techniques.

INDEX OF A COMMUNICATION NET

Another interesting problem in the synthesis of nonoriented communication nets is the case where the terminal matrix is not given. Situations like this occur when not enough information is available, as in the case of military warning systems or other types of systems for emergency purposes. Under these circumstances, an average measure of the system's capability may be obtained in terms of the index of the communication net. The index of a communication net, I, is defined as[9]

$$I = \sum_{i>j} t_{ij}, \qquad \text{for } i, j = 1, 2, \ldots n. \tag{3.9}$$

It is seen that the index is a function of the realization. Suppose one is given only the total edge-capacity and not the terminal matrix. Then one would be interested in obtaining a net such that the index is maximized, assuming we have a fixed number of nodes.

Before investigating possible ways of maximizing the index, let us first observe a very special type of communication net. Suppose we have a completely connected graph with edges of equal capacity e; then all terminal-capacities would also be equal to each other. In fact, for all terminal-capacities,

$$t_{ij} = t = (n-1)e. \qquad (3.10)$$

In the meantime, the total edge-capacity of the net is equal to

$$E_T = [n(n-1)/2]e = nt/2 \qquad (3.11)$$

$$I = (n-1)E_T. \qquad (3.12)$$

Now suppose one submatrix of a terminal matrix has the following form:

$$\mathbf{T_0} = \begin{bmatrix} \mathbf{A} & \mathbf{T_1} \\ \mathbf{T_1}^t & \mathbf{B} \end{bmatrix}, \qquad (3.13)$$

where $\mathbf{A}$ is a uniform matrix of value t_A and has u nodes, $\mathbf{B}$ is a uniform matrix of value t_B and has v nodes, and $\mathbf{T_1}$ is a matrix which is uniform and of value t_1. Without loss of generality we may assume

$$t_1 \leqslant t_A \leqslant t_B. \qquad (3.14)$$

The minimum total edge-capacity for this submatrix is

$$E_{T_0} = \tfrac{1}{2}(ut_A + vt_B), \qquad (3.15)$$

while the index of this submatrix is

$$I_0 = [u(u-1)/2]t_A + [v(v-1)/2]t_B + uvt_1. \qquad (3.16)$$

Solving Eq. (3.15) for t_B and substituting it into Eq. (3.16), one obtains

$$I_0 = (v-1)E_{T_0} + (ut_A/2)(u+v). \qquad (3.17)$$

This means that if we increase t_A, the index will always increase with the total edge-capacity held constant. The highest possible value for t_A is reached when it is equal to t_B, and this value is

$$2E_{T_0}/(u+v). \qquad (3.18)$$

Thus if the total edge-capacity is held constant, the maximum value for the index is

$$(I_0)_{max} = (v-1)E_{T_0} + u(u+v)[2E_{T_0}/(u+v)] = E_{T_0}(u+v-1). \quad (3.19)$$

By smoothing out the terminal-capacities we have succeeded in increasing the index to a maximum while making all terminal-capacities equal in the meantime. Due to the fact that in any terminal matrix the smallest partitions are always uniform in value, the above procedure may be applied repetitively so that any terminal matrix can eventually have the maximum possible index for the given total edge-capacity; in the meantime all the terminal-capacities will have the same value.

SIMULTANEOUS FLOWS IN NONORIENTED COMMUNICATION NETS

Up to now, it was assumed that at any instant there is in the communication net only one source and one destination. In reality, this is a rather rare case. In general, simultaneous flows occur and their magnitude may vary with time. However, as we shall show in this section, many of the concepts discussed are very useful even in the treatment of simultaneous flows.[11]

The requirements for simultaneous flow are conveniently represented in the form of a set of matrices:

$$\mathbf{R}^{(k)} = \begin{bmatrix} \textcircled{1} & r_{12}^{(k)} \ldots & r_{1n}^{(k)} \\ \cdot & \cdot \cdot \cdot \cdot \cdot & \cdot \cdot \\ \cdot & \cdot \cdot \cdot \cdot \cdot & \cdot \cdot \\ r_{n1}^{(k)} & r_{n2}^{(k)} \ldots & \textcircled{m} \end{bmatrix} \qquad k = 1, 2, \ldots, m, \quad (3.20)$$

where $r_{ij}^{(k)} = r_{ji}^{(k)}$ is a constant for each k, and m is a positive integer. This set of R-matrices may be considered as the sampled values of the requirements at m different time instants.

A network which satisfies the simultaneous flow requirements must provide, for each fixed k, enough capacities for all $r_{ij}^{(k)}$ and $i, j = 1, 2, \ldots, n, i \neq j$. This means that the network described by the E-matrix must be such that for each fixed k it can be written as the sum of $n(n-1)/2$ component matrices:

$$\mathbf{E} = \sum_{i<j} \mathbf{E}_{(ij)}^{(k)}, \qquad (3.21)$$

where each component matrix $\mathbf{E}_{(ij)}^{(k)}$ yields a corresponding network which provides a minimum cut-set between nodes i and j of a value no less than $r_{ij}^{(k)}$. That is, in the graph of each $E_{(ij)}^{(k)}$,

$$t_{ij}^{(k)} \geqslant r_{ij}^{(k)}. \tag{3.22}$$

We shall first show a network which, satisfying the simultaneous flow requirements, gives an upper bound on the minimum total edge-capacity.

Let the set of matrices

$$\mathbf{R}^{(k)} = [r_{ij}^{(k)}], \qquad \text{for } k = 1, 2, \ldots m \tag{3.23}$$

describe the given simultaneous flow requirements. If we consider only one matrix from the set, an obvious realization is a network with

$$\mathbf{E}^{(k)} = \overline{\mathbf{R}}^{(k)}, \tag{3.24}$$

which may be considered as the "requirement network." Following the same line of thought, we can obtain an immediate realization satisfying the given set of R-matrices by choosing

$$\mathbf{E} = \overline{\mathbf{R}} = [\bar{r}_{ij}], \tag{3.25}$$

where

$$\bar{r}_{ij} = \max_k r_{ij}^{(k)}. \tag{3.26}$$

This is clear since we can always write

$$\mathbf{E} = \overline{\mathbf{R}} = [\bar{r}_{ij}] = \sum_{i<j} \mathbf{E}_{ij}^{(k)}, \tag{3.27}$$

if each component matrix $\mathbf{E}_{(ij)}^{(k)}$ contains exactly one nonzero pair

$$e_{ij}^{(k)} = e_{ji}^{(k)} = \bar{r}_{ij} \geqslant r_{ij}^{(k)} \tag{3.28}$$

and zero everywhere else. This decomposition is clearly good for all $k = 1, 2, \ldots, m$.

The total edge-capacity of this realization is an upper bound of the minimum total edge-capacity with no constraint on the configuration.

We now analyze the cut-set requirements of a network which satisfies the given simultaneous flow requirements.

Let γ be a subset of N, the set of all nodes. Let $C(\gamma)$ denote the capacity of the cut-set which separates it from its complement $(N-\gamma)$. Then,

$$C(\gamma) = \sum_{\substack{i \in \gamma \\ j \in N-\gamma}} e_{ij}. \tag{3.29}$$

Let the requirement matrices $\mathbf{R}^{(k)} = [r_{ij}^{(k)}]$, $k = 1, 2, \ldots, m$, for the simultaneous flow be represented by the requirement networks satisfying $\mathbf{E}^{(k)} = \mathbf{R}^{(k)}$. For each $\mathbf{E}^{(k)}$ we denote the capacity of $(2^{n-1} - 1)$ possible cut-sets by $C^{(k)}(\alpha)$, $C^{(k)}(\beta)$, ..., where $\alpha, \beta, \ldots$, are $(2^{n-1} - 1)$ distinct proper subsets of N, none of which is the complement of the other. For any γ we define

$$\overline{C}(\gamma) = \max_k C^{(k)}(\gamma). \tag{3.30}$$

Theorem 18. If a network of $\mathbf{E}$ satisfies the requirement matrices $\mathbf{R}^{(k)}$ $k = 1, 2, \ldots, m$, for simultaneous flow, then for every collection of nodes γ, $C(\gamma)$ of $\mathbf{E}$ satisfies

$$C(\gamma) \geqslant \overline{C}(\gamma). \tag{3.31}$$

Proof: Assume the theorem is false. Then

$$C(\gamma) < \overline{C}(\gamma) \tag{3.32}$$

for some γ. From Eq. (3.30) it follows that

$$C(\gamma) \leqslant C^{(k)}(\gamma) \tag{3.33}$$

for some k. However, we can write for this k

$$\mathbf{E} = \sum_{i<j} \mathbf{E}_{ij}^{(k)} \tag{3.34}$$

such that

$$t_{ij}^{(k)} \geqslant r_{ij}^{(k)}. \tag{3.35}$$

This implies

$$C(\gamma) \geqslant \sum_{\substack{i \in \gamma \\ j \in N - \gamma}} e_{ij}^{(k)} \geqslant C^{(k)}(\gamma), \tag{3.36}$$

which is a contradiction.

Theorem 19.

$$\overline{C}(\alpha) + \overline{C}(\beta) \geqslant \overline{C}(\alpha \oplus \beta). \tag{3.37}$$

Proof: For any k it is straightforward to show that

$$C^{(k)}(\alpha) + C^{(k)}(\beta) = C^{(k)}(\alpha \oplus \beta) + 2 \sum_{\substack{i \in \alpha, j \in \beta \\ i, j \notin \alpha \cap \beta}} r_{ij}^{(k)} + 2 \sum_{\substack{i \in \alpha \cap \beta \\ j \in N - (\alpha \cup \beta)}} r_{ij}^{(k)}$$

$$\geqslant C^{(k)}(\alpha \oplus \beta). \tag{3.38}$$

Therefore

$$\max_k \{C^{(k)}(\alpha)\} + \max_k \{C^{(k)}(\beta)\} \geqslant \max_k \{C^{(k)}(\alpha \oplus \beta)\}. \tag{3.39}$$

Theorem 20. A tree-network satisfies the given set of simultaneous flow requirement matrices if and only if the $(n-1)$ simple cut-sets satisfy the inequality of (3.32).

Proof: The "only if" follows directly from Theorem 18. For the "if," let the $(n-1)$ simple cut-sets of a given tree satisfy the inequality of (3.31). For each k write

$$\mathbf{E} = \sum_{i<j} \mathbf{F}_{(ij)}^{(k)}, \tag{3.40}$$

such that each $\mathbf{E}_{(ij)}^{(k)}$ contains exactly a path which is the same as the path connecting node i and node j in the given tree. Let the capacity of this path be $r_{ij}^{(k)}$.

First of all, the corresponding network $\mathbf{E}$ thus obtained has the same tree configuration. Also we see that any edge of this tree $e_{(xy)}^{(k)}$ has the capacity

$$e_{xy}^{(k)} = \sum_{\substack{x \in \alpha \\ y \in N-\alpha}} r_{xy}^{(k)}, \tag{3.41}$$

where $e_{xy}^{(k)}$ separates the tree into subgraphs α and $(N-\alpha)$. Since it was assumed that all the simple cut-sets of the given tree satisfy the inequality of (3.32), we thus have

$$e_{xy}^{(k)} = \sum_{\substack{x \in \alpha \\ y \in N-\alpha}} r_{xy}^{(k)} = C^{(k)}(\alpha) \leqslant \bar{C}(\alpha) \leqslant e_{xy}, \tag{3.42}$$

for all k and all x, y.

NONORIENTED COMMUNICATION NETS AND n-PORT RESISTIVE NETWORKS

From the graph theory point of view, the only difference between a nonoriented communication net and a resistive network is the unit of the weight associated with each edge. For communication nets the unit is the unit of flow while the unit used in a resistive network is the unit of resistance or conductance. Although the problems considered in each case are quite different in general, due to this similarity in graphs some interesting relationships have been discovered.[12]

Theorem 21. If a nonoriented communication net N_c and a resistive network N_r have the same weighted graph, then, for any pair of terminals a and b,

$$t_{ab} \geqslant Y_{ab}, \tag{3.43}$$

where Y_{ab} is the driving-point admittance at nodes a and b.

Proof: Suppose S is a cut-set which cuts the graph G into subgraphs G_1 and G_2, and S is a minimal cut-set for nodes a and b. Then

$$C(S) = t_{ab}. \tag{3.44}$$

When all nodes in G_1 and all nodes in G_2 are identified respectively, the maximum flow does not change. However, if we calculate the

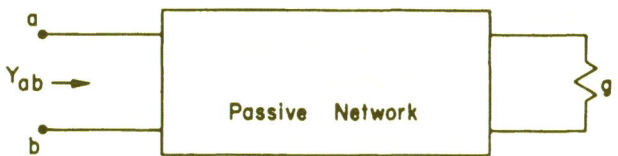

Fig. 3.2. Four-terminal passive network

effect on Y_{ab} due to any change of conductance g (see Fig. 3.2), we see that

$$Y_{ab} = y_{11} - [y_{12}^2/(y_{22}+g)]. \tag{3.45}$$

Therefore

$$dY_{ab}/dg = y_{12}^2/(y_{22}+g)^2 = \text{positive number.} \tag{3.46}$$

The identification of nodes can only increase the value of Y_{ab}. This leads to the conclusion that

$$t_{ab} \geqslant Y_{ab}. \tag{3.47}$$

Another analogy may be made between the realizability of resistive networks and communication nets. If we remember our study of the realization of a Y-matrix in terms of a resistive multiport network in Part III, we see that the following two conditions are necessary and sufficient for the realization of a symmetric matrix $\mathbf{Y} = [y_{ij}]$, with positive real elements, possibly after rows and columns are interchanged, as the short-circuit admittance matrix of an n-port resistive network with $(n+1)$ nodes and a linear-tree port-structure. Namely, $\mathbf{Y}$ must be symmetric and be uniformly tapered. That is,

$$y_{ii} \geqslant y_{i,i+1} \geqslant y_{i,i+2} \geqslant \ldots \geqslant y_{i,i+n} \tag{3.48a}$$

$$y_{ii} \geqslant y_{i-1,i} \geqslant y_{i-2,i} \geqslant \ldots \geqslant y_{i-n,i}, \qquad \text{for all } i \tag{3.48b}$$

$$y_{ij} + y_{i-1,j+1} \geqslant y_{i-1,j} + y_{i,j+1} \qquad \text{for all } i \text{ and } j. \dagger \tag{3.49}$$

† See Reference 5b of Part III.

We now relate the realization method of a *Y*-matrix of a resistive network to that of a terminal matrix of a communication net.

Theorem 22. With proper assignment of diagonal elements, every real matrix which can be realized as the terminal matrix of a non-oriented communication net of *n* nodes is also realizable as the short-circuit conductance matrix of an *n*-port resistive network of $(n+1)$ nodes in the series-parallel form.

The proof of this theorem is rather lengthy and is therefore beyond the scope of this presentation. Interested readers are urged to read Reference 12. We shall be content to illustrate this property with a simple example.

Definition 7. In an *n*-port resistive network, the edge-conductance matrix $\mathbf{G} = [g_{ij}]$ is a symmetric matrix in which element g_{ij} is the conductance of the edge between *i* and *j*. Element g_{ii} is the conductance of the edge(s) connected directly across port *i*.

Example 8: We are given the terminal matrix

$$\mathbf{T} = \begin{bmatrix} ① & 5 & 5 & 4 & 4 \\ 5 & ② & 6 & 4 & 4 \\ 5 & 6 & ③ & 4 & 4 \\ 4 & 4 & 4 & ④ & 6 \\ 4 & 4 & 4 & 6 & ⑤ \end{bmatrix}. \tag{3.50}$$

The conductance matrix **Y** is identical with **T** except for the diagonal elements. Obtain *G* by Guillemin's method. We have

$$\mathbf{G} = \begin{bmatrix} ① & -5 & 0 & 1 & 0 & 4 \\ 0 & ② & -6 & 1 & 0 & 0 \\ 1 & 1 & ③ & -6 & 0 & 0 \\ 0 & 0 & 0 & ④ & -6 & 2 \\ 4 & 0 & 0 & 2 & ⑤ & -6 \end{bmatrix}. \tag{3.51}$$

The network, *N*, in series-parallel form, corresponding to **G**, is shown in Fig. 3.3.

With regard to the analysis of communication nets, Ford and Fulkerson first formulated the problem mathematically and gave a clear-cut solution in terms of the max-flow min-cut theorem. Evidently this result is the most basic result and the key to the theory of communication nets. Elias, Feinstein, and Shannon[13] were able to give a shorter proof and pointed out that the extension to the case of the multisource multisink single commodity is indeed fairly straightforward. There exists a wealth of further investigations along the same line and along the lines of applications to other combinatorial problems. But these are of particular interest to mathematicians, rather than engineers. From the engineering point of view, the analysis of one-terminal-pair flow in communication nets is completely solved. The major engineering problem today in the

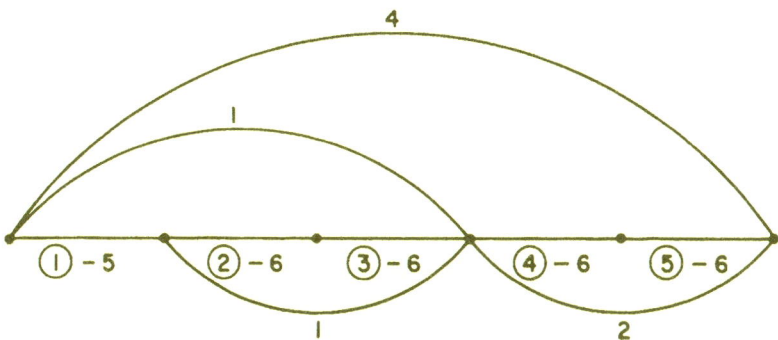

Fig. 3.3. *A resistive network realization of the terminal matrix of Eq. (3.50)*

analysis of communication nets is the simultaneous flow problem of multicommodity. Some interesting results have been obtained concerning necessary and sufficient conditions for realizability. The simultaneous flow problem is important since it is very closely connected to the regulation of traffic in communication and other systems.

In the synthesis of communication nets, most investigations up to now have treated the problem of realizing a terminal matrix. Mayeda first found a necessary and sufficient condition for the realization of terminal matrices of nonoriented communication nets, and was able to show that a terminal matrix can always be realized in terms of a tree network. As in the oriented case, the realizability conditions are still open, although partial results have allowed us to devise many useful synthesis techniques.

As in many engineering problems, one is always interested not only in realization but in realization with minimum cost. Many methods have been suggested for the synthesis of nonoriented communication nets with minimum total edge-capacity. In the case of a linear cost function, linear programming techniques have been applied to great advantage. But more work in this area is still necessary in order to handle problems of practical interest. Unfortunately very little is known with regard to nonlinear cost functions. Any significant result in nonlinear functions will have many practical implications.

From the systems point of view, the formulation of the synthesis problem in terms of a single-terminal matrix is not an adequate representation of the actual problem. A more realistic model would treat the terminal-capacity requirements as random variables which are functions of time. The communication net is then realized to give a certain expected value of the waiting time for any message. Such a network, if realizable with minimum nonlinear cost, would certainly provide a satisfactory solution to the problem. However, a solution of this order and magnitude would most likely be approached in small steps, and the synthesis of a single terminal matrix is probably the first necessary step in this direction. The synthesis of stationary simultaneous flow communication nets seems to be a second logical step. We can see already that the notion of a terminal matrix is rather handy in the treatment of simultaneous flow.

Extensions of the theory of communication nets have also been made in other directions. Synthesis of communication nets with discrete sets of capacities has been treated by Kim, Ash, and Frisch.[14,15] Yau has extended the model to include radio communication nets.[16]

REFERENCES

1a. L. R. Ford and D. R. Fulkerson, "Flows in networks," Rand Corporation Report No. R-375, December, 1960.

1b. L. R. Ford and D. R. Fulkerson, "Maximal flow through a network," *Can. J. Math.*, 8 (1956), 399–404.

2. G. B. Dantzig and D. R. Fulkerson, "On the max-flow min-cut theorem of networks," in *Linear Inequalities and Related Systems*, pp. 215–21. Annals of Mathematics, Study 38. Princeton, New Jersey, Princeton University Press, 1956.

3. W. Mayeda, "Terminal and branch capacity matrices of a communication net," *IRE Trans.* CT-7 (1960), 261–70.

4. D. T. Tang and R. T. Chien, "Analysis and synthesis techniques of oriented communication nets," *IRE Trans.* CT-8 (1961), 39–43.

5. W. Mayeda, "Maximum flow through a communications network," Interim Tech. Report No. 13, Contract DA-11-022-ORD-1983, University of Illinois, Urbana, Illinois, February, 1959.

6. R. T. Chien, "A Method for Computing Maximum Flows through a Communication Net," in *Proceedings of the Sixth National Symposium on Communications Systems*, pp. 282–85, 1960.

7. O. Wing and R. T. Chien, "Optimal synthesis of a communication net," *IRE Trans. CT*-8 (1961), 44–49.

8. R. E. Gomory and T. C. Hu, "Multi-terminal network flows," IBM Research Report, 1960.

9. R. T. Chien, "Synthesis of a communication net," *IBM J. Research & Develop.*, 4 (1960), 311–20.

10. R. E. Gomory and T. C. Hu, "An application of generalized linear programming to network flows," IBM Research Report, 1960.

11. D. T. Tang, "Communication networks with simultaneous flow requirements," IBM Research Report, 1961.

12. I. T. Frisch and W. H. Kim, "N-port resistive networks and communication nets," *IRE Trans. CT*-8 (1961), 493–96.

13. P. Elias, A. Feinstein, and C. E. Shannon, "A note on the maximal flow through a network," *IRE Trans. IT*-2 (1956), 117–19.

14. R. B. Ash and W. H. Kim, "On the Synthesis of Information Networks," delivered at the URSI Conference, Pennsylvania State University, October, 1958.

15. I. T. Frisch and W. H. Kim, "Realization of Communication Nets with Maximum Information Flow," in *Proceedings of the Seventh National Symposium on Communications Systems*, pp. 254–61, 1961.

16. S. S. Yau, "A Generalization of the Cut Sets for Application to Communication Nets," Ph.D. thesis, Department of Electrical Engineering, University of Illinois, Urbana, Illinois, June, 1961.

Index

Errata

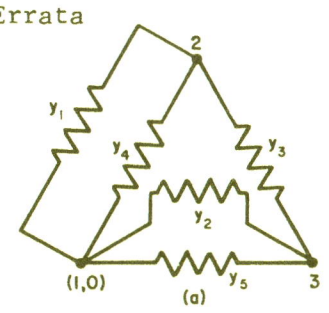

Page 57:
 For Fig. 2.6a
 substitute

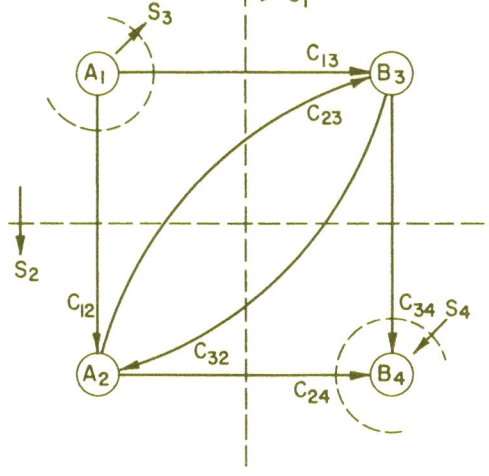

Page 260:
 For Fig. 1.4
 substitute

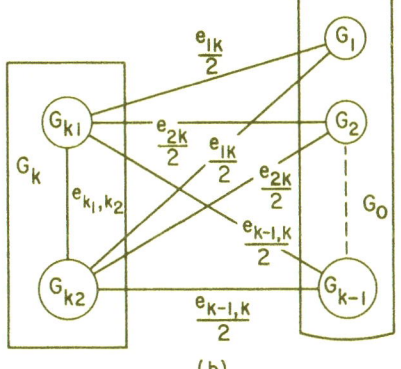

Page 286:
 For Fig. 2.10b
 substitute

Errata

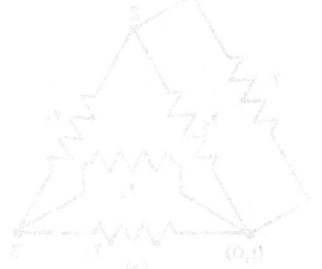

(a)

Page 57:
For Fig. 2.2
substitute

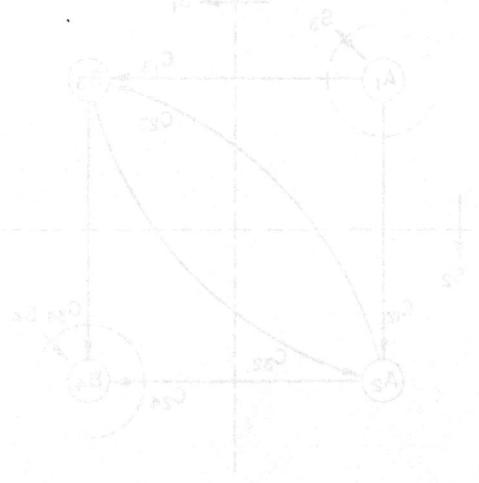
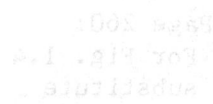

Page 260:
For Fig. 1.4
substitute

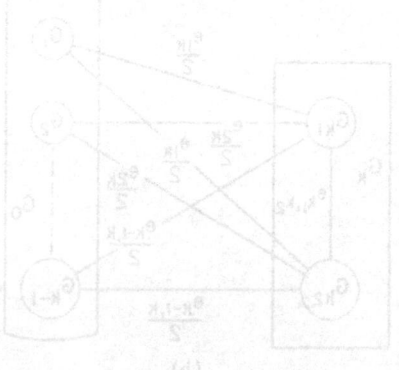

(b)

Page 286:
For Fig. 2.10b
substitute

Bei Fragen zur Produktsicherheit wenden Sie sich bitte an:
If you have any questions regarding product safety,
please contact:

Walter de Gruyter GmbH
Genthiner Straße 13
10785 Berlin
productsafety@degruyterbrill.com